Acoustics and Psychoacoustics

Music
TECHNOLOGY
S e r i e s

Titles in the series

Acoustics and Psychoacoustics, 2nd edition
(with accompanying website:
http://www-users.york.ac.uk/~dmh8/AcPsych/acpsyc.htm)
David M. Howard and James Angus

The Audio Workstation Handbook
Francis Rumsey

Composing Music with Computers (with CD-ROM)
Eduardo Reck Miranda

Computer Sound Synthesis for the Electronic Musician
(with CD-ROM)
Eduardo Reck Miranda

Digital Audio CD and Resource Pack
Markus Erne
(Digital Audio CD also available separately)

Network Technology for Digital Audio
Andy Bailey

Digital Sound Processing for Music and Multimedia
(with accompanying website:
http://www.York.ac.uk/inst/mustech/dspmm.htm)
Ross Kirk and Andy Hunt

MIDI Systems and Control, 2nd edition
Francis Rumsey

Sound and Recording: An introduction, 3rd edition
Francis Rumsey and Tim McCormick

Sound Synthesis and Sampling
Martin Russ

Sound Synthesis and Sampling CD-ROM
Martin Russ

Acoustics and Psychoacoustics

Second edition

David M. Howard and
James Angus

Focal Press
OXFORD   AUCKLAND   BOSTON   JOHANNESBURG   MELBOURNE   NEW DELHI

Focal Press
An imprint of Butterworth-Heinemann
Linacre House, Jordan Hill, Oxford OX2 8DP
225 Wildwood Avenue, Woburn, MA 01801-2041
A division of Reed Educational and Professional Publishing Ltd

A member of the Reed Elsevier plc group

First published 1996
Reprinted 1998, 1999, 2000
Second edition 2001
Reprinted 2001

© David M. Howard and James A.S. Angus 2001

British Library Cataloguing in Publication Data
Howard, David M. (David Martin), 1956–
   Acoustics and psychoacoustics. – 2nd ed. – (Music
   technology series)
   1. Acoustical engineering   2. Psychoacoustics   3. Music –
   Acoustics and physics
   I. Title   II.. Angus, James
   620.2

Library of Congress Cataloguing in Publication Data
Howard, David M. (David Martin), 1956–
   Acoustics and psychoacoustics/David M. Howard and James Angus – 2nd ed.
   p. cm.
   Includes index.
   ISBN 0–240–51609–5 (alk. paper)
   1. Music – Acoustics and physics.   2. Psychoacoustics.   3. Sound –
   Recording and reproducing – Digital techniques.   I. Angus, J. A. S. (James
   A. S.)   II. Title.

   ML3805 .H77 2000
   152.1'5–dc21   00-049494

ISBN 0 240 51609 5

For more information on all Butterworth-Heinemann publications
please visit our website at www.bh.com

Composition by Scribe Design, Gillingham, Kent, UK
Printed and bound in Great Britain

Contents

Series introduction   vii

Preface to the second edition   ix

1   **Introduction to sound   1**
   1.1   Pressure waves and sound transmission   1
   1.2   Sound intensity, power and pressure level   14
   1.3   Adding sounds together   20
   1.4   The inverse square law   28
   1.5   Sound interactions   33
   1.6   Time and frequency domains   51
   1.7   Analysing spectra   56

2   **Introduction to hearing   65**
   2.1   The anatomy of the hearing system   66
   2.2   Critical bands   74
   2.3   Frequency and pressure sensitivity ranges   79
   2.4   Loudness perception   82
   2.5   Noise-induced hearing loss   91
   2.6   Perception of sound source direction   96

3   **Notes and harmony   109**
   3.1   Musical notes   109
   3.2   Hearing pitch   119
   3.3   Hearing notes   136
   3.4   Tuning systems   144

4   **Acoustic model for musical instruments   152**
   4.1   A 'black box' model of musical instruments   152
   4.2   Stringed instruments   155

4.3   Wind instruments   166
4.4   Percussion instruments   194
4.5   The speaking and singing voice   198

5   **Hearing timbre and deceiving the ear   210**
5.1   What is timbre?   210
5.2   Acoustics of timbre   211
5.3   Psychoacoustics of timbre   220
5.4   The pipe organ as a timbral synthesiser   226
5.5   Deceiving the ear   228

6   **Hearing music in different environments   247**
6.1   Acoustics of enclosed spaces   247
6.2   Room modes and standing waves   288
6.3   Absorption materials   305
6.4   Diffusion materials   311
6.5   Sound isolation   316
6.6   Energy–time considerations   321

7   **Processing sound electronically   327**
7.1   Filtering   327
7.2   Equalisation and tone controls   331
7.3   Artificial reverberation   335
7.4   Chorus, automatic double tracking (ADT), phasing and flanging effects   343
7.5   Pitch processing and time modification   347
7.6   Sound morphing and vocoding   354
7.7   Spatial processing   356
7.8   Loudness processing   361
7.9   Summary   364

Appendix 1
Solving the ERB equation   367
Appendix 2
Converting between frequency ratios and cents   369
Appendix 3
Deriving the reverberation time equation   371
Appendix 4
Deriving the reverberation time equation for different frequencies and surfaces   375

Index   379

Series introduction

The Focal Press Music Technology Series is intended to fill a growing need for authoritative books to support college and university courses in music technology, sound recording, multimedia and their related fields. The books will also be of value to professionals already working in these areas and who want either to update their knowledge or to familiarise themselves with topics that have not been part of their mainstream occupations.

Information technology and digital systems are now widely used in the production of sound and in the composition of music for a wide range of end uses. Those working in these fields need to understand the principles of sound, musical acoustics, sound synthesis, digital audio, video and computer systems. This is a tall order, but people with this breadth of knowledge are increasingly sought after by employers. The series will explain the technology and techniques in a manner which is both readable and factually concise, avoiding the chattiness, informality and technical woolliness of many books on music technology. The authors are all experts in their fields and many come from teaching and research backgrounds.

Dr Francis Rumsey
Series Editor

Preface to the second edition

The first edition has proved to be something of a success, but inevitably a number of minor corrections have been pointed out to us during its lifetime. We still believe in this musical application-related manner of presenting acoustics and psycho-acoustics to our intended readership. The opportunity to prepare a second edition has enabled us to attend to those minor corrections that have been brought to our attention; we are particularly grateful to readers who have done so.

There are also additions made, some following comments from readers and reviewers and others from our own teaching experiences, which we hope make the work more complete. In particular there are margin notes in Chapter 1 to provide some additional background to terms presented, the sections in Chapter 4 relating to wind instruments have been expanded, an introduction is given in Chapter 5 to audio coding based on psychoacoustic principles, and in Chapter 6 there is additional material on diffusion and more example room designs.

We are very pleased to know that many readers are finding this form of presentation of the material useful and fervently hope that this will continue to be the case. We are always keen to receive comments and suggestions via the publisher. This field is developing rapidly as the desire to listen to music anywhere is matched by technological advances that enable this to happen. The importance of a proper understanding of the underlying principles behind acoustics and psychoacoustics cannot be underestimated if this is to be carried out satisfactorily, effectively and professionally.

David M. Howard and James A.S. Angus

1 Introduction to sound

Sound is something most people take for granted. Our environment is full of noises, which we have been exposed to from before birth. What is sound, how does it propagate, and how can it be quantified? The purpose of this chapter is to introduce the reader to the basic elements of sound, the way it propagates, and related topics. This will help us to understand both the nature of sound, and its behaviour in a variety of acoustic contexts and allow us to understand both the operation of musical instruments and the interaction of sound with our hearing.

1.1 Pressure waves and sound transmission

At a physical level sound is simply a mechanical disturbance of the medium, which may be air, or a solid, liquid or other gas. However, such a simplistic description is not very useful as it provides no information about the way this disturbance travels, or any of its characteristics other than the requirement for a medium in order for it to propagate. What is required is a more accurate description which can be used to make predictions of the behaviour of sound in a variety of contexts.

1.1.1 The nature of sound waves

Consider the simple mechanical model of the propagation of sound through some physical medium, shown in Figure 1.1. This

Figure 1.1 Golf ball and spring model of a sound propagating material.

shows a simple one-dimensional model of a physical medium, such as air, which we call the golf ball and spring model because it consists of a series of masses, e.g. golf balls, connected together by springs. The golf balls represent the point masses of the molecules in a real material, and the springs represent the inter-molecular forces between them. If the golf ball at the end is pushed toward the others then the spring linking it to the next golf ball will be compressed and will push at the next golf ball in the line which will compress the next spring, and so on. Because of the mass of the golf balls there will be a time lag before they start moving from the action of the connecting springs. This means that the disturbance caused by moving the first golf ball will take some time to travel down to the other end. If the golf ball at the beginning is returned to its original position the whole process just described will happen again, except that the golf balls will be pulled rather than pushed and the connecting springs will have to expand rather than compress. At the end of all this the system will end up with the golf balls having the same average spacing that they had before they were pushed and pulled.

The region where the golf balls are pushed together is known as a compression whereas the region where they are pulled apart is known as a rarefaction, and the golf balls themselves are the propagating medium. In a real propagating medium, such as air, a disturbance would naturally consist of either a compression followed by a rarefaction or a rarefaction followed by a compression in order to allow the medium to return to its normal state. A picture of what happens is shown in Figure 1.2. Because of the way the disturbance moves—the golf balls are pushed and pulled in the direction of the disturbance's travel—this type of propagation is known as a longitudinal wave. Sound waves are therefore longitudinal waves which propagate via a series of compressions and rarefactions in a medium, usually air.

There is an alternative way that a disturbance could be propagated down the golf ball and spring system. If, instead of being pushed and pulled toward each other, the golf balls were moved from side to side then a lateral disturbance would be propagated, due to the forces exerted by the springs on the golf balls as described earlier. This type of wave is known as a transverse wave and is often found in the vibrations of parts of musical instruments, such as strings or membranes.

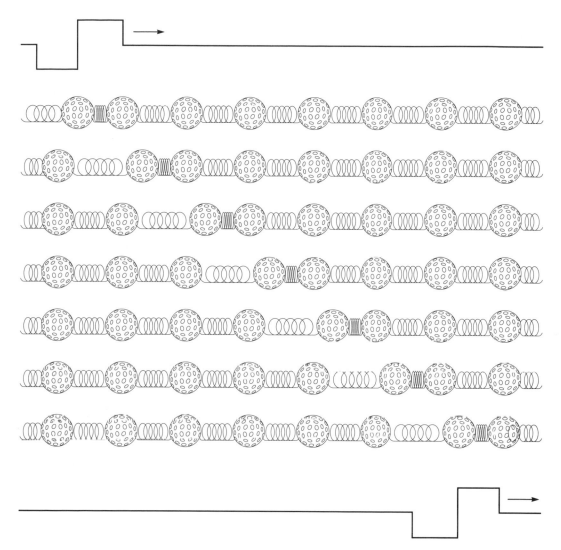

Figure 1.2 Golf ball and spring model of a sound pulse propagating in a material.

1.1.2 The velocity of sound waves

The speed at which a disturbance, of either kind, moves down the 'string' of connected golf balls will depend on two things:

- The mass of the golf balls: the mass affects the speed of disturbance propagation because a golf ball with more mass will take longer to start and stop moving. In real materials the density of the material determines the effective mass of the golf balls. A higher density gives a higher effective mass and so the propagation will travel more slowly.
- The strength of the springs: the strength of the springs connecting the golf balls together will also affect the speed of disturbance propagation because a stronger spring will be

Young's modulus is a measure of the 'springiness' of a material. A high Young's modulus means the material needs more force to compress it. It is measured in newtons per square metre (N m^{-2}).

able to push harder on the next golf ball and so accelerate it faster. In real materials the strength of the springs is equivalent to the elastic modulus of the material, which is also known as the Young's modulus of the material. A higher elastic modulus in the material implies a stiffer spring and therefore a faster speed of disturbance propagation.

For longitudinal waves in solids, the speed of propagation is only affected by the density and Young's modulus of the material and this can be simply calculated from the following equation:

A newton (N) is a measure of force.

$$v = \sqrt{\frac{E}{\rho}} \qquad (1.1)$$

where v = the speed in metres per second (ms^{-1})
ρ = the density of the material (in kg m^{-3})
and E = the Young's modulus of the material (in N m^{-2})

Density is the mass per unit volume. It is measured in kilograms per cubic metre (kg m^{-3}).

However, although the density of a solid is independent of the direction of propagation in a solid, the Young's modulus may not be. For example, brass will have a Young's modulus which is independent of direction because it is homogeneous whereas wood will have a different Young's modulus depending on whether it is measured across the grain or with the grain. Thus brass will propagate a disturbance with a velocity which is independent of direction but in wood the velocity will depend on whether the disturbance is travelling with the grain or across it. To make this clearer let us consider an example.

$\sqrt{}$ is the square root symbol. It means take the square root of whatever is inside it.

Example 1.1 Calculate the speed of sound in steel and in beech wood.

The density of steel is 7800 kg m^{-3}, and its Young's modulus is 2.1×10^{11} N m^{-2}, so the speed of sound in steel is given by:

s^{-1} means per second.

$$v_{\text{steel}} = \sqrt{\frac{2.1 \times 10^{11}}{7800}} = 5189 \text{ ms}^{-1}$$

The density of beech wood is 680 kg m^{-3}, and its Young's modulus is 14×10^9 N m^{-2} along the grain and 0.88×10^9 N m^{-2} across the grain. This means that the speed of sound is different in the two directions and they are given by:

$$v_{\text{beech along the grain}} = \sqrt{\frac{14 \times 10^9}{680}} = 4537 \text{ ms}^{-1}$$

and

$$v_{\text{beech across the grain}} = \sqrt{\frac{0.88 \times 10^9}{680}} = 1138 \text{ ms}^{-1}$$

Thus the speed of sound in beech is four times faster along the grain than across the grain.

This variation of the speed of sound in materials such as wood can affect the acoustics of musical instruments made of wood and has particular implications for the design of loudspeaker cabinets, which are often made of wood. In general, loudspeaker manufacturers choose processed woods, such as plywood or MDF (medium density fibreboard), which have a Young's modulus which is independent of direction.

1.1.3 The velocity of sound in air

So far the speed of sound in solids has been considered. However, sound is more usually considered as something that propagates through air, and for music this is the normal medium for sound propagation. Unfortunately air does not have a Young's modulus so Equation 1.1 cannot be applied directly, even though the same mechanisms for sound propagation are involved. Air is springy, as any one who has held their finger over a bicycle pump and pushed the plunger will tell you, so a means of obtaining something equivalent to Young's modulus for air is required. This can be done by considering the adiabatic, meaning no heat transfer, gas law given by:

$$PV^{\gamma} = \text{constant} \tag{1.2}$$

where P = the pressure of the gas (in N m^{-2})
V = the volume of the gas (in m^3)
and γ = is a constant which depends on the gas (1.4 for air)

The adiabatic gas law equation is used because the disturbance moves so quickly that there is no time for heat to transfer from the compressions or rarefactions. Equation 1.2 gives a relationship between the pressure and volume of a gas and this can be used to determine the strength of the air spring, or the equivalent to Young's modulus for air, which is given by:

$$E_{gas} = \gamma P \tag{1.3}$$

The density of a gas is given by:

$$\rho_{gas} = \frac{m}{V} = \frac{PM}{RT} \tag{1.4}$$

where m = the mass of the gas (in kg)
M = the molecular mass of the gas (in kg mole^{-1})
R = the gas constant (8.31 J K^{-1} mole^{-1})
and T = the absolute temperature (in K)

Pressure is the force, in newtons, exerted by a gas on a surface. This arises because the gas molecules 'bounce' off the surface. It is measured in newtons per square metre (N m^{-2}).

The molecular mass of a gas is approximately equal to the total number of protons and neutrons in the molecule expressed in grams (g). Molecular mass expressed in this way always contains the same number of molecules (6.022 × 10^{23}). This number of molecules is known as a mole (mol).

Equations 1.3 and 1.4 can be used to give the equation for the speed of sound in air, which is:

$$v_{\text{gas}} = \sqrt{\frac{E_{\text{gas}}}{\rho_{\text{gas}}}} = \sqrt{\frac{\gamma P}{\left(\dfrac{PM}{RT}\right)}} = \sqrt{\frac{\gamma RT}{M}} \tag{1.5}$$

Equation 1.5 is important because it shows that the speed of sound in a gas is not affected by pressure. Instead, the speed of sound is strongly affected by the absolute temperature and the molecular weight of the gas. Thus we would expect the speed of sound in a light gas, such as helium, to be faster than that of a heavy gas, such as carbon dioxide, and air to be somewhere in between. For air we can calculate the speed of sound as follows.

Example 1.2 Calculate the speed of sound in air at 0°C and 20°C.

The composition of air is 21% oxygen (O_2), 78% nitrogen (N_2), 1% argon (Ar), and minute traces of other gases. This gives the molecular weight of air as:

$$M = 21\% \times 16 \times 2 + 78\% \times 14 \times 2 + 1\% \times 18$$
$$= 2.87 \times 10^{-2} \text{ kg mole}^{-1}$$

and

$$\gamma = 1.4$$

$$R = 8.31 \text{ J K}^{-1} \text{ mole}^{-1}$$

which gives the speed of sound as:

$$v = \sqrt{\frac{1.4 \times 8.31}{2.87 \times 10^{-2}} T}$$

$$v = 20.1 \sqrt{T}$$

Thus the speed of sound in air is dependent only on the square root of the absolute temperature, which can be obtained by adding 273 to the Celsius temperature; thus the speed of sound in air at 0°C and 20°C is:

$$v_{0°C} = 20.1 \sqrt{(273 + 0)} = 332 \text{ ms}^{-1}$$

$$v_{20°C} = 20.1 \sqrt{(273 + 20)} = 344 \text{ ms}^{-1}$$

The reason for the increase in the speed of sound as a function of temperature is two fold. Firstly, as shown by Equation 1.4 which describes the density of an ideal gas, as the temperature rises the volume increases and providing the pressure remains constant, the density decreases. Secondly, if the pressure does alter, its effect on the density is compensated for by an increase in the effective Young's modulus for air, as given by Equation 1.3. In fact the dominant factor other than temperature on the speed of sound in a gas is the molecular weight of the gas. This is clearly different if the gas is different from air but the effective molecular weight can also be altered by the presence of water vapour, that is humidity, and this also can alter the speed of sound compared with dry air.

Although the speed of sound in air is proportional to the square root of absolute temperature we can approximate this change over our normal temperature range by the linear equation:

$$v \approx 33.1 + 0.6t \text{ ms}^{-1} \tag{1.6}$$

where t = the temperature of the air in °C

Therefore we can see that sound increases by about 0.6 ms⁻¹ for each °C rise in ambient temperature and this can have important consequences for the way in which sound propagates.

Table 1.1 gives the density, Young's modulus and corresponding velocity of longitudinal waves, for a variety of materials.

Table 1.1 Young's modulus, densities and speeds of sound for some common materials

Material	Young's modulus (N m⁻²)	Density (kg m⁻³)	Speed of sound (ms⁻¹)
Steel	2.10×10^{11}	7800	5189
Aluminium	6.90×10^{10}	2720	5037
Lead	1.70×10^{10}	11400	1221
Glass	6.00×10^{10}	2400	5000
Concrete	3.00×10^{10}	2400	3536
Water	2.30×10^{9}	1000	1517
Air (at 20°C)	1.43×10^{5}	1.21	344
Beech wood (along the grain)	1.40×10^{10}	680	4537
Beech wood (across the grain)	8.80×10^{8}	680	1138

1.1.4 The velocity of transverse waves

The velocities of transverse vibrations are affected by other factors. For example, the static spring tension will have a significant effect on the acceleration of the golf balls in the golf ball and spring model. If the tension is low then the force which restores the golf balls back to their original position will be lower and so the wave will propagate more slowly than when the

Figure 1.3 Some different forms of transverse wave.

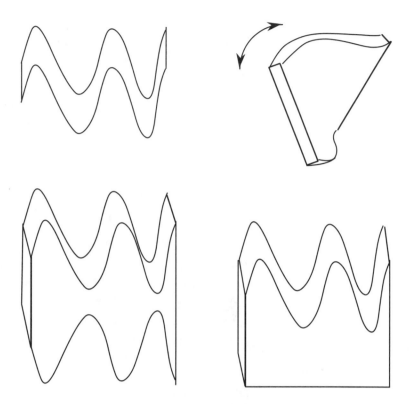

tension is higher. Also there are several different possible transverse waves in three-dimensional objects. For example, there are different directions of vibration and in addition there are different forms, depending on whether opposing surfaces are vibrating in similar or contrary motion, such as transverse, torsional and others, see Figure 1.3. As all of these different ways of moving will have different spring constants, and will be affected differently by external factors such as shape, this means that for any shape more complicated than a thin string the velocity of propagation of transverse modes of vibration become extremely complicated. This becomes important when one considers the operation of percussion instruments.

For transverse waves, calculating the velocity is more complex because for anything larger than a—in principle infinitely thin—string the speed is affected by the geometry of the propagating medium and the type of wave, as mentioned earlier. However, the transverse vibration of strings are quite important for a number of musical instruments and the velocity of a transverse wave in a piece of string can be calculated by the following equation:

$$v_{\text{transverse}} = \sqrt{\frac{T}{\mu}} \qquad (1.7)$$

where μ = the mass per unit length (in kg m^{-1})
and T = the tension of the string (in N)

This equation, although it is derived assuming an infinitely thin string, is applicable to most strings that one is likely to meet in practice. But it is applicable only to pure transverse vibration: it does not apply to torsional or other modes of vibration. However, this is the dominant form of vibration for thin strings. Its main error is due to the inherent stiffness in real materials which results in a slight increase in velocity with frequency. This effect does alter the timbre of percussive stringed instruments, like the piano, and gets stronger for thicker pieces of wire. However, Equation 1.7 can be used for most practical purposes. Let us calculate the speed of a transverse vibration on a steel string.

Example 1.3 Calculate the speed of a transverse vibration on a steel wire which is 0.8 mm in diameter (this could be a steel guitar string), and under 627 N of tension.

The mass per unit length is given by:

$$\mu_{\text{steel}} - \rho_{\text{steel}}(\pi r^2) - 7800 \times 3.14 \times \left(\frac{0.8 \times 10^{-3}}{2}\right)^2 - 3.92 \times 10^{-3} \text{ kg m}^{-1}$$

The speed of the transverse wave is thus:

$$v_{\text{steel transverse}} = \sqrt{\frac{627}{3.92 \times 10^{-3}}} = 400 \text{ ms}^{-1}$$

This is considerably slower than a longitudinal wave in the same material and generally transverse waves propagate more slowly than longitudinal ones in a given material.

1.1.5 The wavelength and frequency of sound waves

So far we have only considered the propagation of a single disturbance through the golf ball and spring model and we have seen that the disturbance travels at a constant velocity which is dependent only on the characteristics of the medium. Thus any other type of disturbance, such as a periodic one, would also travel at a constant speed. Figure 1.4 shows the golf ball and spring model being excited by a pin attached to a wheel rotating at a constant rate of rotation. This will produce a pressure variation as a function of time which is proportional to the sine of the

Figure 1.4 Golf ball and spring model of a sine wave propagating in a material.

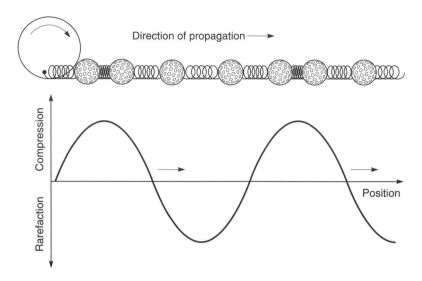

angle of rotation. This is known as a sinusoidal excitation and produces a sine wave. It is important because it represents the simplest form of periodic excitation. As we shall see later in the chapter, more complicated wave forms can always be described in terms of these simpler sine waves. Sine waves have three parameters: their amplitude, rate of rotation or frequency, and their starting position or phase. The frequency used to be expressed in units of cycles per second, reflecting the origin of the waveform, but now it is measured in the equivalent units of

Figure 1.5 The wavelength of propagating sine wave.

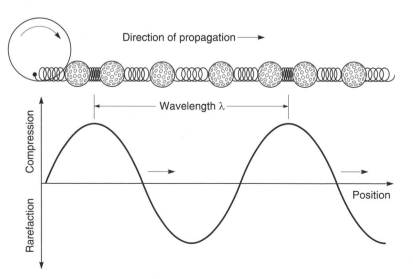

Hertz (Hz). This type of excitation generates a travelling sine wave disturbance down the model, where the compressions and rarefactions are periodic. Because the sine wave propagates at a given velocity a length can be assigned to the distance between the repeats of the compressions or rarefactions, as shown in Figure 1.5. Furthermore, because the velocity is constant, the distance between these repeats will be inversely proportional to the rate of variation of the sine wave, known as its frequency. The distance between the repeats is an important acoustical quantity and is called the wavelength (λ). Because the wavelength and frequency are linked together by the velocity, it is possible to calculate one of the quantities given the knowledge of two others using the following equation.

$$v = f\lambda \tag{1.8}$$

where v = the velocity of sound in the medium (in ms^{-1})
f = the frequency of the sound (in Hz, 1 Hz = 1 cycle per second)
and λ = the wavelength of the sound in the medium (in m)

This equation can be used to calculate the frequency given the wavelength, the wavelength given the frequency, and even the speed of sound in the medium given the frequency and wavelength, and is applicable to both longitudinal and transverse waves.

Example 1.4 Calculate the wavelength of sound, being propagated in air at 20 °C, at 20 Hz and 20 kHz.

For air the speed of sound at 20 °C is 344 ms^{-1} (see Example 1.2), thus the wavelengths at the two frequencies are given by:

$$\lambda = \frac{v}{f}$$

which gives:

$$\lambda = \frac{344}{20} = 17.2 \text{ m for 20 Hz}$$

and

$$\lambda = \frac{344}{20 \times 10^3} = 1.72 \text{ cm for 20 kHz}$$

These two frequencies correspond to the extremes of the audio frequency range so one can see that the range of wavelength sizes involved is very large!

Example 1.5 Calculate the frequency of sound with a wavelength of 34 cm in air at 20 °C.

The frequency is given by:

$$f = \frac{v}{\lambda} = \frac{344}{0.34} = 1012 \text{ Hz}$$

In acoustics the wavelength is often used as the 'ruler' for measuring length, rather than metres, feet or furlongs, because many of the effects of real objects, such as rooms or obstacles, on sound waves are dependent on the wavelength.

1.1.6 The relationship between pressure, velocity and impedance in sound waves

Another aspect of a propagating wave to consider is the movement of the molecules in the medium which is carrying it. The wave can be seen as a series of compressions and rarefactions which are travelling through the medium. The force required to effect the displacement, a combination of both compression and acceleration, forms the pressure component of the wave.

In order for the compressions and rarefactions to occur, the molecules must move closer together or further apart. Movement implies velocity, so there must be a velocity component which is associated with the displacement component of the sound wave. This behaviour can be observed in the golf ball model for sound propagation described earlier. In order for the golf balls to get closer for compression they have some velocity to move towards each other. This velocity will become zero when the compression has reached its peak, because at this point the molecules will be stationary. Then the golf balls will start moving with a velocity away from each other in order to get to the rarefacted state. Again the velocity between the golf balls will become zero at the trough of the rarefaction. The velocity does not switch instantly from one direction to another, due to the inertia of the molecules involved, instead it accelerates smoothly from a stationary to a moving state and back again. The velocity component reaches its peak in between the compressions and rarefactions, and for a sine wave pressure component the associated velocity component is a cosine. The force required to accelerate the molecules forms the pressure component of the wave. As this is associated with the velocity component of the velocity of the wave it is in phase with it. That is, if the velocity component is a cosine then the pressure component will also be a cosine. Figure 1.6 shows a sine wave propagating in the golf ball model with plots of the associated

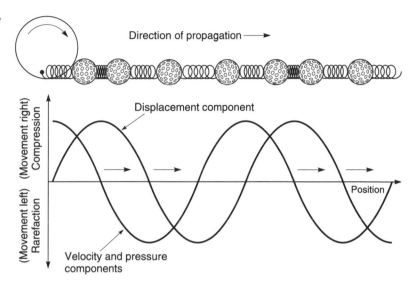

Figure 1.6 Pressure, velocity and displacement components of a sine wave propagating in a material.

components. Therefore a sound wave has both pressure and velocity components that travel through the medium at the same speed.

Pressure is a scalar quantity and therefore has no direction; we talk about pressure at a point and not in a particular direction. Velocity on the other hand must have direction; things move from one position to another. It is the velocity component which gives a sound wave its direction.

The velocity and pressure components of a sound wave are also related to each other in terms of the density and springiness of the propagating medium. A propagating medium which has a low density and weak springs would have a higher amplitude in its velocity component for a given pressure amplitude compared with a medium which is denser and has stronger springs. This relationship can be expressed, for a wave some distance away from the source and any boundaries, using the following equation:

$$\frac{\text{Pressure component amplitude}}{\text{Velocity component amplitude}} = \text{Constant} = Z_{\text{acoustic}} = \frac{p}{U}$$

where p = the pressure component amplitude
U = the volume velocity component amplitude
and Z_{acoustic} = the acoustic impedance

This constant is known as the acoustic impedance and is analogous to the resistance (or impedance) of an electrical circuit.

The amplitude of the pressure component is a function of the springiness (Young's modulus) of the material and the volume velocity component is a function of the density. This allows us to calculate the acoustic impedance using the Young's modulus and density with the following equation:

$$Z_{acoustic} = \sqrt{\rho E}$$

However the velocity of sound in the medium, usually referred to as c, is also dependent on the Young's modulus and density so the above equation is often expressed as:

$$Z_{acoustic} = \sqrt{\rho E} = \sqrt{\rho^2 \left(\frac{E}{\rho}\right)} = \rho c = 1.21 \times 344$$

$$= 416 \text{ kg m}^{-2}\text{s}^{-1} \text{ in air at 20 °C} \tag{1.9}$$

Note that the acoustic impedance for a wave in free space is also dependent only on the characteristics of the propagating medium.

However, if the wave is travelling down a tube whose dimensions are smaller than a wavelength, then the impedance predicted by Equation 1.9 is modified by the tube's area to give:

$$Z_{acoustic\ tube} = \frac{\rho c}{S_{tube}}$$

where S_{tube} = the tube area

This means that for bound waves the impedance depends on the surface area within the bounding structure and so will change as the area changes. As we shall see later, changes in impedance can cause reflections. This effect is important in the design and function of many musical instruments as discussed in Chapter 4.

1.2 Sound intensity, power and pressure level

The energy of a sound wave is a measure of the amount of sound present. However, in general we are more interested in the rate of energy transfer, instead of the total energy transferred. Therefore we are interested in the amount of energy transferred per unit of time, that is the number of joules per second (watts) that propagate. Sound is also a three-dimensional quantity and so a sound wave will occupy space. Because of this it is helpful to characterise the rate of energy transfer with respect to area, that is, in terms watts per unit area. This gives a quantity known as the sound intensity which is a measure of the power density of a sound wave propagating in a particular direction, as shown in Figure 1.7.

Figure 1.7 Sound intensity.

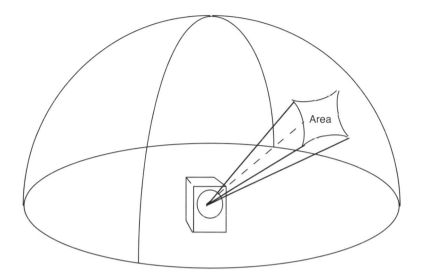

1.2.1 Sound intensity level

The sound intensity represents the flow of energy through a unit area. In other words it represents the watts per unit area from a sound source and this means that it can be related to the sound power level by dividing it by the radiating area of the sound source. As discussed earlier, sound intensity has a direction which is perpendicular to the area that the energy is flowing through, see Figure 1.7. The sound intensity of real sound sources can vary over a range which is greater than one million-million (10^{12}). Because of this, and because of the way we perceive the loudness of a sound, the sound intensity level is usually expressed on a logarithmic scale. This scale is based on the ratio of the actual power density to a reference intensity of 1 picowatt per square metre (10^{-12} W m^{-2}). Thus the sound intensity level (SIL) is defined as:

The symbol for power in watts is W.

$$SIL = 10 \log_{10}\left(\frac{I_{\text{actual}}}{I_{\text{ref}}}\right) \tag{1.10}$$

where I_{actual} = the actual sound power flux level (in W m^{-2})
and I_{ref} = the reference sound power flux level (10^{-12} W m^{-2})

The factor of 10 arises because this makes the result a number in which an integer change is approximately equal to the smallest change that can be perceived by the human ear. A factor of 10 change in the power density ratio is called the bel; in Equation 1.10 this would result in change of 10 in the outcome. The integer unit that results from Equation 1.10 is therefore called the decibel

15

(dB). It represents a $^{10}\sqrt{10}$ change in the power density ratio, that is a ratio of about 1.26.

Example 1.6 A loudspeaker with an effective diameter of 25 cm radiates 20 mW. What is the sound intensity level at the loudspeaker?

Sound intensity is the power per unit area. First we must work out the radiating area of the loudspeaker which is:

$$A_{\text{speaker}} = \pi r^2 = \pi \left(\frac{0.25 \text{ m}}{2}\right)^2 = 0.049 \text{ m}^2$$

Then we can work out the sound intensity as:

$$I = \left(\frac{W}{A_{\text{speaker}}}\right) = \left(\frac{20 \times 10^{-3} \text{ W}}{0.049 \text{ m}^2}\right) = 0.41 \text{ W m}^{-2}$$

This result can be substituted into Equation 1.12 to give the sound intensity level, which is:

$$SIL = 10 \log_{10}\left(\frac{I_{\text{actual}}}{I_{\text{ref}}}\right) = 10 \log_{10}\left(\frac{0.41 \text{ W m}^{-2}}{10^{-12} \text{ W m}^{-2}}\right) = 116 \text{ dB}$$

1.2.2 Sound power level

The sound power level is a measure of the total power radiated in all directions by a source of sound and it is often given the abbreviation SWL, or sometimes PWL. The sound power level is also expressed as the logarithm of a ratio in decibels and can be calculated from the ratio of the actual power level to a reference level of 1 picowatt (10^{-12} W) as follows:

$$SWL = 10 \log_{10}\left(\frac{w_{\text{actual}}}{w_{\text{ref}}}\right) \tag{1.11}$$

where w_{actual} = the actual sound power level (in watts)
and w_{ref} = the reference sound power level (10^{-12} W)

The sound power level is useful for comparing the total acoustic power radiated by objects, for example ones which generate unwanted noises. It has the advantage of not depending on the acoustic context, as we shall see in Chapter 6. Note that, unlike the sound intensity, the sound power has no particular direction.

Example 1.7 Calculate the SWL for a source which radiates a total of 1 watt.

Substituting into Equation 1.11 gives:

$$SWL = 10 \log_{10}\left(\frac{w_{actual}}{w_{ref}}\right) = 10 \log_{10}\left(\frac{1 \text{ watt}}{1 \times 10^{-12} \text{ watts}}\right)$$

$$= 10 \log_{10}(1 \times 10^{12}) = 120 \text{ dB}$$

One watt would be a very loud sound, if you were to receive all the power. However in most situations the listener would only be subjected to a small proportion of this power.

1.2.3 Sound pressure level

The sound intensity is one way of measuring and describing the amplitude of a sound wave at a particular point. However, although it is useful theoretically, and can be measured, it is not the usual quantity used when describing the amplitude of a sound. Other measures could be either the amplitude of the pressure, or the associated velocity component of the sound wave. Because human ears are sensitive to pressure, which will be described in Chapter 2, and because it is easier to measure, pressure is used as a measure of the amplitude of the sound wave. This gives a quantity which is known as the sound pressure, which is the root mean square (rms) pressure of a sound wave at a particular point. The sound pressure for real sound sources can vary from less than 20 microPascals (20 μPa or 20×10^{-6} Pa) to greater than 20 Pascals (20 Pa). Note that 1 Pa equals a pressure of 1 N m^{-2}. These two pressures broadly correspond to the threshold of hearing (20 μPa) and the threshold of pain (20 Pa) for a human being, at a frequency of 1 kHz, respectively. Thus real sounds can vary over a range of pressure amplitudes which is greater than a million. Because of this, and because of the way we perceive sound, the sound pressure level is also usually expressed on a logarithmic scale. This scale is based on the ratio of the actual sound pressure to the notional threshold of hearing at 1 kHz of 20 μPa. Thus the sound pressure level (SPL) is defined as:

The pascal (Pa) is a measure of pressure; 1 pascal (1 Pa) is equal to 1 newton per square metre (1 Nm^{-2}).

$$SPL = 20 \log_{10}\left(\frac{p_{actual}}{p_{ref}}\right) \qquad (1.12)$$

where p_{actual} = the actual pressure level (in Pa)
and p_{ref} = the reference pressure level (20 μPa)

The multiplier of 20 has a twofold purpose. The first is to make the result a number in which an integer change is approximately equal to the smallest change that can be perceived by the human ear. The second is to provide some equivalence to intensity measures of sound level as follows.

The intensity of an acoustic wave is given by the product of the volume velocity and pressure amplitude as:

$$I_{acoustic} = Up$$

where p = the pressure component amplitude
and U = the volume velocity component amplitude

However the pressure and velocity component amplitudes are linked via the acoustic impedance (Equation 1.10) so the intensity can be calculated in terms of just the sound pressure and acoustic impedance by:

$$I_{acoustic} = Up = \left(\frac{p}{Z_{acoustic}}\right)p = \frac{p^2}{Z_{acoustic}}$$

Therefore the sound intensity level could be calculated using the pressure component amplitude and the acoustic impedance using:

$$SIL = 10\log_{10}\left(\frac{I_{acoustic}}{I_{ref}}\right) = 10\log_{10}\left(\frac{\frac{p^2}{Z_{acoustic}}}{I_{ref}}\right) = 10\log_{10}\left(\frac{p^2}{Z_{acoustic}\,I_{ref}}\right)$$

This shows that the sound intensity is proportional to the square of pressure, in the same way that electrical power is proportional to the square of voltage. The operation of squaring the pressure can be converted into multiplication of the logarithm by a factor of two, which gives:

$$SIL = 20\log_{10}\left(\frac{p}{\sqrt{Z_{acoustic}\,I_{ref}}}\right)$$

This equation is similar to Equation 1.12 except that the reference level is expressed differently. In fact, this equation shows that if the pressure reference level was calculated as:

$$p_{ref} = \sqrt{Z_{acoustic}\,I_{ref}} = \sqrt{416 \times 10^{-12}} = 20.4 \times 10^{-6}$$

then the two ratios would be equivalent. The actual pressure reference level of 20 µPa is close enough to say that the two measures of sound level are broadly equivalent. That is, $SIL \approx SPL$ for a single sound wave a reasonable distance from the source and any boundaries. They can be equivalent because the sound pressure level is calculated at a single point and sound intensity is the power density from a sound source at the measurement point. However, whereas the sound intensity level is the power density from a sound source at the measurement point, the sound pressure level is the sum of the sound pressure waves at the measurement point. If there is only a single pressure wave from the sound source at the measurement point, that is there are no extra pressure waves due to reflections, the sound pressure level and the sound intensity level are approximately equivalent, $SIL \approx SPL$. This will be the case for sound waves in the atmosphere well away from any reflecting surfaces. It will not be true when there are additional pressure waves due to reflections, as might arise in any room or if the acoustic impedance changes. However, changes in level for both SIL and SPL will be equivalent because if the sound intensity increases then the sound pressure at a point will also increase by the same proportion, provided that nothing alters the number and proportions of the sound pressure waves arriving at the point at which the sound pressure is measured. Thus a 10 dB change in SIL will result in a 10 dB change in SPL.

These different means of describing and measuring sound amplitudes can be confusing and one must be careful to ascertain which one is being used in a given context. In general a reference to sound level implies that the SPL is being used because the pressure component can be measured easily and corresponds most closely to what we hear.

Let us calculate the SPLs for a variety of pressure levels.

Example 1.8 Calculate the SPL for sound waves with rms pressure amplitudes of 1 Pa, 2 Pa and 2 µPa.

Substituting the above values of pressure into Equation 1.12 gives:

$$SPL_{1\,Pa} = 20 \log_{10}\left(\frac{p_{actual}}{p_{ref}}\right) = 20 \log_{10}\left(\frac{1\,Pa}{20\,\mu Pa}\right)$$

$$= 20 \times \log_{10}(5 \times 10^4) = 94\,dB$$

1 Pa is often used as a standard level for specifying microphone sensitivity and, as the above calculation shows, represents a loud sound.

$$SPL_{2\,Pa} = 20 \log_{10}\left(\frac{p_{actual}}{p_{ref}}\right) = 20 \log_{10}\left(\frac{2\,Pa}{20\,\mu Pa}\right)$$

$$= 20 \times \log_{10}(1 \times 10^5) = 100\,dB$$

Doubling the pressure level results in a 6 dB increase in sound pressure level, and a tenfold increase in pressure level results in a 20 dB increase in SPL.

$$SPL_{2\ \mu Pa} = 20 \log_{10}\left(\frac{p_{\text{actual}}}{p_{\text{ref}}}\right) = 20 \log_{10}\left(\frac{2\ \mu Pa}{20\ \mu Pa}\right)$$

$$= 20 \times \log_{10}(1 \times 10^{-1}) = -20\ dB$$

If the actual level is less than the reference level then the result is a negative SPL. The decibel concept can also be applied to both sound intensity and the sound power of a source.

1.3 Adding sounds together

So far we have only considered the amplitude of single sources of sound. However, in most practical situations where more than one source of sound is present, these may result from other musical instruments or reflections from surfaces in a room. There are two different situations which must be considered when adding sound levels together.

- *Correlated sound sources:* in this situation the sound comes from several sources which are related. In order for this to happen the extra sources must be derived from a single source. This can happen in two ways. Firstly, the different sources may be related by a simple reflection, such as might arise from a simple reflection from a nearby surface. If the delay is short then the delayed sound will be similar to the original and so it will be correlated with the primary sound source. Secondly, the sound may be derived from a common electrical source, such as a recording or a microphone, and then may be reproduced using several loudspeakers. Because the speakers are being fed the same signal, but are spatially disparate, they act like several related sources and so are correlated. Figure 1.8 shows two different situations.
- *Uncorrelated sound sources:* in this situation the sound comes from several sources which are unrelated. For example, it may come from two different instruments, or from the same source but with a considerable delay due to reflections. In the first case the different instruments will be generating different waveforms and at different frequencies. Even when the same instruments play in unison, these differences will occur. In the second case, although the additional sound source comes from the primary one and so could be expected to be related to it, the delay will mean that the waveform from the additional source will no longer be the same. This is because in the intervening time, due to the delay, the primary source of the sound will have changed in pitch, amplitude and waveshape. Because the delayed

Example 1.9 The sound at a particular point consists of a main loudspeaker signal and a reflection at the same amplitude that has been delayed by 1 millisecond. What is the pressure amplitude at this point at 250 Hz, 500 Hz and 1 kHz?

The equation for pressure at a point due to a single frequency is given by the equation:

$$P_{\text{at a point}} = P_{\text{sound amplitude}} \sin(2\pi ft) \text{ or } P_{\text{sound amplitude}} \sin(360°ft)$$

where f = the frequency (in Hz)
and t = the time (in s)

Note the multiplier of 2π, or $360°$, within the sine function is required to express accurately the position of the wave within the cycle. Because a complete rotation occurs every cycle, one cycle corresponds to a rotation of 360 degrees, or, more usually, 2π radians. This representation of frequency is called angular frequency (1 Hz (cycle per second) = 2π radians per second).

The effect of the delay to the difference in path lengths alters the time of arrival of one of the waves, and so the pressure at a point due to a single frequency delayed by some time, τ, is given by the equation:

$$P_{\text{at a point}} = P_{\text{sound amplitude}} \sin(2\pi f(t + \tau))$$

$$\text{or } P_{\text{sound amplitude}} \sin(360°f(t + \tau))$$

where τ = the delay (in s)

Add the delayed and undelayed sine waves together to give:

$$P_{\text{total}} = P_{\text{delayed}} \sin(360°f(t + \tau)) + P_{\text{undelayed}} \sin(360°ft)$$

Assuming that the delayed and undelayed signals are the same amplitude this can be rearranged to give:

$$P_{\text{total}} = 2P \cos\left(360°f\left(\frac{\tau}{2}\right)\right) \sin\left(360°f\left(t + \frac{\tau}{2}\right)\right)$$

The cosine term in this equation is determined by the delay and frequency and the sine term represents the original wave slightly delayed. Thus we can express the combined pressure amplitude of the two waves as:

$$P_{\text{total}} = 2P \cos\left(360°f\left(\frac{\tau}{2}\right)\right)$$

Using the above equation we can calculate the effect of the delay on the pressure amplitude at the three different frequencies as:

$$P_{\text{total 250 Hz}} = 2P \, \cos\left(360°f\left(\frac{\tau}{2}\right)\right)$$

$$= 2P \, \cos\left(360° \times 250 \text{ Hz} \times \left(\frac{1 \times 10^{-3} \text{ s}}{2}\right)\right) = 1.41 \, P$$

$$P_{\text{total 500 Hz}} = 2P \, \cos\left(360°f\left(\frac{\tau}{2}\right)\right)$$

$$= 2P \, \cos\left(360° \times 500 \text{ Hz} \times \left(\frac{1 \times 10^{-3} \text{ s}}{2}\right)\right) = 0$$

$$P_{\text{total 1 kHz}} = 2P \, \cos\left(360°f\left(\frac{\tau}{2}\right)\right)$$

$$= 2P \, \cos\left(360° \times 1 \text{ kHz} \times \left(\frac{1 \times 10^{-3} \text{ s}}{2}\right)\right) = 2 \, P$$

These calculations show that the summation of correlated sources can be strongly frequency dependent and can vary between zero and two times the wave pressure amplitude.

1.3.2 The level when uncorrelated sounds add

On the other hand, if the sound waves are uncorrelated then they do not add algebraically, like correlated waves; instead we must add the powers of the individual waves together. As stated earlier, the power in a waveform is proportional to the square of the pressure levels so in order to sum the powers of the waves we must square the pressure amplitudes before adding them together. If we want the result as a pressure then we must take the square root of the result. This can be expressed in the following equation:

$$P_{\text{total uncorrelated}} = \sqrt{(P_1^2 + P_2^2 + \dots + P_N^2)} \tag{1.14}$$

Adding uncorrelated sources is different from adding correlated sources in several respects. Firstly, the resulting total is related to the power of the signals combined and so is not dependent on their relative phases. This means that the result of combining uncorrelated sources is always an increase in level. The second difference is that the level increase is lower because powers rather than pressures are being added. Recall that the maximum increase for two equal correlated sources was a factor of two increase in pressure amplitude. However, for uncorrelated sources the powers of the sources are added and, as the power is proportional to the square of the pressure, this means that the maximum amplitude increase for two uncorrelated sources is only $\sqrt{2}$. However, the addition of uncorrelated components always results in an increase in level without any of the cancellation effects that correlated sources suffer. Because of the lack of cancellation effects, the spatial variation in the sum of uncorrelated sources is usually much less than that of correlated ones, as the result only depends on the amplitude of the sources. As an example let us consider the effect of adding together several uncorrelated sources of the same amplitude.

Example 1.10 Calculate the increase in signal level when two vocalists sing together at the same level and when a choir of N vocalists sing together, also at the same level.

The total level from combining several uncorrelated sources together is given by Equation 1.14 as:

$$P_{\text{total uncorrelated}} = \sqrt{(P_1^2 + P_2^2 + ... + P_N^2)}$$

For N sources of the same amplitude this can be simplified to:

$$P_{\text{N uncorrelated}} = \sqrt{(P^2 + P^2 + ... + P^2)} = \sqrt{NP^2} = P\sqrt{N}$$

Thus the increase in level, for uncorrelated sources of equal amplitude, is proportional to the square root of the number of sources. In the case of just two sources this gives:

$$P_{\text{two uncorrelated}} = P\sqrt{N} = P\sqrt{2} = 1.41 \ P$$

How does the addition of sources affect the sound pressure level (SPL), the sound power level (SWL), and the sound intensity level (SIL)? For the SWL and SIL, because we are adding powers, the results will be the same whether the sources are correlated or not. However, for SPL, there will be a difference between the correlated and uncorrelated results. The main difficulty that arises when these measures are used during the calculation of the effect of combining sound sources, is confusion over the correct use of decibels during the calculation.

1.3.3 Adding decibels together

Decibels are a logarithmic scale and this means that *adding decibels together is not the same as adding the sources' amplitudes together*. This is because adding logarithms together is equivalent to the logarithm of the multiplication of the quantities that the logarithms represent. Clearly this is not the same as a simple summation!

When decibel quantities are added together it is important to convert them back to their original ratios before adding them together. If a decibel result of the summation is required then it must be converted back to decibels after the summation has taken place. To make this clearer let us look at Example 1.11.

Example 1.11 Calculate the increase in signal level when two vocalists sing together, one at 69 dB and the other at 71 dB SPL.

From Equation 1.12 the SPL of a single source is:

$$SPL = 20 \log_{10}\left(\frac{p_{\text{actual}}}{p_{\text{ref}}}\right)$$

For multiple, uncorrelated, sources this will become:

$$SPL = 20 \log_{10}\left(\frac{\sqrt{(P_1^2 + P_2^2 + \ldots + P_N^2)}}{p_{\text{ref}}}\right) =$$

$$10 \log_{10}\left(\frac{P_1^2 + P_2^2 + \ldots + P_N^2}{p_{\text{ref}}^2}\right) \tag{1.15}$$

We must substitute the pressure squared values that the singer's SPLs represent. These can be obtained with the following equation:

$$P^2 = 10^{\left(\frac{SPL}{10}\right)} p_{\text{ref}}^2$$

where $p_{\text{ref}}^2 = 4 \times 10^{-10}$ N^2 m^{-4}

Substituting in our two SPL values gives:

$$P^2_{69\text{ dB}} = 10^{\left(\frac{69}{10}\right)} \times 4 \times 10^{-10}\text{ N}^2\text{ m}^{-4} = 3.18 \times 10^{-3}\text{ N}^2\text{ m}^{-4}$$

and

$$P^2_{71\text{ dB}} = 10^{\left(\frac{71}{10}\right)} \times 4 \times 10^{-10}\text{ N}^2\text{ m}^{-4} = 5.04 \times 10^{-3}\text{ N}^2\text{ m}^{-4}$$

substituting these two values into equation 1.15 gives the result as:

$$SPL = 10 \log_{10}\left(\frac{P^2_{69\text{ dB}} + P^2_{71\text{ dB}}}{p_{\text{ref}}^2}\right)$$

$$= 10 \log_{10} \left(\frac{3.18 \times 10^{-3} + 5.04 \times 10^{-3}}{4 \times 10^{-10}} \right) = 73.1 \text{ dB}$$

Note that the combined sound level is only about 2 dB more than the louder of the two sounds and *not* 69 dB greater, which is the result that would be obtained if the SPLs were added directly.

There are some areas of sound level calculation where the fact that the addition of decibels represents multiplication is an advantage. In these situations the result can be expressed as a multiplication, and so can be expressed as a summation of decibel values. In other words decibels can be added when the underlying sound level calculation is a multiplication. In this context the decibel representation of sound level is very useful, as there are many acoustic situations in which the effect on the sound wave is multiplicative, for example the attenuation of sound through walls or their absorption by a surface. To make the use of decibels in this context let us consider Example 1.12.

Example 1.12 Calculate the increase in the sound pressure level (SPL) when two vocalists sing together at the same level and when a choir of N vocalists sing together, also at the same level. The total level from combining several uncorrelated single sources is given by:

$$P_{\text{N uncorrelated}} = P\sqrt{N}$$

This can be expressed in terms of the SPL as:

$$SPL_{\text{N uncorrelated}} = 20 \log_{10} \left(\frac{P\sqrt{N}}{p_{\text{ref}}} \right) = 20 \log_{10} \left(\frac{P}{p_{\text{ref}}} \right) + 20 \log_{10}(\sqrt{N})$$

In this equation the first term simply represents the SPL of a single source and the addition of the decibel equivalent of the square root of the number of source represents the increase in level due to the multiple sources. So this equation can be also expressed as:

$$SPL_{\text{N uncorrelated}} = SPL_{\text{single source}} + 10 \log_{10}(N)$$

This equation will give the total SPL for N uncorrelated sources of equal level. For example ten sources will raise the SPL by 10 dB, since $10 \log(10) = 10$.

In the case of two singers the above equation becomes:

$$SPL_{\text{N uncorrelated}} = SPL_{\text{single source}} + 10 \log_{10}(2) = SPL_{\text{single source}} + 3 \text{ dB}$$

27

So the summation of two uncorrelated sources increases the sound pressure level by 3 dB.

1.4 The inverse square law

So far we have only considered sound as a disturbance that propagates in one direction. However, in reality sound propagates in three dimensions. This means that the sound from a source does not travel on a constant beam, instead it spreads out as it travels away from the radiating source, as shown in Figure 1.7.

As the sound spreads out from a source it gets weaker. This is not due to it being absorbed but due to its energy being spread more thinly. Figure 1.11 gives a picture of what happens. Consider a half blown up spherical balloon, which is coated with honey to a certain thickness. If the balloon is blown up to double its radius, the surface area of the balloon would have increased four fold. As the amount of honey has not changed it

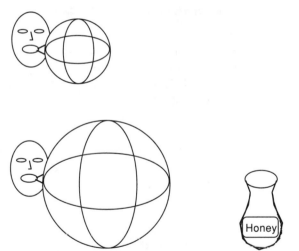

Figure 1.11 The honey and balloon model of the inverse square law for sound.

must therefore have a quarter of the thickness that it had before. The sound intensity from a source behaves in an analogous fashion in that every time the distance from a sound source is doubled the intensity reduces by a factor of four, that is there is an inverse square relationship between sound intensity and the distance from the sound source. The area of a sphere is give by the equation:

$$A_{sphere} = 4\pi r^2$$

The sound intensity is defined as the power per unit area. Therefore the sound intensity as a function of distance from a sound source is given by:

$$I = \frac{W_{\text{source}}}{A_{\text{sphere}}} = \frac{W_{\text{source}}}{4\pi r^2} \qquad (1.16)$$

where I = the sound intensity (in W m⁻²)
W_{source} = the power of the source (in W)
and r = the distance from the source (in m)

Equation 1.16 shows that the sound intensity for a sound wave that spreads out in all directions from a source reduces as the square of the distance. Furthermore this reduction in intensity is purely a function of geometry and is not due to any physical absorption process. In practice there are additional sources of absorption in air, for example impurities and water molecules, or smog and humidity. These extra sources of absorption have more effect at high frequencies and, as a result sound not only gets quieter, but also gets duller, as one moves away from a source, due to the extra attenuation these cause at high frequencies. The amount of excess attenuation is dependent on the level of impurities and humidity and is therefore variable.

A milliwatt is one thousandth of a watt (10^{-3} watts).

Example 1.13 An omnidirectional loudspeaker radiates one hundred milliwatts (100 mW). What is the sound intensity level (SIL) at a distance of 1 m, 2 m and 4 m from the loudspeaker? How does this compare with the sound power level (SWL) at the loudspeaker?

The sound power level can be calculated from Equation 1.11 and is given by:

$$SWL = 10 \log_{10}\left(\frac{w_{\text{actual}}}{w_{\text{ref}}}\right) = 10 \log_{10}\left(\frac{100 \text{ mW}}{1 \times 10^{-12} \text{ W}}\right)$$

$$= 10 \log_{10}(1 \times 10^{11}) = 110 \text{ dB}$$

The sound intensity at a given distance can be calculated using Equations 1.10 and 1.16 as:

$$SIL = 10 \log_{10}\left(\frac{I_{\text{actual}}}{I_{\text{ref}}}\right) = 10 \log_{10}\left(\frac{\dfrac{W_{\text{source}}}{4\pi r^2}}{I_{\text{ref}}}\right)$$

this can be simplified to give:

$$SIL = 10 \log_{10}\left(\frac{W_{\text{source}}}{W_{\text{ref}}}\right) - 10 \log_{10}(4\pi) - 10 \log_{10}(r^2)$$

29

which can be simplified further to:

$$SIL = 10 \log_{10}\left(\frac{W_{\text{source}}}{W_{\text{ref}}}\right) - 20 \log_{10}(r) - 11 \text{ dB} \qquad (1.17)$$

This equation can then be used to calculate the intensity level at the three distances as:

$$SIL_{1 \text{ m}} = 10 \log_{10}\left(\frac{100 \text{ mW}}{10^{-12} \text{ W}}\right) - 20 \log_{10}(1) - 11 \text{ dB}$$

$$= 110 \text{ dB} - 0 \text{ dB} - 11 \text{ dB} = 99 \text{ dB}$$

$$SIL_{2 \text{ m}} = 10 \log_{10}\left(\frac{100 \text{ mW}}{10^{-12} \text{ W}}\right) - 20 \log_{10}(2) - 11 \text{ dB}$$

$$= 110 \text{ dB} - 6 \text{ dB} - 11 \text{ dB} = 93 \text{ dB}$$

$$SIL_{4 \text{ m}} = 10 \log_{10}\left(\frac{100 \text{ mW}}{10^{-12} \text{ W}}\right) - 20 \log_{10}(4) - 11 \text{ dB}$$

$$= 110 \text{ dB} - 12 \text{ dB} - 11 \text{ dB} = 87 \text{ dB}$$

From these results we can see that the sound at 1 m from a source is 11 dB less than the sound power level at the source. Note that the sound intensity level at the source is, in theory, infinite because the area for a point source is zero. In practice, all real sources have a finite area so the intensity at the source is always finite. We can also see that the sound intensity level reduces by 6 dB every time we double the distance; this is a direct consequence of the inverse square law and is a convenient rule of thumb. The reduction in intensity of a source with respect to the logarithm of distance is plotted in Figure 1.12 and shows the 6 dB per doubling of distance relationship as straight line except when one is very close to the source. In this situation the fact that the source is finite in extent renders Equation 1.16 invalid. As an approximate rule the nearfield region occurs within the radius described by the physical size of the source. In this region the sound field can vary wildly depending on the local variation of the vibration amplitudes of the source.

Equation 1.16 describes the reduction in sound intensity for a source which radiates in all directions. However, this is only possible when the sound source is well away from any surfaces that might reflect the propagating wave. Sound radiation in this type of propagating environment is often called the free field radiation, because there are no boundaries to restrict wave propagation.

Figure 1.12 Sound intensity as a function of distance from the source.

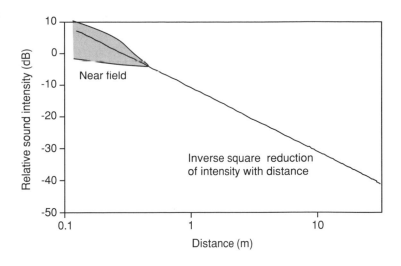

1.4.1 *The effect of boundaries*

But how does a boundary affect Equation 1.16? Clearly many acoustic contexts involve the presence of boundaries near acoustic sources, or even all the way round them in the case of rooms and some of these effects will be considered in Chapter 7. However, in many cases a sound source is placed on a boundary, such as a floor. In these situations the sound is radiating into a restricted space, as shown in Figure 1.13. However, despite the restriction of the radiating space, the surface area of the sound wave still increases in proportion to the square of the distance, as shown in Figure 1.13. The effect of the boundaries is to merely concentrate the sound power of the source into a smaller range of angles. This concentration can be expressed as an extra multiplication factor in Equation 1.16. Therefore the equation can be rewritten as:

$$I_{\text{directive source}} = \frac{QW_{\text{source}}}{4\pi r^2} \tag{1.18}$$

where $I_{\text{directvie source}}$ = the sound intensity (in W m^{-2})
Q = the directivity of the source (compared to a sphere)
W_{source} = the power of the source (in W)
and r = the distance from the source (in m)

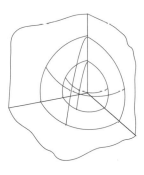

Figure 1.13 The inverse square law for sound at boundaries.

Equation 1.18 can be applied to any source of sound which directs its sound energy into a restricted set of angles which are less than a sphere. Obviously the presence of boundaries is one means of restriction, but other techniques can also achieve the same effect. For example the horn structure of brass instruments results in the same effect. However, it is important to remember that the sound intensity of a source reduces in proportion to the square of the distance, irrespective of the directivity.

The effect of having the source on a boundary can also be calculated; as an example let us examine the effect of boundaries on the sound intensity from a loudspeaker.

Example 1.14   A loudspeaker radiates 100 mW. Calculate the sound intensity level (SIL) at a distance of 2 m from the loudspeaker when it is mounted on 1, 2 and 3 boundaries.

The sound intensity at a given distance can be calculated using Equations 1.10 and 1.18 as:

$$SIL = 10 \log_{10}\left(\frac{I_{actual}}{I_{ref}}\right) = 10 \log_{10}\left(\frac{\dfrac{QW_{source}}{4\pi r^2}}{W_{ref}}\right)$$

which can be simplified to give:

$$SIL = 10 \log_{10}\left(\frac{W_{source}}{W_{ref}}\right) + 10 \log_{10}(Q) - 10 \log_{10}(4\pi) - 20 \log_{10}(r)$$

This is similar to Equation 1.18 except for the addition of the term for the directivity, Q. The presence of 1, 2 and 3 boundaries reduces the sphere to a hemisphere, half hemisphere and quarter hemisphere, which corresponds to a Q of 2, 4 and 8, respectively. As the only difference between the result with the boundaries is the term in Q, the sound intensity level at 2 m can be calculated as:

$$SIL_{1\ boundary} = SIL_{2\ m} + 10 \log_{10}(Q) = 93\ dB + 10 \log_{10}(2)$$
$$= 93\ dB + 3\ dB = 96\ dB$$

$$SIL_{2\ boundaries} = SIL_{2\ m} + 10 \log_{10}(Q) = 93\ dB + 10 \log_{10}(4)$$
$$= 93\ dB + 6\ dB = 99\ dB$$

$$SIL_{3\ boundaries} = SIL_{2\ m} + 10 \log_{10}(Q) = 93\ dB + 10 \log_{10}(8)$$
$$= 93\ dB + 9\ dB = 102\ dB$$

From these calculations we can see that each boundary increases the sound intensity at a point by 3 dB, due to the increased directivity. Note that one cannot use the above equations on more than three boundaries because then the sound can no longer expand without bumping into something. We shall examine this subject in more detail in Chapter 6. However, it is possible to have directivities of greater than 8 using other techniques. For example, horn loudspeakers with a directivity of 50 are readily available as a standard product from public address loudspeaker manufacturers.

1.5 Sound interactions

So far we have only considered sound in isolation and we have seen that sound has velocity, frequency, wavelength and reduces in intensity in proportion to the square of the distance from the source. However, sound also interacts with physical objects and other sound waves, and is affected by changes in the propagating medium. The purpose of this section is to examine some of these interactions as an understanding of them is necessary in order to understand both how musical instruments work and how sound propagates in buildings.

1.5.1 Superposition

When sounds destructively interfere with each other they do not disappear. Instead they travel through each other. Similarly, when they constructively interfere they do not grow but simply pass through each other. This is because, although the total pressure, or velocity component, may be either equal to zero or the sum of the amplitude of the individual waves, the energy flow of the sound wave is still preserved and so the wave continues to propagate. Thus the pressure or velocity at a given point in space is simply the sum, or superposition, of the individual waves that are propagating through that point, as shown in Figure 1.14. This characteristic of sound waves is called linear superposition and is very useful as it allows us to describe, and therefore analyse, the sound wave at a given point in space as the linear sum of individual components.

1.5.2 Sound refraction

This is analogous to the refraction of light at the boundary of different materials. In the optical case refraction arises because the speed of light is different in different materials, for example it is slower in water than it is in air. In the acoustic case refraction arises for the same reasons, because the velocity of sound in air is dependent on the temperature, as shown in Equation 1.5.

Figure 1.14 Superposition of a sound wave in the golf ball and spring model.

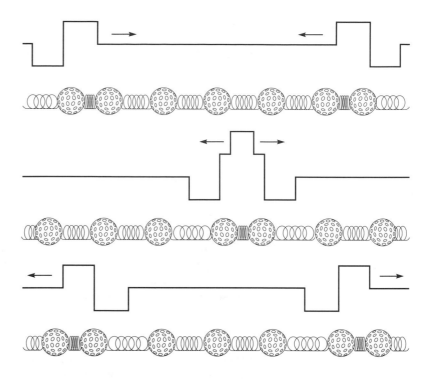

Consider the situation shown in Figure 1.15 where there is a boundary between air at two different temperatures. When a sound wave approaches this boundary at an angle, then the direction of propagation will alter according to Snell's law, that is, using Equation 1.5:

$$\frac{\sin \theta_1}{\sin \theta_2} = \frac{v_{T1}}{v_{T2}} = \frac{20.1\sqrt{T_1}}{20.1\sqrt{T_2}} = \sqrt{\frac{T_1}{T_2}} \tag{1.19}$$

where θ_1, θ_2 = the propagation angles in the two media

v_{T1}, v_{T2} = the velocities of the sound wave in the two media

and T_1, T_2 = the absolute temperatures of the two media

Thus the change in direction is a function of the square root of the ratio of the absolute temperatures of the air on either side of the boundary. As the speed of sound increases with temperature one would expect to observe that when sound moves from colder to hotter air that it would be refracted away from the normal direction and that it would refract towards the normal when moving from hotter to colder air. This effect has some interesting consequences for outdoor sound propagation. Normally the temperature of air reduces as a function of height and this results in the sound wave being bent upwards as it moves away from

Figure 1.15 Refraction of a sound wave.

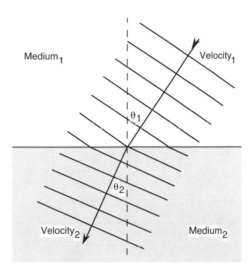

a sound source, as shown in Figure 1.16. This means that listeners on the ground will experience a reduction in sound level, as they move away from the sound, which reduces more quickly than the inverse square law would predict. This is a helpful effect for reducing the effect of environmental noise nuisance. However, if the temperature gradient increases with height then instead of being bent up the sound waves are bent down, as shown in Figure 1.17. This effect can often happen on summer evenings and results in a greater sound level at a given distance than predicted by the inverse square law. This behaviour is often responsible for the pop concert effect where people some distance away from the concert experience noise disturbance whereas people living nearer the concert do not experience the same level of noise.

Figure 1.16 Refraction of a sound wave due to a normal temperature gradient.

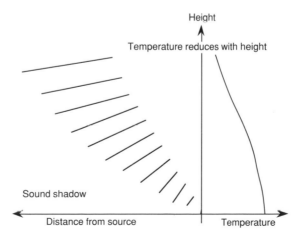

Figure 1.17 Refraction of a sound wave due to an inverted temperature gradient.

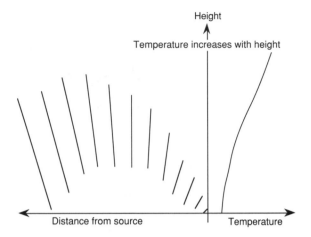

Refraction can also occur at the boundaries between liquids at different temperatures, such as water, and in some cases the level of refraction can result in total internal reflection. This effect is sometimes used by submarines to hide from the sonar of other ships; it can also cause the sound to be ducted between two boundaries and in these cases sound can cover large distances. It is thought that these mechanisms allow whales and dolphins to communicate over long distances in the ocean.

Wind can also cause refraction effects because the velocity of sound is affected by the velocity of the medium. The velocity of a sound wave in a moving medium is the sum of the two velocities, so that it is increased when the sound is moving with the wind and is reduced when it is moving against the wind. As the velocity of air is generally less at ground level compared with the velocity higher up (due to the effect of the friction of the ground), sound

Figure 1.18 Refraction of a sound wave due to a wind velocity gradient.

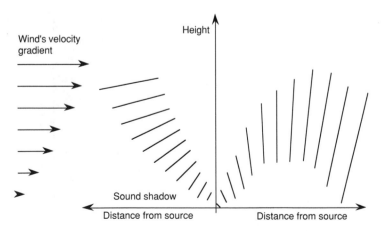

waves are bent upwards or downwards depending on their direction relative to the wind. The degree of direction change depends on the rate of change in wind velocity as a function of height; a faster rate of change results in a greater direction change. Figure 1.18 shows the effect of wind on sound propagation.

1.5.3 Sound absorption

Sound is absorbed when it interacts with any physical object. One reason is the fact that when a sound wave hits an object then that object will vibrate, unless it is infinitely rigid. This means that energy is transferred from the sound wave to the object that has been hit as vibration. Some of this energy will be absorbed because of the internal frictional losses in the material that the object is made of. Another form of energy loss occurs when the sound wave hits, or travels through, some porous material. In this case there is a very large surface area of interaction in the material, due to all the fibres and holes. There are frictional losses at the surface of any material due to the interaction of the velocity component of the sound wave with the surface. A larger surface area will have a higher loss which is why porous materials such as cloth or rockwool absorb sound waves strongly.

1.5.4 Sound reflection from hard boundaries

Sound is also reflected when it strikes objects and we have all experienced the effect as an echo when we are near a large hard object such as a cliff, or large building. There are two main situations in which reflection can occur.

In the first case the sound wave strikes an immovable object, or hard boundary, as shown in Figure 1.19. At the boundary between the object and the air the sound wave must have zero velocity, because it can't move the wall. This means that at that point all the energy in the sound is in the compression of the air, or pressure. As the energy stored in the pressure cannot transfer in the direction of the propagating wave, it bounces back in the reverse direction, which results in a change of phase in the velocity component of the wave. Figure 1.19 shows this effect using our golf ball and spring model. One interesting effect occurs due to the fact that the wave has to change direction and that is that the spring connected to the immovable boundary is compressed twice as much compared to a spring well away from the boundary. This occurs because the velocity components associated with the reflected (bounced back) wave are moving in contrary motion to the velocity components of the incoming wave, due to the

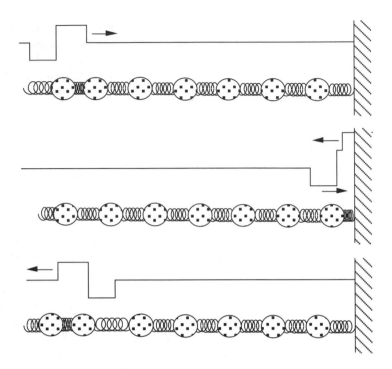

Figure 1.19 Reflection of a sound wave due to a rigid barrier.

change of phase in the reflected velocity components. In acoustic terms this means that while the velocity component at the reflecting boundary is zero, the pressure component is twice as large.

1.5.5 Sound reflection from bounded to unbounded boundaries

In the second case the wave moves from a bounded region, for example a tube, into an unbounded region, for example free space, as shown in Figure 1.20. At the boundary between the bounded and unbounded regions the molecules in the unbounded region find it a lot easier to move than in the bounded region. The result is that, at the boundary, the sound wave has a pressure component which is close to zero and a large velocity component. Therefore at this point all the energy in the sound is in the kinetic energy of the moving air molecules, in other words, the velocity component. Because there is less resistance to movement in the unbounded region the energy stored in the velocity component cannot transfer in the direction of the propagating wave, due to there being less 'springiness' to act on. Therefore the momentum of the molecules is transferred back to the 'springs' in the bounded region which pushed them in the first place, by stretching them

Figure 1.20 Reflection of a sound wave due to bounded–unbounded transition.

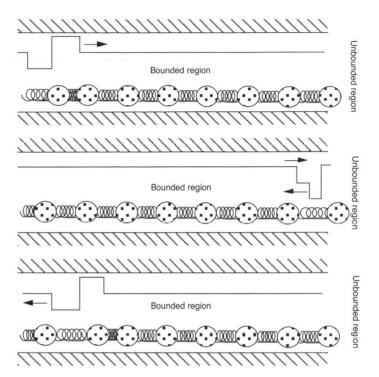

still further. This is equivalent to the reflection of a wave in the reverse direction in which the phase of the wave is reversed, because it has started as a stretching of the 'springs', or rarefaction, as opposed to a compression. Figure 1.20 shows this effect using the golf ball and spring model in which the unbounded region is modelled as having no springs at all. An interesting effect also occurs in this case, due to the fact that the wave has to change direction. That is, the mass that is connected to the unbounded boundary is moving with twice the velocity compared to masses well away from the boundary. This occurs because the pressure components associated with the reflected (bounced back) wave are moving in contrary motion to the pressure components of the incoming wave, due to the change of phase in the reflected pressure components. In acoustic terms this means that while the pressure component at the reflecting boundary is zero, the velocity component is twice as large.

To summarise, reflection from a solid boundary results in a reflected pressure component that is in phase with the incoming wave whereas reflection from a bounded to unbounded boundary results in a reflected pressure component which is in antiphase with the incoming wave. This arises due to the difference in

acoustic impedance at the boundary. In the first case the impedance of the boundary is greater than the propagating medium and
in the second case it is smaller. For angles of incidence on the
boundary, away from the normal, the usual laws of reflection
apply.

1.5.6 Sound interference

We saw earlier that when sound waves come from correlated
sources then their pressure and associated velocity components
simply add. This meant that the pressure amplitude could vary
between zero and the sum of the pressure amplitudes of the
waves that are being added together, as shown in Example 1.9.
Whether the waves add together constructively or destructively
depends on their relative phases and this will depend on the
distance each one has had to travel. Because waves vary in space
over their wavelength then the phase will also spatially vary.
This means that the constructive or destructive addition will also
vary in space. Consider the situation shown in Figure 1.21, which
shows two correlated sources feeding sound into a room. When
the listening point is equidistant from the two sources (P1), the
two sources add in constructively because they are in phase. If
one moves to another point (P2) which is not equidistant, the
waves no longer necessarily add constructively. In fact if the
relative delays between the two paths is equal to half a
wavelength then the two waves will add destructively and there
will be no net pressure amplitude at that point. This effect is
called interference, because correlated waves interfere with each
other; note that this effect does not occur for uncorrelated
sources. The relative phases of the waves depends on their

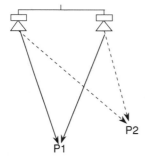

Figure 1.21 Interference
from correlated sources.

Figure 1.22 Effect of
position on interference at a
given frequency.

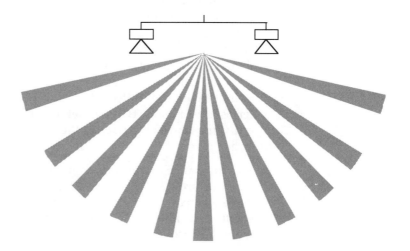

Figure 1.23 Effect of frequency, or wavelength, on interference at a given position.

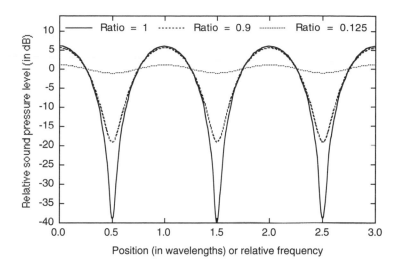

relative delays, and this depends on the relative distances to the listening point from the sources. Because of this the pattern of constructive and destructive interferences depends strongly on position, as shown in Figure 1.22. Less obviously the interference is also strongly dependent on frequency. This is because the factor that determines whether or not the waves add constructively or destructively is the relative distances from the listening point to the sources measured in wavelengths. As the wavelength is inversely proportional to frequency one would expect to see the pattern of interference vary directly with frequency, and this is indeed the case. Figure 1.23 shows the amplitude that results when two sources of equal amplitude but different relative distances are combined. The amplitude is plotted as a function of the relative distance measured in wavelengths (λ). Figure 1.23 shows that the waves constructively interfere when the relative delay is equal to a multiple of a wavelength, and that they interfere destructively at multiples of an odd number of half wavelengths. As the number of wavelengths for a fixed distance increases with frequency, this figure shows that the interference at a particular point varies with frequency. If the two waves are not of equal amplitude then the interference effect is reduced, as shown in Figure 1.23. In fact once the amplitude interfering wave is less than one eighth of the other wave then the peak variation in sound pressure level is less than 1 dB.

There are several acoustical situations which can cause interference effects. The obvious ones are when two loudspeakers radiate the same sound into a room or when the same sound is coupled into a room via two openings which are separated. A

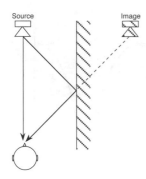

Figure 1.24 Interference arising from reflections from a boundary.

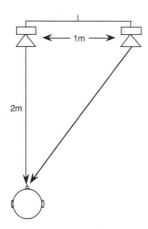

Figure 1.25 Interference at a point due to two loudspeakers.

less obvious situation is when there is a single sound source spaced away from a reflecting boundary, of either type. In this situation an image source is formed by the reflection and thus there are effectively two sources available to cause interference, as shown in Figure 1.24. This latter situation can often cause problems for recording or sound reinforcement due to a microphone picking up a direct and reflected sound component and so suffering interference.

Example 1.15 Two loudspeakers are one metre apart and radiate the same sound pressure level. A listener is two metres away from one speaker on a line which is perpendicular to the line joining the two loudspeakers, see Figure 1.25. What are the first two frequencies at which destructive interference occurs? When does the listener first experience constructive interference, other than at very low frequencies?

First work out the path length difference using Pythagoras' theorem:

$$\Delta_{\text{path length}} = \sqrt{(1 \text{ m}^2 + 2 \text{ m}^2)} - 2 \text{ m} = 0.24 \text{ m}$$

The frequencies at which destructive interference will occur will be at $\lambda/2$ and $3\lambda/2$. The frequencies at which this will happen will be when these wavelengths equal the path length difference. Thus the first frequency can be calculated using:

$$\frac{\lambda}{2} = \Delta_{\text{path length}} \qquad \text{i.e. } \lambda = 2\Delta_{\text{path length}}$$

$$f_{\lambda/2} = \frac{v}{\lambda} = \frac{v}{2\Delta_{\text{path length}}} = \frac{344 \text{ ms}^{-1}}{2 \times 0.24\text{m}} = 717 \text{ Hz}$$

The second frequency will occur at 3 times the first and so can be given by:

$$f_{3\lambda/2} = 3 \times f_{\lambda/2} = 3 \times 717 \text{ Hz} = 2150 \text{ Hz}$$

The frequency at which the first constructive interference happens will occur at 2 times the frequency of the first destructive interference which will be:

$$f_{\lambda} = 2 \times f_{\lambda/2} = 2 \times 717 \text{ Hz} = 1434 \text{ Hz}$$

If the listener moved closer to the centre line of the speakers then the relative delays would reduce and the frequencies at which destructive interference occurs would get higher. In the limit when the listener was equidistant the interference frequencies would be infinite, that is, there would be no destructive interference.

1.5.7 Standing waves at hard boundaries

The linear superposition of sound can also be used to explain a wave phenomenon known as standing waves, which is applicable to any form of sound wave. Standing waves occur when sound waves bounce between reflecting surfaces. The simplest system in which this can occur consists of two reflecting boundaries as shown in Figure 1.26. In this system the sound wave shuttles backwards and forwards between the two reflecting surfaces. At

Figure 1.26 Reflection of a sound wave between two parallel surfaces.

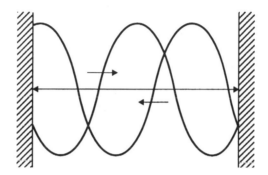

most frequencies the distance between the two boundaries will not be related to the wavelength and so the compression and rarefaction peaks and troughs will occupy all positions between the two boundaries, with equal probability, as shown in Figure 1.27. However, when the wavelength is related to the distance between the two boundaries then, as it travels between the two boundaries, the wave keeps tracing the same path. This means that the compressions and rarefactions always end up in the same position between the boundaries. Thus the sound wave will appear to be stationary between the reflecting boundaries, and so is called a

Figure 1.27 A non-stationary sound wave between two parallel surfaces.

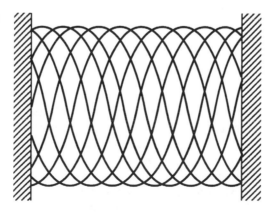

Figure 1.28 The pressure components of a standing wave between two hard boundaries.

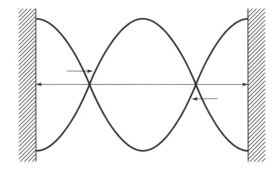

standing wave. It is important to realise that the wave is still moving at its normal speed, it is merely that, like a toy train, the wave endlessly retraces the same positions between the boundaries with respect to the wavelength, as shown in Figure 1.28. Figures 1.28 and 1.29 show the pressure and velocity components respectively of a standing wave between two hard reflecting boundaries. In this situation the pressure component is a maximum and the velocity component is a minimum at the two boundaries. The largest wave that can fit these constraints is a half wavelength and this sets the lowest frequency at which a standing wave can exist for a given distance between reflectors, and can be calculated using the following equation:

$$f_{lowest} \Rightarrow L = \frac{\lambda}{2} \Rightarrow \lambda\ 2L \Rightarrow f_{lowest} = \frac{v}{2L}$$

where f_{lowest} = the standing wave frequency (in Hz)
L = the distance between the boundaries (in m)
λ = the wavelength (in m)
and v = the velocity of sound (in ms^{-1})

Any multiple of half wavelengths will also fit between the two reflectors as well and so there is, in theory, an infinite number of frequencies at which standing waves occur which are all multiples of f_{lowest}. These can be calculated directly using:

$$f_n = \frac{nv}{2L} \tag{1.20}$$

where f_n = the nth standing wave frequency (in Hz)
and n = 1, 2, ..., ∞

An examination of Figures 1.28 and 12.9 also shows that there are points of maximum and minimum amplitude of the pressure and velocity components. For example, in Figure 1.28 the pressure component's amplitude is a maximum at the two boundaries and the velocity component is zero at the midpoint. The point at which the pressure amplitude is zero is called a

Figure 1.29 The velocity components of a standing wave between two hard boundaries.

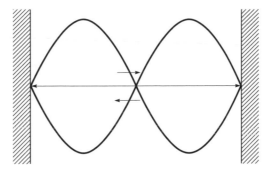

Figure 1.30 The pressure and velocity components of a standing wave between two hard boundaries.

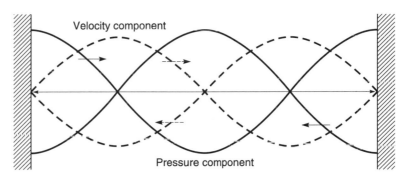

Velocity component

Pressure component

pressure node and the maximum points are called pressure antinodes. Note that as the number of half wavelengths in the standing waves increases then the number of nodes and antinodes increases, and for hard reflecting boundaries the number of pressure nodes is equal to, and the number of pressure antinodes is one more than, the number of half wavelengths. Velocity nodes and antinodes also exist, and they are always in the opposite sense to the pressure nodes, that is, a velocity antinode occurs at a pressure node and vice versa, as shown in Figure 1.30. This happens because the energy in the travelling wave must always exist at a pressure node carried in the velocity component and at a velocity node the energy is carried in the pressure component.

1.5.8 Standing waves at other boundaries

There are two other pairs of boundary arrangements which can support standing waves. The first, shown in Figure 1.32, is the case of a bounded to unbounded boundary at both ends. An example would be a tube or pipe which is open at both ends. In this situation the pressure component is zero at the boundaries whereas the velocity component is at a maximum, as shown in Figures 1.31 and 1.32. Like the hard reflecting boundaries the minimum frequency for a standing wave occurs when there is

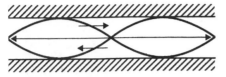

Figure 1.31 The pressure components of a standing wave between two bound–unbound boundaries.

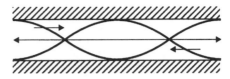

Figure 1.32 The velocity components of a standing wave between two bound–unbound boundaries.

precisely half a wavelength between the two boundaries, and at all subsequent multiples of this frequency. This means that Equation 1.20 can also be used to calculate the standing wave frequencies for this boundary arrangement as well.

Example 1.16 Calculate the first two modal (standing wave) frequencies for a pipe 98.5 cm long, and open at both ends.

As this is a system with identical boundaries we can use Equation 1.20 to calculate the two frequencies as:

$$f_{1 \text{ open tube}} = \frac{nv}{2L} = \frac{1 \times 344 \text{ ms}^{-1}}{2 \times 0.985 \text{ m}} = 174.6 \text{ Hz}$$

$$f_{2 \text{ open tube}} = \frac{nv}{2L} = \frac{2 \times 344 \text{ ms}^{-1}}{2 \times 0.985 \text{ m}} = 349.2 \text{ Hz}$$

These two frequencies correspond to the notes F3 and F4, which differ by an octave.

The second is more interesting and consists of one hard boundary and one bound–unbound boundary, and is shown in Figures 1.33 and 1.34. In this situation there is a pressure node at the bound–unbound boundary and a pressure antinode at the hard boundary. The effect of this is to allow a standing wave to exist when there is only a quarter of a wavelength between the two

Figure 1.33 The pressure components of a standing wave between mixed boundaries.

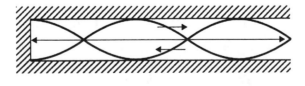

Figure 1.34 The velocity components of a standing wave between two mixed boundaries.

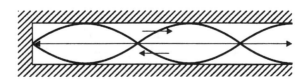

boundaries. However, a standing wave cannot exist at twice this frequency as this would require a pressure node or antinode at both ends as shown in Figures 1.28 and 1.31. Instead the next standing wave that can be supported occurs at a frequency which is three times the lowest frequency. The frequencies which can support standing waves continue in this vein at odd multiples of the base frequency. This can be expressed in the following equation as:

$$f_n = \frac{(2n + 1)v}{4L} \tag{1.21}$$

where f_n = the nth standing wave frequency (in Hz)
    and n = 0, 1, 2, ..., ∞

Standing waves can also occur for any type of wave propagation. A transverse wave on a string which is clamped at the ends has standing waves which can be predicted using Equation 1.20, provided one uses the propagation velocity of the transverse wave for v.

Standing waves in an acoustic context are often called the modes of a given system; the lowest frequency standing wave is known as the first order mode, and the multiples of this are higher order modes. So the third order mode of a system is the third lowest frequency standing wave pattern which can occur in it. Standing waves are also not just restricted to situations with two parallel reflecting boundaries. In fact any sequence of reflections or refractions which returns the wave back to the beginning of its phase will support a standing wave or mode. This can happen in one, two and three dimensions and with any form of wave propagation. The essential requirement is a cyclic path in which the time of propagation results in the wave travelling around this path in phase with the previous time round. Figure 1.35 shows an example of a two-dimensional standing wave.

Figure 1.35 A two-dimensional standing wave.

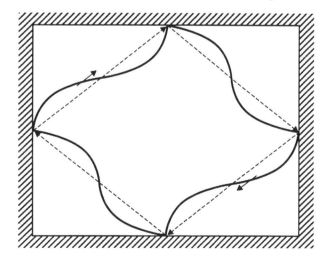

Example 1.17 Calculate the first two modal (standing wave) frequencies for the same pipe as Example 1.16 with one end closed.

As this is a system with non-identical boundaries we must use Equation 1.21 to calculate the two frequencies as:

$$f_{1 \text{ stopped tube}} = \frac{(2n + 1)v}{4L} = \frac{(2 \times 0 + 1) \times 344 \text{ ms}^{-1}}{4 \times 0.985 \text{ m}} = 87.3 \text{ Hz}$$

$$f_{2 \text{ stopped tube}} = \frac{(2n + 1)v}{4L} = \frac{(2 \times 1 + 1) \times 344 \text{ ms}^{-1}}{4 \times 0.985 \text{ m}} = 261.9 \text{ Hz}$$

In this case the first mode is at half the frequency of the pipe which is open at both ends, an octave below on the musical scale which is F2. The second mode is now at three times the lowest mode, which is approximately equal to C4 on the musical scale.

1.5.9 Sound diffraction

We have all experienced the ability of sound to travel around the corners of building or other objects. This is due to a process, known as diffraction, in which the sound bends round objects, as shown in Figure 1.36. Diffraction occurs because the variations in air pressure, due to the compressions and rarefactions in the sound wave, cannot go abruptly to zero after passing the edge of an object. This is because there is interaction between the adjacent molecules that are propagating the wave. In order to allow the compressions and rarefactions to die out gracefully in the boundary between the wave and the shadow there must be a region in which part of the propagating wave changes direction and it is this bent part of the wave that forms the diffracted component.

Figure 1.36 Diffraction around an object.

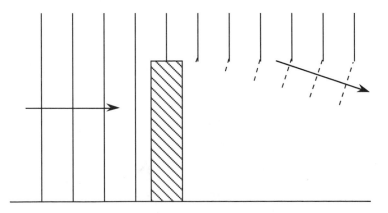

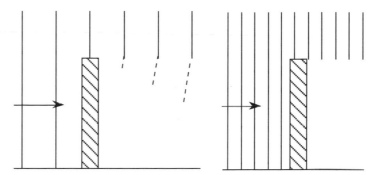

Figure 1.37 Diffraction around an edge at low frequencies.

Figure 1.38 Diffraction around an edge at high frequencies.

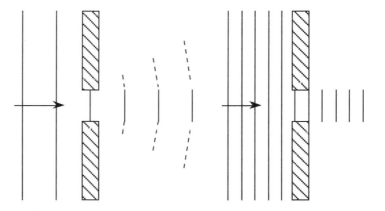

Figure 1.39 Diffraction through an opening at low frequencies.

Figure 1.40 Diffraction through an opening at high frequencies.

The degree of diffraction depends on wavelength because it effectively takes a certain number of wavelengths for the edge of the wave to make the transition to shadow. Thus the amount of diffraction around an edge, such as a building or wall, will be greater at low and less at high frequencies. This effect is shown in Figures 1.37 and 1.38. Similar effects occur when sound has to pass through an opening, as shown in Figures 1.39 and 1.40. Here the sound wave is diffracted away from the edges of the opening. The amount of diffraction depends on the size of the opening relative to the wavelength. When the size of the wavelength is large relative to the opening the wave is diffracted strongly and when the wavelength is small relative to the opening then the diffraction is low, see Figures 1.39 and 1.40. The transition between the wavelength being small and large with respect to the opening occurs when the opening size is about two thirds of the wavelength ($\frac{2}{3}\lambda$).

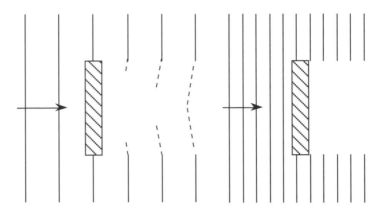

Figure 1.41 Diffraction around an object at low frequencies.

Figure 1.42 Diffraction around an object at high frequencies.

As well as occurring through openings, diffraction also happens around solid objects; one could consider them to be anti-openings, as shown in Figures 1.41 and 1.42. Here the wave diffracts around the object and meets behind it. The effect is also a function of the size of the object with respect to the wavelength. When the size of the wavelength is large relative to the object then the wave is diffracted strongly around it and the object has little influence on the wave propagation. On the other hand when the wavelength is small relative to the object then the diffraction is less, and the object has a significant shadow behind it; these effects are shown in Figure 1.42. The size at which an object becomes significant with respect to a wavelength is when its size is again about two thirds of a wavelength (⅔ λ).

1.5.10 Sound scattering

Sound which is incident on an object is not just diffracted around it—some of the incident energy will be reflected, or scattered from the side facing the incident wave, as shown in Figure 1.43. As in the case of diffraction, it is the size of the object with respect to the wavelength that determines how much, and in what way, the energy is scattered. When the object is large with respect to the wavelength then most of the sound incident on the object is scattered, according to the laws of reflection, and there is very little spreading or diffraction of the reflected wave, as shown in Figure 1.44. However, when the object is small with respect to the wavelength only a small proportion of the incident energy is scattered and there is a large amount of spreading or diffraction of the reflected wave, as shown in Figure 1.43. As in the case of diffraction, the size at which the object becomes significant is

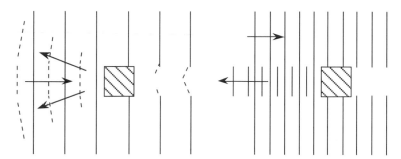

Figure 1.43 Scattering from an object at low frequencies.

Figure 1.44 Scattering from an object at high frequencies.

when it is about two thirds of a wavelength ($\frac{2}{3}\lambda$). Thus objects smaller than this will tend to scatter the energy in all directions whereas objects bigger than this will be more directional.

In addition to the scattering effects, interference effects happen when the scattering object is about two thirds of a wavelength ($\frac{2}{3}\lambda$) in size. This is because at this frequency both reflection from, and diffraction around, the object is occurring. In this situation the waves on the two sides of the object can influence or interact with each other. This results in enhanced reflection from or diffraction around the object at particular wavelengths. The precise nature of these effects depends on the interaction between the front and back of the scattering object and so will be significantly affected by its shape. Thus the variation in scattering will be different for a sphere compared to rectangular plate. As the sum of energy diffracted around the object and the scattered energy must be constant, the reflection and diffraction responses will be complementary.

1.6 Time and frequency domains

So far we have mainly considered a sound wave to be a sinusoidal wave at a particular frequency. This is useful as it allows us to consider aspects of sound propagation in terms of the wavelength. However most musical sounds have a waveform which is more complex than a simple sine wave and a selection is shown in Figure 1.45. How can we analyse real sound waveforms, and make sense of them in acoustical terms? The answer is based on the concept of superposition and a technique called Fourier analysis.

1.6.1 The spectrum of periodic sound waves

Fourier analysis states that any waveform can be built up by using an appropriate set of sine waves of different frequencies,

Figure 1.45 Waveforms from musical instruments.

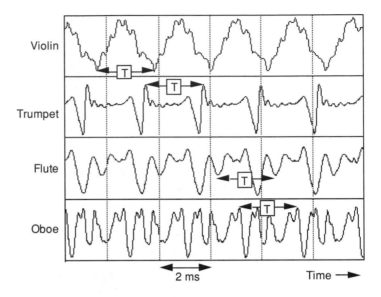

Figure 1.46 The effect of adding several harmonically related sine waves together.

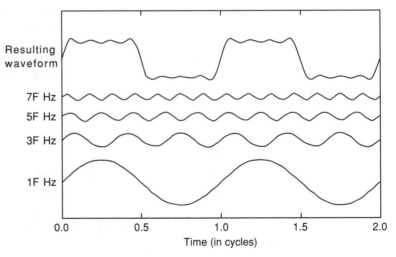

amplitudes and phases. To see how this might work consider the situation shown in Figure 1.46. This shows four sine waves whose frequencies are 1F Hz, 3F Hz, 5F Hz, and 7F Hz, whose phase is zero (that is, they all start from the same value) and whose amplitude is inversely proportional to the frequency. This means that the 3F Hz component is ⅓ the amplitude of the component at 1F Hz and so on. When these sine waves are added together, as shown in Figure 1.46, the result approximates a square wave, and, if more high frequency components were added, it would become progressively closer to an ideal square wave. The higher frequency components are needed in order to provide the fast rise, and sharp corners, of the square wave. In

Figure 1.47 The frequency domain representation, or spectrum, of a square wave.

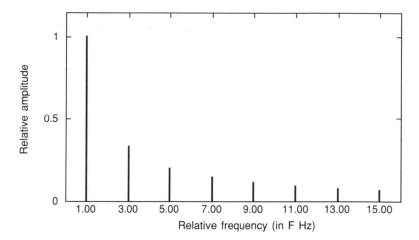

general, as the rise time gets faster, and/or the corners get sharper, then more high frequency sine waves are required to represent the waveform accurately. In other words we can look at a square wave as a waveform that is formed by summing together sine waves which are odd multiples of its fundamental frequency and whose amplitudes are inversely proportional to frequency. A sine wave represents a single frequency and there fore a sine wave of a given amplitude can be plotted as a single line on a graph of amplitude versus frequency. The components of a square wave plotted in this form are shown in Figure 1.47. Figure 1.47 clearly shows that the square wave consists of a set of progressively reducing discrete sine wave components at odd multiples of the lowest frequency. This representation is called the frequency domain representation, or spectrum, of a waveform and the waveform's amplitude versus time plot is called its time domain representation. The individual sine wave components of the waveform are often called the partials of the waveform. If they are integer related, as in the square wave, then they can be called harmonics. The lowest frequency is called the fundamental, or first harmonic, and the higher frequency harmonics are labelled according to their frequency multiple relative to the fundamental. Thus the second harmonic is twice the frequency of the fundamental and so on. Partials on the other hand need not be harmonically related to the fundamental and are numbered in their order of appearance with frequency. So for the square wave the second partial is the third harmonic and the third partial is the fifth harmonic. Other waveforms have different frequency domain representations, because they are made up of sine waves of different amplitudes and frequencies. Some examples of other waveforms in both the time and frequency domains are shown in Chapter 3.

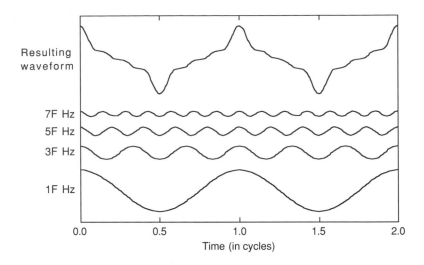

Figure 1.48 The effect of adding harmonically related sine waves together with different phase shifts.

1.6.2 The effect of phase

The phase, which expresses the starting value of the individual sine wave components, also affects the waveshape. Figure 1.48 shows what happens to the square wave if alternate partials are subtracted rather than added and this is equivalent to changing the phase of these components by 180°. That is, alternate frequency components start from half way around the circle compared with the other components. However, although the time domain waveform is radically different the frequency domain is very similar, as the amplitudes are identical, only the phase of some of the harmonics have changed. Interestingly, in many cases the resulting wave is perceived as sounding the same, even though the waveform is different. This is because the ear, as we will see later, appears to be less sensitive to the phase of the individual frequency compared to the relative amplitudes. Because of this, often only the amplitude of the frequency components are plotted in the spectrum and, in order to handle the range of possible amplitudes and because of the way we perceive sound, the amplitudes are usually plotted as decibels. For example, Figure 4.24 in Chapter 4 shows the waveform and spectrum plotted in this fashion for middle C played on a clarinet and tenor saxophone.

1.6.3 The spectrum of non-periodic sound waves

So far only the spectrum of waveforms which are periodic, that is, have pitch, has been considered. However some instruments, especially percussion, do not have pitch and hence are

Figure 1.49 The effect of adding several non-harmonically related sine waves together.

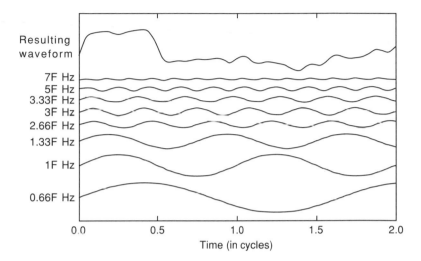

non-periodic, or aperiodic. How can we analyse these instruments in terms of a basic waveform, such as a sine wave, which is inherently periodic? The answer is shown in Figure 1.49. Here the square wave example discussed earlier has had four more sine waves added to it. However, these sine waves are between the harmonics of the square wave and so are unrelated to the period, but they do start off in phase with the harmonics. The effect of these other components is to start cancelling out the repeat periods of the square waves, because they are not related in frequency to them. By adding more components which sit in between the harmonics, this cancellation of the repeats becomes more effective so that when in the limit, the whole space between the harmonics is filled with sine wave components of the appropriate amplitude and phase. These extra components will add constructively only at the beginning of the waveform and will interfere with successive cycles due to their different frequencies. Therefore, in this case, only one square wave will exist. Thus the main difference between the spectrum of periodic and aperiodic waveforms is that periodic waveforms have discrete partials, which can be represented as lines in the spectrum, with a spacing which is inversely proportional to the period of the waveform. Aperiodic waveforms by contrast will have a spectrum which is continuous and therefore does not have any discrete components. However, the envelope of the component amplitudes as a function of frequency will be the same for both periodic and aperiodic waves of the same shape, as shown in Figure 1.50. Figure 3.6 in Chapter 3 shows the aperiodic waveform and spectrum of a brushed snare.

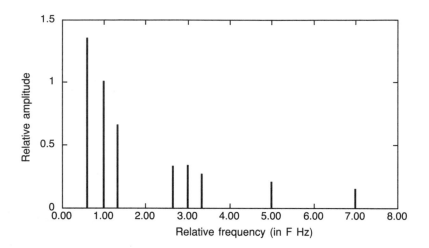

Figure 1.50 The frequency domain representation, or spectrum, of an aperiodic square wave.

1.7 Analysing spectra

Because the spectrum of a sound is an important part of the way we perceive it there is often a need to look at the spectrum of a real signal. The way this is achieved is to use a bank of filters, as shown in Figure 1.51.

Figure 1.51 A filter bank for analysing spectra.

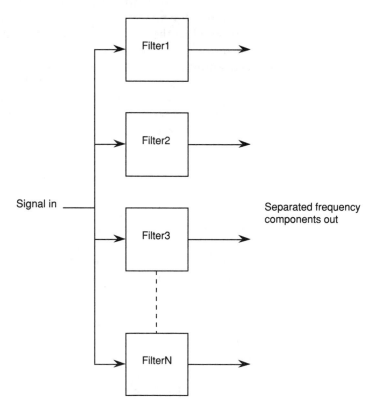

Figure 1.52 The effect of different filter types.

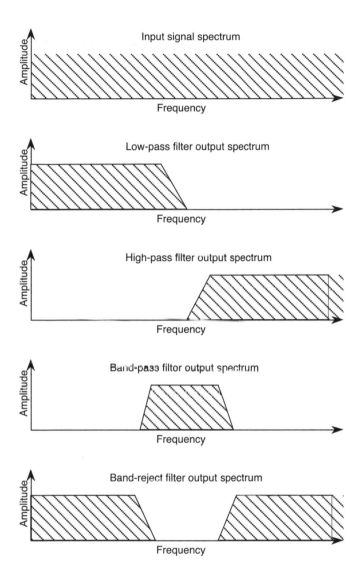

1.7.1 Filters and filter types

A filter is a device which separates out a portion of the frequency spectrum of a sound signal from the total, this is shown in Figure 1.52. There are four basic types of filters, which are classified in terms of their effect as a function of signal frequency. This is known as the filters frequency response. These basic types of filter are as follows:

- *Low-pass:* the filter only passes frequencies below a frequency known as the filter's cut-off frequency.

- *High-pass:* the filter only passes frequencies above the cut-off frequency.
- *Band-pass:* the filter passes a range of frequencies between two cut-off frequencies. The frequency range between the cut-off frequencies is known as the filter's bandwidth.
- *Band-reject:* the filter rejects a range of frequencies between two cut-off frequencies.

The effects of these four types of filter are shown in Figure 1.52 and one can see that a bank of band-pass filters are the most appropriate for analysing the spectrum of a sound wave. Note that although practical filters have specified cut-off frequencies which are determined by their design, they do not cut off instantly as the frequency changes. Instead they take a finite frequency range to attenuate the signal.

The effect of a filter on a spectrum is multiplicative in that the output spectrum after filtering is the product of the filter's frequency response with the input signal's spectrum. Thus we can easily determine the effect of a given filter on a signal's spectrum. This is a useful technique which can be applied to the analysis of musical instruments, as we shall see later, by treating some of their characteristics as a form of filtering. In fact, filtering can be carried out using mechanical, acoustical and electrical means, and many instruments perform some form of filtering on the sounds they generate (see Chapter 4).

1.7.2 Filter time responses

There is a problem, however, with filtering a signal in order to derive the spectrum and this is the effect of the filter on the time

Figure 1.53 The effect of low-pass filtering a square wave on the waveform.

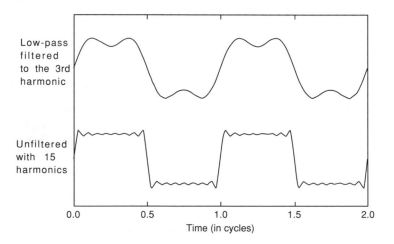

Low-pass filtered to the 3rd harmonic

Unfiltered with 15 harmonics

Time (in cycles)

Figure 1.54 The spectral effect of low-pass filtering a square wave.

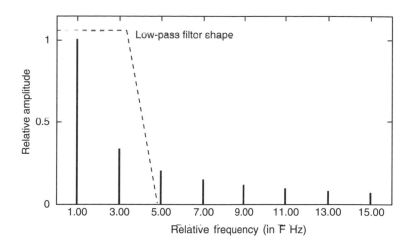

response of the signal. Filters have a time response due to the fact that they do not allow all frequencies to pass through. Why do filters have a time response? The answer can be obtained by reconsidering the Fourier analysis approach to analysing a sound signal. For example, if a signal is low-pass filtered such that the maximum frequency is F_{max}, then there can be no sine waves with a frequency greater than F_{max} in the output. As a sine wave has a slope which is a function of its frequency the maximum rate of change in the output will be determined by the frequency of F_{max}. Any faster rate of change would require higher frequency sine waves that are no longer present, as shown in Figure 1.53 which shows the effect of low-pass filtering a square wave such that only the first two partials are passed, as shown in Figure 1.54.

A similar argument can used for band-pass filters. In this case there is a maximum range of frequencies that can be passed around a given frequency. Although the sine waves corresponding to the band-pass frequencies will be passed, and they may be at quite a high frequency, their amplitude envelope cannot vary quickly, as shown in Figure 1.55. This is because the speed of variation of the envelope depends on the number of, and total frequency occupied by, the sine wave components in the output of the filter, as shown in Figure 1.56. As it is the amplitude variation of the output of a band-pass filter that carries the information, it too has an inherent time response which is a function of its bandwidth.

Thus all filters have a time response which is a function of the frequency range or bandwidth of the signals that they pass. Note that this is an inherent problem that cannot be solved by technology. The time response of a filter is inversely proportional to the

Figure 1.55 The effect of band-pass filtering on a square wave envelope.

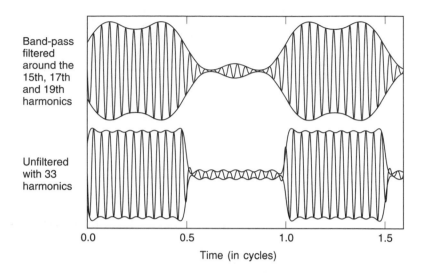

Band-pass filtered around the 15th, 17th and 19th harmonics

Unfiltered with 33 harmonics

Time (in cycles)

Figure 1.56 The spectral effect of band-pass filtering a square wave envelope.

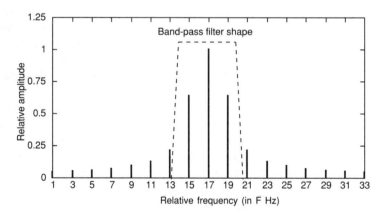

bandwidth so a narrow-band filter has a slow rise and fall time whereas a wide-band filter has faster rise and fall times. In practice this effect means that if the frequency resolution of the spectral analysis is good, implying the use of narrow-band analysis filters, then the time response is poor and there is significant smearing of the signal. On the other hand if the time response is good then the frequency resolution will be poor, because the filter bandwidths will have to be wider.

1.7.3 Time responses of acoustic systems

This argument can be reversed to show that when the output of an acoustic system such as a musical instrument changes slowly

Figure 1.57 The effect of an increasing number of harmonics on a square wave.

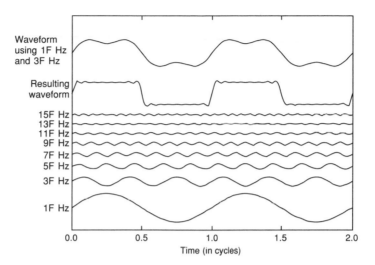

then its spectrum occupies a narrow bandwidth, whereas when it changes quickly then it must occupy a larger bandwidth. This can be either due to an increase in the number of harmonics present in the spectrum, as would be the case if the waveform becomes sharper or more spiky. Figure 1.57 shows the effect of an increasing number of harmonics on a square wave. Or, as in the band-pass case described earlier, it would be due to an increase in the bandwidths occupied by the individual partials, or harmonics, of the sound. This would be the case if the envelope of the sound changed more rapidly. Figures 1.58 and 1.59 show this effect, which in this case compares two similar systems, one with a slow and one with a fast rate of amplitude

Figure 1.58 The decay rate of two systems with different bandwidths.

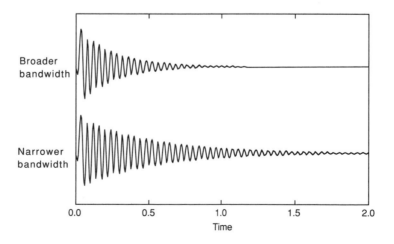

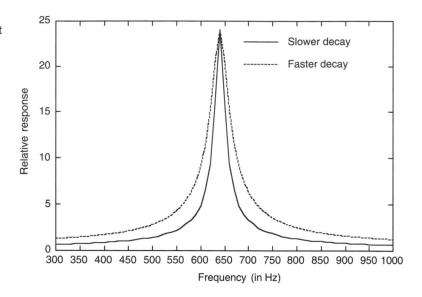

Figure 1.59 The response of two systems with different decay times.

decay, the latter being due to a higher loss of energy from the system. The figure clearly shows that in the system which decays more rapidly there are more harmonics and there is a higher bandwidth due to the more rapid change in the sound's envelope.

1.7.4 Time and frequency representations of sounds

Figure 1.60 is a useful way of showing both the time and frequency characteristics of a sound signal at the same time, called a spectrogram. In this representation the spectrum of a signal is plotted as a function of time, with frequency and time forming the main axes. The decibel amplitude of the signal is plotted as a grey scale with black representing a high amplitude and white representing a low amplitude. This representation is an excellent way of seeing the time evolution of the spectral components of a musical sound. However it does suffer from the time smearing effects discussed earlier. Figure 1.60 shows both narrow-band and broad-band analysis of the beginning of a harpsichord note. In the narrow-band version, although the harmonics are clearly resolved, the start of the signal is smeared due to the slow time response of the narrow-band filters. On the other hand the broad-band version has excellent time response and even shows the periodic variation in amplitude due to the pitch of the sound, shown as vertical striations in the spectrogram; however it is unable to resolve the harmonics of the signal.

Figure 1.60 Acoustic pressure waveform (a) narrow band (40 Hz analysis filter) and (b) wide band (300 Hz analysis filter) spectrograms for middle C played on a harpsichord.

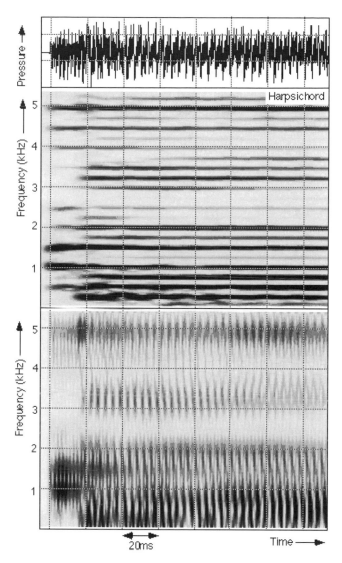

Ideally one would like a filter which has a fast time response and a narrow bandwidth, but this is impossible. However, as we shall see later, the human hearing system is designed to provide a compromise which gives both good frequency and time resolution.

In this chapter we have examined many aspects of sound waveforms and their characteristics. However, sound by itself is useless unless it is heard and therefore we must consider the way in which we hear sound in order to fully understand the nature of musical instruments and musical signal processing. This is the subject of the next chapter.

References

Beranek, L.L. (1986). *Acoustics*. New York: Acoustical Society of America.

Everest, F.A. (1994). *The Master Handbook of Acoustics*. New York: Tab Books.

Kuttruff, H. (2000). *Room Acoustics*. London: E&FN Spon.

2 Introduction to hearing

Psychoacoustics is the study of how humans perceive sound. To begin our exploration of psychoacoustics it is first necessary to become familiar with the basic anatomy of the human hearing system to facilitate understanding of:

- the effect the normal hearing system has on sounds entering the ear,
- the origin of fundamental psychoacoustic findings relevant to the perception of music,
- how listening to very loud sounds can cause hearing damage, and
- some of the listening problems faced by the hearing impaired.

This chapter introduces the main anatomical structures of the human hearing system which form the path along which incoming music signals travel up to the point where the signal is carried by nerve fibres from the ear(s) to the brain. It also introduces the concept of 'critical bands' which is the single most important psychoacoustic principle for an understanding of the perception of music and other sounds in terms of pitch, loudness and timbre.

It should be noted that many of the psychoacoustic effects have been observed experimentally, mainly as a result of playing sounds that are carefully controlled in terms of their acoustic nature to panels of listeners from whom responses are monitored. These responses are often the result of comparing two sounds and indicating, for example, which is louder or higher in pitch or 'brighter'. Many of the results from such experiments cannot as yet be described in terms of either where anatomically, or by what physical means they occur. Psychoacoustics is a developing field of research. However, the results from such experiments give a firm foundation for understanding the nature of human perception of musical sounds, and knowledge of minimum changes that are perceived provide useful guideline bounds for those exploring the subtleties of sound synthesis.

2.1 The anatomy of the hearing system

The anatomy of the human hearing system is illustrated in Figure 2.1. It consists of three sections:

- the outer ear,
- the middle ear, and
- the inner ear.

The anatomical structure of each of these is discussed below along with the effect that each has on the incoming acoustic signal.

2.1.1 Outer ear function

The outer ear (see Figure 2.1) consists of the external flap of tissue known as the pinna with its many grooves, ridges and depressions. The depression at the entrance to the auditory canal is known as the concha. The auditory canal is approximately 25 mm to 35 mm long from the concha to the 'tympanic membrane', more commonly known as the eardrum . The outer ear has an acoustic effect on sounds entering the ear in that it helps us both to locate sound sources and it enhances some frequencies with respect to others.

Sound localisation is helped mainly by the acoustic effect of the pinna and the concha. The concha acts as an acoustic resonant cavity. The combined acoustic effects of the pinna and concha are particularly useful for determining whether a sound source is in front or behind and to a lesser extent as to whether it is above or below.

The acoustic effect of the outer ear as a whole serves to modify the frequency response of incoming sounds due to resonance effects, primarily of the auditory canal whose main resonance frequency is in the region around 4 kHz.

Figure 2.1 The main structures of the human ear showing the outer, middle and inner ears.

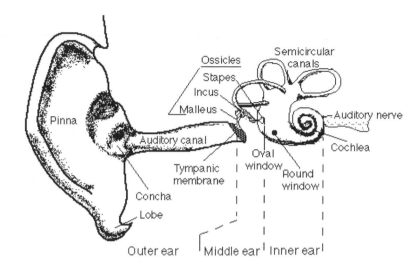

The tympanic membrane is a light, thin highly elastic structure which forms the boundary between the outer and middle ears. It consists of three layers: the outside layer which is a continuation of the skin lining of the auditory canal, the inside layer which is continuous with the mucous lining of the middle ear, and the layer in between these which is a fibrous structure which gives the tympanic membrane its strength and elasticity. The tympanic membrane converts acoustic pressure variations from the outside world into mechanical vibrations to the middle ear.

2.1.2 Middle ear function

The mechanical movements of the tympanic membrane are transmitted through three small bones known as ossicles, comprising the malleus, incus and stapes—more commonly known as the hammer, anvil and stirrup—to the oval window of the cochlea (see Figure 2.1). The oval window forms the boundary between the middle and inner ears.

The malleus is fixed to the middle fibrous layer of the tympanic membrane in such a way that when the membrane is at rest, it is pulled inwards. Thus the tympanic membrane when viewed down the auditory canal from outside appears concave and conical in shape. One end of the stapes, the stapes footplate, is attached to the oval window of the cochlea. The malleus and incus are joined quite firmly such that at normal intensity levels they act as a single unit, rotating together as the tympanic membrane vibrates to move the stapes via a ball and socket joint in a piston-like manner. Thus acoustic vibrations are transmitted via the tympanic membrane and ossicles as mechanical movements to the cochlea of the inner ear.

The function of the middle ear is two fold: (1) to transmit the movements of the tympanic membrane to the fluid which fills the cochlea without significant loss of energy, and (2) to protect the hearing system to some extent from the effects of loud sounds, whether from external sources or the individual concerned.

In order to achieve efficient transfer of energy from the tympanic membrane to the oval window, the effective pressure acting on the oval window is arranged by mechanical means to be greater than that acting on the tympanic membrane. This is to overcome the higher resistance to movement of the cochlear fluid compared to that of air at the input to the ear. Resistance to movement can be thought of as 'impedance' to movement and the impedance of fluid to movement is high compared to that of air. The ossicles act as a mechanical 'impedance converter' or 'impedance transformer' and this is achieved essentially by two means:

- the lever effect of the malleus and incus, and
- the area difference between the tympanic membrane and the stapes footplate.

The lever effect of the malleus and incus arises as a direct result of the difference in their lengths. Figure 2.2 shows this effect. The force at the stapes footplate relates to the force at the tympanic membrane by the ratio of the lengths of the malleus and incus as follows:

$$F1 \times L1 = F2 \times L2$$

where F1 = force at tympanic membrane
F2 = force at stapes footplate
L1 = length of malleus
L2 = length of incus

Therefore:

$$F2 = F1 \times \frac{L1}{L2} \tag{2.1}$$

The area difference has a direct effect on the pressure applied at the stapes footplate compared with the incoming pressure at the tympanic membrane since pressure is expressed as force per unit area as follows:

$$\text{Pressure} = \frac{\text{Force}}{\text{Area}} \tag{2.2}$$

Figure 2.2 The function of the ossicles of the middle ear.

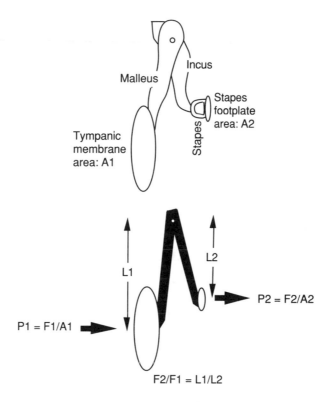

The areas of the tympanic membrane and the stapes footplate in humans are represented in Figure 2.2 as A1 and A2 respectively. The pressure at the tympanic membrane (P1) and the pressure at the stapes footplate (P2) can therefore be expressed as follows:

$$P1 = \frac{F1}{A1}$$

$$P2 = \frac{F2}{A2}$$

The forces can be therefore be expressed in terms of pressures:

$$F1 = (P1 \times A1) \tag{2.3}$$

$$F2 = (P2 \times A2) \tag{2.4}$$

Substituting Equations 2.3 and 2.4 into Equation 2.1 gives:

$$(P2 \times A2) = (P1 \times A1) \times \frac{L1}{L2}$$

Therefore:

$$\frac{P2}{P1} = \frac{A1 \times L1}{A2 \times L2} \tag{2.5}$$

Pickles (1982) describes a third aspect of the middle ear which appears relevant to the impedance conversion process. This relates to a buckling motion of the tympanic membrane itself as it moves, resulting in a twofold increase in the force applied to the malleus.

In humans, the area of the tympanic membrane (A1) is approximately 13 times larger than the area of the stapes footplate (A2), and the malleus is approximately 1.3 times the length of the incus. The buckling effect of the tympanic membrane provides a force increase by a factor of 2. Thus the pressure at the stapes footplate (P2) is about (13 × 1.3 × 2 = 33.8) times larger than the pressure at the tympanic membrane (P1).

Example 2.1 Express the pressure ratio between the stapes footplate and the tympanic membrane in decibels.

The pressure ratio is 33.8:1. Equation (1.12) is used to convert from pressure ratio to decibels:

$$\mathrm{dB(SPL)} = 20 \, \log_{10} \frac{P2}{P1}$$

Substituting 33.8 as the pressure ratio gives:

$$20 \, \log_{10} [33.8] = 30.6 \, \mathrm{dB}$$

The second function of the middle ear is to provide some protection for the hearing system from the effects of loud sounds, whether from external sources or the individual concerned. This occurs as a result of the action of two muscles in the middle ear: the tensor tympani and the stapedius muscle. These muscles contract automatically in response to sounds with levels greater than approximately 75 dB(SPL) and they have the effect of increasing the impedance of the middle ear by stiffening the ossicular chain. This reduces the efficiency with which vibrations are transmitted from the tympanic membrane to the inner ear and thus protects the inner ear to some extent from loud sounds. Approximately 12 to 14 dB of attenuation is provided by this protection mechanism, but this is for frequencies below 1 kHz only. The names of these muscles derive from where they connect with the ossicular chain: the tensor tympani is attached near the tympanic membrane and the stapedius muscles is attached to the stapes.

This effect is known as the 'acoustic reflex'. It takes some 60 ms to 120 ms for the muscles to contract in response to a loud sound. In the case of a loud impulsive sound such as the firing of a large gun, it has been suggested that the acoustic reflex is too slow to protect the hearing system. In gunnery situations, a sound loud enough to trigger the acoustic reflex, but not so loud as to damage the hearing systems, is often played at least 120 ms before the gun is fired.

2.1.3 Inner ear function

The inner ear consists of the snail-like structure known as the cochlea. The function of the cochlea is to convert mechanical vibrations into nerve firings to be processed eventually by the brain. Mechanical vibrations reach the cochlea at the oval window via the stapes footplate of the middle ear.

The cochlea consists of a tube coiled into a spiral with approximately 2.75 turns (see Figure 2.3(a)). The end with the oval and round windows is the 'base' and the other end is the 'apex'. Figure 2.3(c) illustrates the effect of slicing through the spiral vertically, and it can be seen in part (d) that the tube is divided into three sections by Reissner's membrane and the basilar membrane. The outer channels, the scala vestibuli (V) and scala tympani (T), are filled with an incompressible fluid known as perilymph, and the inner channel is the scala media (M). The scala vestibuli terminates at the oval window and the scala tympani at the round window. An idealised unrolled cochlea is shown in part (b). There is a small hole at the apex known as the helicotrema through which the perilymph fluid can flow.

Input acoustic vibrations result in a piston-like movement of the stapes footplate at the oval window which moves the perilymph fluid within the cochlea. The membrane covering the round window moves to compensate for oval window movements since the perilymph fluid is essentially incompressible. Inward movements of the stapes footplate at the oval window cause the round window to move outwards and outward movement of the stapes footplate cause the round window to move inwards. These movements cause travelling waves to be set up in the scala vestibuli which displace both Reissner's membrane and the basilar membrane.

The basilar membrane is responsible for carrying out a frequency analysis of input sounds. In shape, the basilar membrane is both narrow and thin at the base end of the cochlea, becoming both wider and thicker along its length to the apex, as illustrated in Figure 2.4. Small structures respond best to higher frequencies than large structures (compare, for

71

Figure 2.3 (a) The spiral nature of the cochlea. (b) The cochlea 'unrolled'. (c) Vertical cross-section through the cochlea. (d) Detailed view of the cochlear tube.

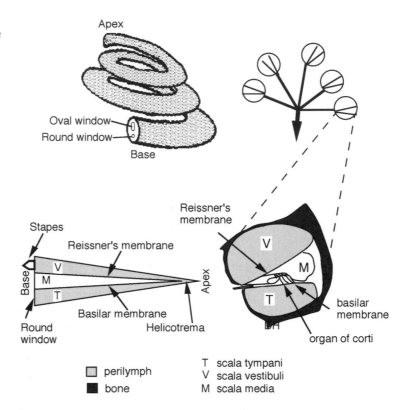

example, the sizes of a violin and a double bass or the strings at the treble and bass ends of a piano). The basilar membrane therefore, responds best to high frequencies where it is narrow and thin (at the base) and to low frequencies where it is wide and thick (at the apex). Since its thickness and width change gradually along its length, input pure tones at different frequency will produce a maximum basilar membrane movement at different positions or 'places' along its length. This is illustrated in Figure 2.5 for a section of the length of the membrane. This is the basis of the 'place' analysis of sound by the hearing system. The extent, or 'envelope', of basilar membrane movement is plotted

Figure 2.4 Idealised shape of unrolled basilar membrane.

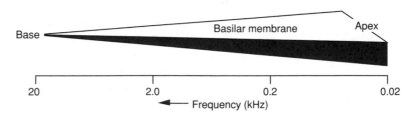

Figure 2.5 Idealised envelope of basilar membrane movement to sounds at five different frequencies.

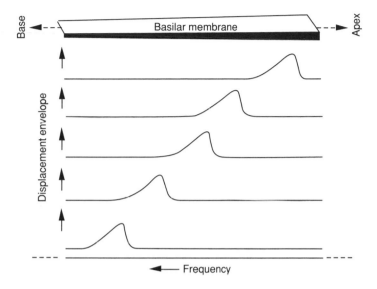

against frequency in an idealised manner for five input pure tones of different frequencies. If the input sound were a complex tone consisting of many components, the overall basilar membrane response is effectively the sum of the responses for each individual component. The basilar membrane is stimulated from the base end (see Figure 2.3) which responds best to high frequencies, and it is important to note that its envelope of movement for a pure tone (or individual component of a complex sound) is not symmetrical, but that it tails off less rapidly towards high frequencies than towards low frequencies. This point will be taken up again in Chapter 5.

The movement of the basilar membrane for input sine waves at different frequencies has been observed by a number of researchers following the pioneering work of von Békésy (1960). They have confirmed that the point of maximum displacement along the basilar membrane changes as the frequency of the input is altered. It has also been shown that the linear distance measured from the apex to the point of maximum basilar membrane displacement is directly proportional to the logarithm of the input frequency. The frequency axis in Figure 2.5 is therefore logarithmic. It is illustrated in the figure as being 'back-to-front' (i.e. with increasing frequency changing from right to left, low frequency at the apex and high at the base) to maintain the left to right sense of flow of the input acoustic signal and to reinforce understanding of the anatomical nature of the inner ear. The section of the inner ear which is responsible for the analysis of low frequency sounds is the end farthest away from the oval window, coiled into the centre of the cochlear spiral.

In order that the movements of the basilar membrane can be transmitted to the brain for further processing, they have to be converted into nerve firings. This is the function of the organ of corti which consists of a number of hair cells that trigger nerve firings when they are bent. These hair cells are distributed along the basilar membrane and they are bent when it is displaced by input sounds. The nerves from the hair cells form a spiral bundle known as the auditory nerve. The auditory nerve leaves the cochlea as indicated in Figure 2.1.

2.2 Critical bands

Section 2.1 describes how the inner ear carries out a frequency analysis of sound due to the mechanical properties of the basilar membrane and how this provides the basis behind the 'place' theory of hearing. The next important aspect of the place theory to consider is how well the hearing system can discriminate between individual frequency components of an input sound. This will provide the basis for understanding the resolution of the hearing system and it will underpin discussions relating to the psychoacoustics of how we hear music, speech and other sounds.

Each component of an input sound will give rise to a displacement of the basilar membrane at a particular place, as illustrated in Figure 2.5. The displacement due to each individual component is spread to some extent on either side of the peak. Whether or not two components that are of similar amplitude and close together in frequency can be discriminated depends on the extent to which the basilar membrane displacements due to each of the two components are clearly separated or not.

Suppose two pure tones, or sine waves, with amplitudes A_1 and A_2 and frequencies F_1 and F_2 respectively are sounded together. If F_1 is fixed and F_2 is changed slowly from being equal to or in unison with F_1 either upwards (downwards) in frequency, the following is generally heard (see Figure 2.6). When F_1 is equal to F_2, a single note is heard. As soon as F_2 is moved higher (lower) than F_1 a sound with a clearly undulating amplitude variations known as beats is heard. The frequency of the beats is equal to $(F_2 - F_1)$, or $(F_1 - F_2)$ if F_1 is greater than F_2, and the amplitude varies between $(A_1 + A_2)$ and $(A_1 - A_2)$, or $(A_1 + A_2)$ and $(A_2 - A_1)$ if A_2 is greater than A_1. Note that when the amplitudes are equal $(A_1 = A_2)$ the amplitude of the beats varies between $(2 \times A_1)$ and 0.

Figure 2.6 An illustration of the perceptual changes which occur when a pure tone fixed at frequency F_1 is heard combined with a pure tone of variable frequency F_2.

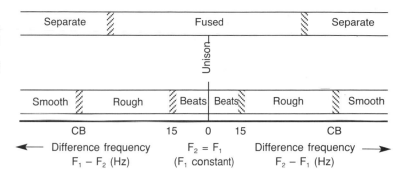

For the majority of listeners beats are usually heard when the frequency difference between the tones is less than about 12.5 Hz, and the sensation of beats generally gives way to one of a 'fused' tone which sounds 'rough' when the frequency difference is increased above 15 Hz. As the frequency difference is increased further there is a point where the fused tone gives way to two separate tones but still with the sensation of roughness, and a further increase in frequency difference is needed for the rough sensation to become smooth. The smooth separate sensation persists while the two tones remain within the frequency range of the listener's hearing.

The changes from fused to separate and from beats to rough to smooth are shown hashed in Figure 2.6 to indicate that there is no exact frequency difference at which these changes in perception occur for every listener. However, the approximate frequencies and order in which they occur is common to all listeners, and in common with most psychoacoustic effects, average values are quoted which are based on measurements made for a large number of listeners.

The point where the two tones are heard as being separate as opposed to fused when the frequency difference is increased can be thought of as the point where two peak displacements on the basilar membrane begin to emerge from a single maximum displacement on the membrane. However, at this point the underlying motion of the membrane which gives rise to the two peaks causes them to interfere with each other giving the rough sensation, and it is only when the rough sensation becomes smooth that the separation of the places on the membrane is sufficient to fully resolve the two tones. The frequency difference between the pure tones at the point where a listener's perception changes from rough and separate to smooth and separate is known as the 'critical bandwidth', and it is therefore marked CB in the figure. A more formal definition is given by Scharf (1970):

'The critical bandwidth is that bandwidth at which subjective responses rather abruptly change.'

In order to make use of the notion of critical bandwidth practically, an equation relating the effective critical bandwidth to the filter centre frequency has been proposed by Moore and Glasberg (1983). They define a filter with an ideal rectangular frequency response curve which passes the same power as the filter in question, and provide an equation for the 'equivalent rectangular bandwidth', or ERB, of the critical bandwidth as follows:

$$ERB = \{[6.23 \times 10^{-6} \times f_c^2] + [93.39 \times 10^{-3} \times f_c] + 28.52\} \text{ Hz} \quad (2.6)$$

where f_c = filter centre frequency in Hz
ERB = equivalent rectangular bandwidth in Hz
Equation valid for (100 Hz < f_c < 10 000 Hz)

This relationship is plotted in Figure 2.7 and lines representing where the bandwidth is equivalent to 1, 2, 4 and 7 semitones— or a semitone, whole tone, major third and perfect fifth respectively—are also plotted for comparison purposes. A third octave filter is often used in the studio as an approximation to the critical bandwidth, this is shown in the figure as the 4 semitone line since there are 12 semitones per octave. A keyboard is shown on the filter centre frequency axis to show the equivalent fundamental frequency values of notes (middle C is marked with a spot).

Figure 2.7 The variation of equivalent rectangular bandwidth (ERB) with filter centre frequency, and lines indicating where the bandwidth would be equivalent to 1, 2, 4 and 7 semitones. (Middle C marked with spot on keyboard.)

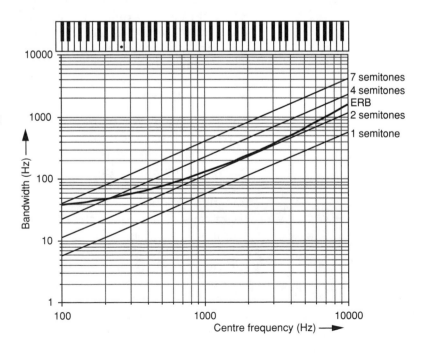

Example 2.2 Calculate the critical bandwidth at 200 Hz and 2000 Hz to three significant figures.

Using Equation (2.8) and substituting 200 Hz and 2000 Hz for f_c gives the critical bandwidth (ERB) as:

$$[6.23 \times 10^{-6} \times 200^2] + [93.39 \times 10^{-3} \times 200] + 28.52 = 47.5 \text{ Hz}$$

$$[6.23 \times 10^{-6} \times 2000^2] + [93.39 \times 10^{-3} \times 2000] + 28.52 = 240 \text{ Hz}$$

The change in critical bandwidth with frequency can be demonstrated if the fixed frequency F_1 in Figure 2.6 is altered to a new value and the new position of CB is found. In practice, critical bandwidth is usually measured by an effect known as 'masking' (see Chapter 5) in which the 'rather abrupt change' is more clearly perceived by listeners.

The response characteristic of an individual filter is illustrated in the bottom curve in Figure 2.8, the vertical axis of which is marked 'filter response' (notice that increasing frequency is plotted from right to left in this figure in keeping with Figure 2.5 relating to basilar membrane displacement). The other curves

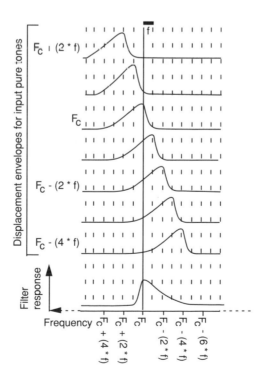

Figure 2.8 Derivation of response of an auditory filter with centre frequency F_c Hz based on idealised envelope of basilar membrane movement to pure tones with frequencies local to the centre frequency of the filter.

Figure 2.9 Idealised response of an auditory filter with centre frequency F_c Hz with increasing frequency plotted in the conventional direction.

in the figure are idealised envelopes of basilar membrane displacement for pure tone inputs spaced by f Hz, where f is the distance between each vertical line as marked. The filter centre frequency F_c Hz is indicated with an unbroken vertical line, which also represents the place on the basilar membrane corresponding to a frequency F_c Hz. The filter response curve is plotted by observing the basilar membrane displacement at the place corresponding to F_c Hz for each input pure tone and plotting this as the filter response at the frequency of the pure tone. This results in the response curve shape illustrated as follows. As the input pure tone is raised to F_c Hz, the membrane displacement gradually increases with the less steep side of the displacement curve. As the frequency is increased above F_c Hz, the membrane displacement falls rapidly with the steeper side of the displacement curve. This results in the filter response curve as shown, which is an exact mirror image about F_c Hz of the basilar membrane displacement curve. Figure 2.9 shows the filter response curve plotted with increasing frequency and plotted more conventionally from left to right in order to facilitate discussion of the psychoacoustic relevance of its asymmetric shape in Chapter 5.

The action of the basilar membrane can be thought of as being equivalent to a large number of overlapping band-pass filters, or a 'bank' of band-pass filters, each responding to a particular band of frequencies (see Chapter 1). Based on the idealised filter response curve shape in Figure 2.9, an illustration of the nature of this bank of filters is given in Figure 2.10. Each filter has an asymmetric shape to its response with a steeper roll-off on the high-frequency side than on the low-frequency side, and the bandwidth of a particular filter is given by the critical bandwidth (see Figure 2.7) for any particular centre frequency. It is not possible to be particularly exact with regard to the extent to

Figure 2.10 Idealised bank of band-pass filters model of the frequency analysis capability of the basilar membrane.

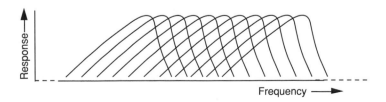

which the filters overlap. A common practical compromise, for example in studio third octave graphic equaliser filter banks, is to overlap adjacent filters at the −3 dB points on their response curves.

In terms of the perception of two pure tones illustrated in Figure 2.6, the 'critical bandwidth' can be thought of as the bandwidth of the band-pass filter in the bank of filters, the centre frequencies of which are exactly half way between the frequencies of the two tones. This ignores the asymmetry of the basilar membrane response (see Figure 2.5) and the consequent asymmetry in the individual filter response curve (see Figure 2.9), but it provides a good working approximation for calculations. Such a filter (and others close to it in centre frequency) would capture both tones while they are perceived as 'beats', 'rough fused' or 'rough separate', and at the point where rough changes to smooth, the two tones are too far apart to be both captured by this *or any other* filter. At this point there is no single filter which captures both tones, but there are filters which capture each of the tones individually and they are therefore resolved and the two tones are perceived as being 'separate and smooth'.

A musical sound can be described by the frequency components which make it up, and an understanding of the application of the critical band mechanism in human hearing in terms of the analysis of the components of musical sounds gives the basis for the study of psychoacoustics. The resolution with which the hearing system can analyse the individual components or sine waves in a sound is important for understanding psychoacoustic discussions relating to, for example, how we perceive:

- melody
- harmony
- chords
- tuning
- intonation
- musical dynamics
- the sounds of different instruments
- blend
- ensemble
- interactions between sounds produced simultaneously by different instruments.

2.3 Frequency and pressure sensitivity ranges

The human hearing system is usually quoted as having an average frequency range of 20 Hz to 20 000 Hz, but there can

Figure 2.11 The general shape of the average human threshold of hearing and the threshold of pain with approximate conversational speech and musically useful ranges.

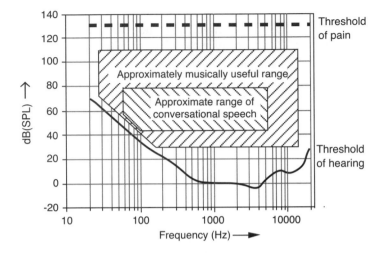

however, be quite marked differences between individuals. This frequency range changes as part of the human ageing process, particularly in terms of the upper limit which tends to reduce. Healthy young children may have a full hearing frequency range up to 20 000 Hz, but by the age of 20, the upper limit may have dropped to 16 000 Hz. From the age of 20, it continues to reduce gradually to approximately 8 000 Hz by retirement age. This is known as 'presbyacusis' or 'presbycusis' and is a function of the normal ageing process. This reduction in the upper frequency limit of the hearing range is accompanied by a decline in hearing sensitivity at all frequencies with age, the decline being less for low frequencies than for high. This natural loss of hearing sensitivity and loss of upper frequencies is more marked for men than for women. Hearing losses can also be induced by other factors such as prolonged exposure to loud sounds (see Section 2.5), particularly with some of the high sound levels now readily available from electronic amplification systems, whether reproduced via loudspeakers or more especially via headphones.

The ear's sensitivity to sounds of different frequencies varies over a vast sound pressure level range. On average, the minimum sound pressure variation which can be detected by the human hearing system around 4 kHz is approximately 10 micropascals (10 µPa), or 10^{-5} Pa (see Figure 2.11). The maximum average sound pressure level which is heard rather than perceived as being painful is 64 Pa. The ratio between the loudest and softest is therefore:

$$\frac{\text{Threshold of pain}}{\text{Threshold of hearing}} = \frac{64}{10^{-5}} = 6\,400\,000 = 6.4 \times 10^6$$

This is a very wide range variation in terms of the numbers involved, and it is not a convenient one to work with. Therefore sound pressure level (SPL), is represented in decibels relative to 20 μPa (see Chapter 1), as dB(SPL) as follows:

$$\mathrm{dB}(SPL) = 20 \log \left(\frac{P_{\mathrm{actual}}}{P_{\mathrm{ref}}} \right)$$

where P_{actual} = measured sound pressure level
P_{ref} = 20 μPa or 2×10^{-5} Pa

Example 2.3 Calculate the threshold of hearing and threshold of pain in dB(SPL).

The threshold of hearing at 1 kHz is, in fact, P_{ref} which in dB(SPL) equals:

$$20 \log \left(\frac{P_{\mathrm{ref}}}{P_{\mathrm{ref}}} \right) = 20 \log \left(\frac{2 \times 10^{-5}}{2 \times 10^{-5}} \right) = 20 \log(1)$$

$$= 20 \times 0 = 0 \, \mathrm{dB}(SPL)$$

and the threshold of pain is 64 Pa which in dB(SPL) equals:

$$20 \log \left(\frac{P_{\mathrm{actual}}}{P_{\mathrm{ref}}} \right) = 20 \log \left(\frac{64}{2 \times 10^{-5}} \right) = 20 \log(6.4 \times 10^{6})$$

$$= 20 \times 6.5 = 130 \, \mathrm{dB}(SPL)$$

Use of the dB(SPL) scale results in a more convenient range of values (0 to 130) to consider, since values in the range of about 0 to 100 are common in everyday dealings. Also, it is a more appropriate basis for expressing acoustic amplitude values, changes in which are primarily perceived as variations in loudness, since loudness perception turns out to be essentially logarithmic in nature.

The threshold of hearing varies with frequency. The ear is far more sensitive in the middle of its frequency range than at the high and low extremes. The lower curve in Figure 2.11 is the general shape of the average threshold of hearing curve for sinusoidal stimuli between 20 Hz and 20 kHz. The upper curve in the figure is the general shape of the threshold of pain, which also varies with frequency but not to such a great extent. It can be seen from the figure that the full 130 dB(SPL) range, or 'dynamic range', between the threshold of hearing and the threshold of pain exists at approximately 4 kHz, but that the

Table 2.1 Typical sound levels in the environment

Example sound	dB(SPL)	Description
Long range gunfire at gunner's ear	140	
Threshold of pain	130	Ouch!
Jet take-off at approximately 100 m	120	
Peak levels on a night club dance floor	110	
Loud shout at 1 m	100	Very noisy
Heavy truck at about 10 m	90	
Heavy car traffic at about 10 m	80	
Car interior	70	Noisy
Normal conversation at 1 m	60	
Office noise level	50	
Living room in a quiet area	40	Quiet
Bedroom at night time	30	
Empty concert hall	20	
Gentle breeze through leaves	10	Just audible
Threshold of hearing for a child	0	

dynamic range available at lower and higher frequencies is considerably less. For reference, the sound level and frequency range for both average normal conversational speech and music are shown in Figure 2.11 and Table 2.1 shows approximate sound levels of everyday sounds for reference.

2.4 Loudness perception

Although the perceived loudness of an acoustic sound is related to its amplitude, there is not a simple one-to-one functional relationship. As a psychoacoustic effect it is affected by both the context and nature of the sound. It is also difficult to measure because it is dependent on the interpretation by listeners of what they hear. It is neither ethically appropriate nor technologically possible to put a probe in the brain to ascertain the loudness of a sound.

In Chapter 1 the concepts of the sound pressure level and sound intensity level were introduced. These were shown to be approximately equivalent in the case of free space propagation in which no interference effects were present. The ear is a pressure sensitive organ that divides the audio spectrum into a set of overlapping frequency bands whose bandwidth increases with frequency. These are both objective descriptions of the amplitude and the function of the ear. However, they tell us nothing about the perception of loudness in relation to the objective measures of sound amplitude level. This will allow us to understand some of the effects that occur when one listens to musical sound sources.

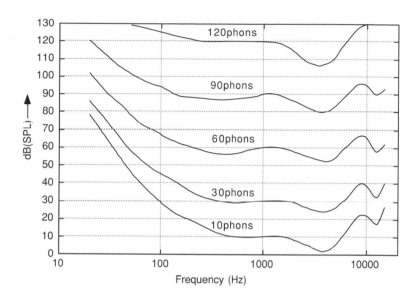

Figure 2.12 Equal loudness contours for the human ear.

The pressure amplitude of a sound wave does not directly relate to its perceived loudness. In fact it is possible for a sound wave with a larger pressure amplitude to sound quieter than a sound wave with a lower pressure amplitude. How can this be so? The answer is that the sounds are at different frequencies and the sensitivity of our hearing varies as the frequency varies. Figure 2.12 shows the equal loudness contours for the human ear. These contours, originally measured by Fletcher and Munson (1933) and by others since, represent the relationship between the measured sound pressure level and the perceived loudness of the sound. The curves show how loud a sound must be in terms of the measured sound pressure level to be perceived as being of the same loudness as a 1 kHz tone of a given level. There are two major features to take note of.

The first is that there are some humps and bumps in the contours above 1 kHz. These are due to the resonances of the outer ear. This is a tube about 25 mm long with one open and one closed end. This will have a first resonance at about 3.4 kHz and, due to its non-uniform shape a second resonance at approximately 13 kHz, as shown in the figure. The effect of these resonances is to enhance the sensitivity of the ear around the resonant frequencies. Note that because this enhancement is due to an acoustic effect in the outer ear it is independent of signal level. The second effect is an amplitude dependence of sensitivity which is due to the way the ear transduces and interprets the sound and, as a result, the frequency response is a function of amplitude. This effect is particularly noticeable at low frequencies but there is also an effect at higher frequencies. The net result of these effects is that the sensitivity of the ear is a function of both frequency and amplitude. In other words the frequency response of the ear is not flat and is also dependent on sound level. Therefore two tones of equal sound pressure level will rarely sound equally loud. For example, a sound at a level which is just audible at 20 Hz would sound much louder if it was at 4 kHz. Tones of different frequencies therefore have to be at different sound pressure levels to sound equally loud and their relative loudness will be also a function of their absolute sound pressure levels.

The loudness of sine wave signals, as a function of frequency and sound pressure levels, is given by the 'phon' scale. The phon scale is a subjective scale of loudness based on the judgements of listeners to match the loudness of tones to reference tones at 1 kHz. The curve for N phons intersects 1 kHz at N dB(SPL) by definition, and it can be seen that the relative shape of the phon curves flattens out at higher sound levels, as shown in Figure 2.12. The relative loudness of different frequencies is not preserved, and therefore the perceived frequency balance of sound varies as the listening level is altered. This is an effect that we have all heard when the volume of a recording is turned down and the bass and treble components appear suppressed relative to the midrange frequencies and the sound becomes 'duller' and 'thinner'. Ideally we should listen to reproduced sound at the level at which it was originally recorded. However, in most cases this would be antisocial, especially as much rock material is mixed at levels in excess of 100 dB(SPL)!

In the early 1970s hi-fi manufacturers provided a 'loudness' button which put in bass and treble boost in order to flatten the Fletcher–Munson curves, and so provide a simple compensation for the reduction in hearing sensitivity at low levels. The action of this control was wrong in two important respects:

- Firstly it directly used the equal loudness contours to perform the compensation, rather than the difference between the curves at two different absolute sound pressure levels, which would be more accurate. The latter approach has been used in professional products to allow night clubs to achieve the equivalent effect of a louder replay level.
- Secondly the curves are a measure of the equal loudness for *sine waves* at a similar level. Real music on the other hand consists of many different frequencies at many different amplitudes and does not directly follow these curves as its level changes. We shall see later how we can analyse the loudness of complex sounds. In fact because the response of

the ear is dependent on both absolute sound pressure level and frequency it cannot be compensated for simply by using treble and bass boost.

2.4.1 Measuring loudness

These effects make it difficult to design a meter which will give a reading which truly relates to the perceived loudness of a sound, and an instrument which gives an approximate result is usually used. This is achieved by using the sound pressure level but frequency weighting it to compensate for the variation of sensitivity as a function of frequency of the ear. Clearly the optimum compensation will depend on the absolute value of the sound pressure level being measured and so some form of compromise is necessary. Figure 2.13 shows two frequency weightings which are commonly used to perform this compensation: termed 'A' and 'C' weightings. The 'A' weighting is most appropriate for low amplitude sounds as it broadly compensates for the low level sensitivity versus frequency curve of the ear. The 'C' weighting on the other hand is more suited to sound at higher absolute sound pressure levels and because of this is more sensitive to low frequency components than the 'A' weighting. The sound levels measured using the 'A' weighting are often given the unit dBA and levels using the 'C' weighting dBC. Despite the fact that it is most appropriate for low sound levels, and is a reasonably good approximation there, the 'A' weighting is now recommended for any sound level, in order to provide a measure of consistency between measurements.

Figure 2.13 The frequency response of 'A' and 'C' weightings.

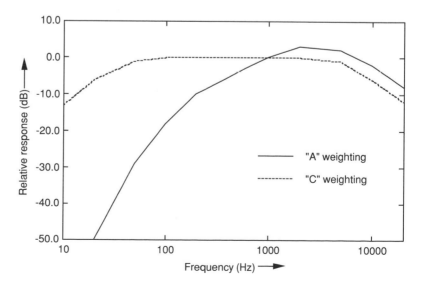

The frequency weighting is not the only factor which must be considered when using a sound level meter. In order to obtain an estimate of the sound pressure level it is necessary to average over at least one cycle, and preferably more, of the sound waveform. Thus most sound level meters have slow and fast time response settings. The slow time response gives an estimate of the average sound level whereas the fast response tracks more rapid variations in the sound pressure level.

Sometimes it is important to be able to calculate an estimate of the equivalent sound level experienced over a period of time. This is especially important when measuring people's noise exposure in order to see if they might suffer noise-induced hearing loss. This cannot be done using the standard fast or slow time responses on a sound level meter; instead a special form of measurement known as the L_{eq} measurement is used. This measure integrates the instantaneous squared pressure over some time interval, such as 15 mins or 8 hrs, and then takes the square root of the result. This provides an estimate of the root mean square level of the signal over the time period of the measurement and so gives the equivalent sound level for the time period. That is the output of the L_{eq} measurement is the constant sound pressure level which is equivalent to the varying sound level over the measurement period. The L_{eq} measurement also provides a means of estimating the total energy in the signal by squaring its output. A series of L_{eq} measurements over short times can also be easily combined to provide a longer time L_{eq} measurement by simply squaring the individual results, adding them together, and then taking the square root of the result, shown in Equation 2.7.

$$L_{eq(total)} = \sqrt{L^2_{eq1} + L^2_{eq2} + ... + L^2_{eqn}} \tag{2.7}$$

where $L_{eq(1-n)}$ = the individual short time L_{eq} measurements

This extendibility makes the L_{eq} measurement a powerful method of noise monitoring. As with a conventional instrument, the 'A' or 'C' weightings can be applied.

2.4.2 Loudness of simple sounds

In Figure 2.11 the two limits of loudness are illustrated: the threshold of hearing and the threshold of pain. As we have already seen, the sensitivity of the ear varies with frequency and therefore so does the threshold of hearing, as shown in Figure 2.14. The peak sensitivities shown in this figure are equivalent to a sound pressure amplitude in the sound wave of 10 µPa or about –6 dB(SPL). Note that this is for monaural listening to a

Figure 2.14 The threshold of hearing as a function of frequency.

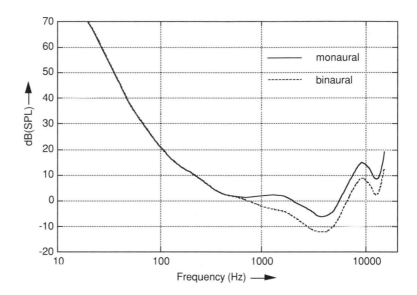

sound presented at the front of the listener. For sounds presented on the listening side of the head there is a rise in peak sensitivity of about 6 dB due to the increase in pressure caused by reflection from the head. There is also some evidence that the effect of hearing with two ears is to increase the sensitivity by between 3 dB and 6 dB. At 4 kHz, which is about the frequency of the sensitivity peak, the pressure amplitude variations caused by the Brownian motion of air molecules, at room temperature and over a critical bandwidth, corresponds to a sound pressure level of about −23 dB. Thus the human hearing system is close to the theoretical physical limits of sensitivity. In other words there would be little point in being much more sensitive to sound, as all we would hear would be a 'hiss' due to the thermal agitation of the air! Many studio and concert hall designers now try to design the building such that environmental noises are less than the threshold of hearing, and so are inaudible.

The second limit is the just noticeable change in amplitude. This is strongly dependent on the nature of the signal, its frequency, and its amplitude. For broad-band noise the just noticeable difference in amplitude is 0.5 to 1 dB when the sound level lies between 20 dB(SPL) and 100 dB(SPL) relative to a threshold of 0 dB(SPL). Below 20 dB(SPL) the ear is less sensitive to changes in sound level. For pure sine waves however, the sensitivity to change is markedly different and is a strong function of both amplitude and frequency. For example at 1 kHz the just noticeable amplitude change varies from 3 dB at 10 dB to 0.3 dB at 80 dB(SPL). This variation occurs at other frequencies as well but in general the just noticeable difference at other frequencies is greater than the values for 1–4 kHz. These different effects make it difficult to judge exactly what difference in amplitude would be noticeable as it is clearly dependent on the precise nature of the sound being listened to. There is some evidence that once more than a few harmonics are present that the just noticeable difference is closer to the broad-band case, of 0.5–1 dB, rather than the pure tone case. As a general rule of thumb the just noticeable difference in sound level is about 1 dB.

The mapping of sound pressure change to loudness variation for larger changes is also dependent on the nature of the sound signal. However, for broad-band noise, or sounds with several harmonics, it generally accepted that a change of about 10 dB in SPL corresponds to a doubling or halving of perceived loudness. However, this scaling factor is dependent on the nature of the sound, and there is some dispute over both its value and its validity.

Example 2.4 Calculate the increase in the number of violinists required to double the loudness of a string section, assuming all the violinists play at the same sound level.

From Chapter 1 the total level from combining several uncorrelated sources is given by:

$$P_{\text{N uncorrelated}} = P\sqrt{N}$$

This can be expressed in terms of the SPL as:

$$SPL_{\text{N uncorrelated}} = SPL_{\text{single source}} + 10\log_{10}(N)$$

In order to double the loudness we need an increase in SPL of 10 dB. Since $10\log(10) = 10$, ten times the number of sources will raise the SPL by 10 dB.

Therefore we must increase the number of violinists in the string section by a factor of ten in order to double their volume.

As well as frequency and amplitude, duration also has an effect on the perception of loudness, as shown in Figure 2.15 for a pure tone. Here we can see that once the sound lasts more than about 200 milliseconds then its perceived level does not change. However, when the tone is shorter than this the perceived amplitude reduces. The perceived amplitude is inversely proportional to the length of the tone burst. This means that when we listen to sounds which vary in amplitude the loudness level is not

Figure 2.15 The effect of tone duration on loudness.

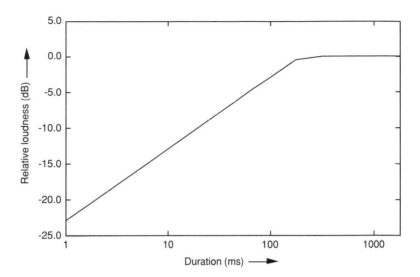

perceived significantly by short amplitude peaks, but more by the sound level averaged over 200 milliseconds.

2.4.3 Loudness of complex sounds

Unlike tones, real sounds occupy more than one frequency. We have already seen that the ear separates sound into frequency bands based on critical bands. The brain seems to treat sounds within a critical band differently to those outside its frequency range and there are consequential effects on the perception of loudness.

The first effect is that the ear seems to lump all the energy within a critical band together and treat it as one item of sound. So when all the sound energy is concentrated within a critical band the loudness is proportional to the total intensity of the sound within the critical band. That is:

$$\text{Loudness} \propto P_1^2 + P_2^2 + \cdots + P_n^2 \tag{2.8}$$

where P_{1-n} = the pressures of the n individual frequency components

As the ear is sensitive to sound pressures, the sound intensity is proportional to the square of the sound pressures, as discussed in Chapter 1. Because the acoustic intensity of the sound is also proportional to the sum of the squared pressure the loudness of a sound within a critical band is independent of the number of frequency components so long as their total acoustic intensity is constant. When the frequency components of the sound extend

Figure 2.16 The effect of tone bandwidth on loudness.

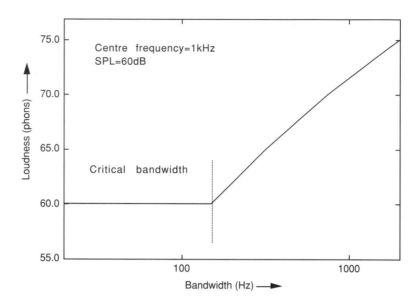

beyond a critical band an additional effect occurs due to the presence of components in other critical bands. In this case more than one critical band is contributing to the perception of loudness and the brain appears to add the individual critical band responses together. The effect is to increase the perceived loudness of the sound even though the total acoustic intensity is unchanged. Figure 2.16 plots the subjective loudness perception of a sound at a constant intensity level as a function of the sound's bandwidth which shows this effect. In the cochlea the critical bands are determined by the place at which the peak of the standing wave occurs, therefore all energy within a critical band will be integrated as one overall effect at that point on the basilar membrane and transduced into nerve impulses as a unit. On the other hand energy which extends beyond a critical band will cause other nerves to fire and it is these extra nerve firings which give rise to an increase in loudness.

The interpretation of complex sounds which cover the whole frequency range is further complicated by psychological effects, in that a listener will attend to particular parts of the sound, such as the soloist, or conversation, and ignore or be less aware of other sounds and will tend to base their perception of loudness on what they have attended to.

Duration also has an effect on the perception of the loudness of complex tones in a similar fashion to that of pure tones. As is the case for pure tones, complex tones have an amplitude which is independent of duration once the sound is longer than about

200 milliseconds and is inversely proportion to duration when the duration is less than this.

2.5 Noise-induced hearing loss

The ear is a sensitive and accurate organ of sound transduction and analysis. However, the ear can be damaged by exposure to excessive levels of sound or noise. This damage can manifest itself in two major forms:

- *A loss of hearing sensitivity:* the effect of noise exposure causes the efficiency of the transduction of sound into nerve impulses to reduce. This is due to damage to the hair cells in each of the organs of corti. Note this is different from the threshold shift due to the acoustic reflex which occurs over a much shorter time period and is a form of built-in hearing protection. This loss of sensitivity manifests itself as a shift in the threshold of hearing that they can hear, as shown in Figure 2.17. This shift in the threshold can be temporary, for short times of exposures, but ultimately it becomes permanent as the hair cells are permanently flattened as a result of the damage, due to long-term exposure, which does not allow them time to recover.
- *A loss of hearing acuity:* this is a more subtle effect but in many ways is more severe than the first effect. We have seen that a crucial part of our ability to hear and analyse sounds is our ability to separate out the sounds into distinct frequency bands, called critical bands. These bands are very

Figure 2.17 The effect of noise exposure on hearing sensitivity (data from Tempest, 1985).

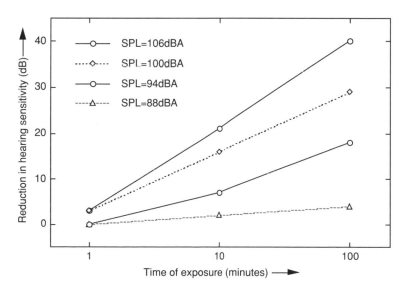

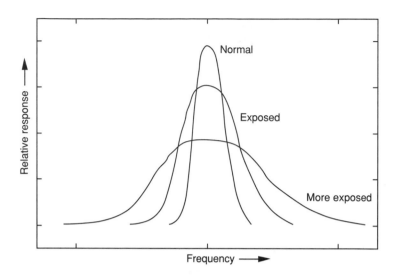

Figure 2.18 Idealised form of the effect of noise exposure on hearing bandwidth.

narrow. Their narrowness is due to an active mechanism of positive feedback in the cochlea which enhances the standing wave effects mentioned earlier. This enhancement mechanism is very easily damaged; it appears to be more sensitive to excessive noise than the main transduction system. The effect of the damage though is not just to reduce the threshold but also to increase the bandwidth of our acoustic filters, as shown in idealised form in Figure 2.18. This has two main effects: firstly, our ability to separate out the different components of the sound is impaired, and this will reduce our ability to understand speech or separate out desired sound from competing noise. Interestingly it may well make musical sounds which were consonant more dissonant because of the presence of more than one frequency harmonic in a critical band, which will be discussed in Chapter 3. The second effect is a reduction in the hearing sensitivity, also shown in Figure 2.18, because the enhancement mechanism also increases the amplitude sensitivity of the ear. This effect is more insidious because the effect is less easy to measure and perceive; it manifests itself as a difficulty in interpreting sounds rather than a mere reduction in their perceived level.

Another related effect due to damage to the hair cells is noise-induced tinnitus. Tinnitus is the name given to a condition in which the cochlea spontaneously generates noise, which can be tonal, random noises, or a mixture of the two. In noise-induced tinnitus exposure to loud noise triggers this, and as well as being disturbing, there is some evidence that people who suffer from

Figure 2.19 Audiograms of normal and damaged hearing (data from Tempest, 1985).

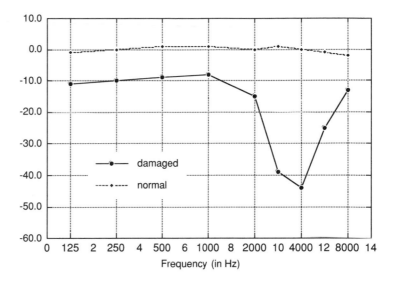

this complaint may be more sensitive to noise induced hearing damage.

Because the damage is caused by excessive noise exposure it is more likely at the frequencies at which the acoustic level at the ear is enhanced. The ear is most sensitive at the first resonance of the ear canal, or about 4 kHz, and this is the frequency at which most hearing damage first shows up. Hearing damage in this region is usually referred to as an audiometric notch because of its distinctive shape on a hearing test, or audiogram, see Figure 2.19. This distinctive pattern is evidence that the hearing loss measured is due to noise exposure rather than some other condition, such as the inevitable high-frequency loss due to ageing.

How much noise exposure is acceptable? There is some evidence that the normal noise in Western society has some long-term effects because measurements on the hearing of other cultures have shown that there is a much lower threshold of hearing at a given age compared with Westerners. However, this may be due to other factors as well; for example, the level of pollution, etc. There is strong evidence, however, that exposure to noises with amplitudes of greater than 90 dB(SPL) will cause permanent hearing damage. This fact is recognised by legislation which requires that the noise exposure of workers be less than this limit. Note that if the work environment has a noise level of greater than this then hearing protection of a sufficient standard should be used to bring the noise level, at the ear, below this figure.

2.5.1 *Integrated noise dose*

In many musical situations the noise level is greater than 90 dB(SPL) for short periods. For example the audience at a concert may well experience brief peaks above this, especially at particular instants in works such as Elgar's *Dream of Gerontius*, or Orff's *Carmina Burana*. Also in many practical industrial and social situations the noise level may be louder than 90 dB(SPL) for only part of the time. How can we relate intermittent periods of noise exposure to continuous noise exposure; for example,

Figure 2.20 Maximum exposure time as a function of sound level plotted on a linear scale (upper) and a logarithmic scale (lower).

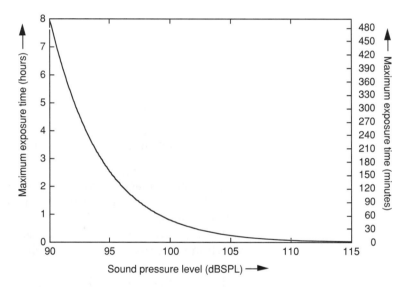

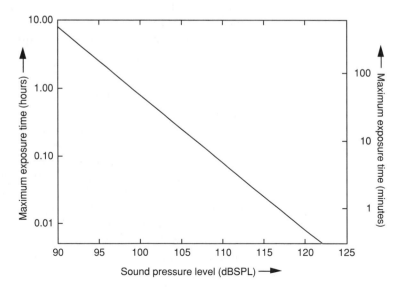

how damaging is a short exposure to a sound of 96 dB(SPL)? The answer is to use a similar technique to that used in assessing the effect of radiation exposure, that of 'integrated dose'. The integrated noise dose is defined as the equivalent level of the sound over a fixed period of time, which is currently 8 hours. In other words the noise exposure can be greater than 90 dB(SPL) providing that it is for an appropriately shorter time which results in a noise dose which is less than that which would result from being exposed to noise at 90 dB(SPL) for 8 hours. The measure used is the L_{eq} mentioned earlier and the maximum dose is 90 dB L_{eq} over eight hours. This means that one can be exposed to 93 dB(SPL) for 4 hours, 96 dB(SPL) for 2 hours, and so on. Figure 2.20 shows how the time of exposure varies with the sound level on linear and logarithmic time scales. It can be seen that exposure to extreme sound levels, greater than 100 dB(SPL), can only be tolerated for a very short period of time, less than half an hour. There is also a limit to how far this concept can be taken because very loud sounds can rupture the eardrum causing instant, and sometimes permanent, loss of hearing.

This approach to measuring the noise dose takes no account of the spectrum of the sound which is causing the noise exposure, because to do so would be difficult in practice. However, it is obvious that the effect of a pure tone at 90 dB(SPL) on the ear is going to be different to the same level spread over the full frequency range. In the former situation there will be a large amount of energy concentrated at a particular point on the basilar membrane and this is likely to be more damaging than the second case in which the energy will be spread out over the full length of the membrane. The specification for noise dose uses 'A' weighting for the measurement which, although it is more appropriate for low rather than high sound levels, weights the sensitive 4 kHz region more strongly.

2.5.2 Protecting your hearing

Hearing loss is insidious and permanent and by the time it is measurable it is too late. Therefore in order to protect hearing sensitivity and acuity one must be proactive. The first strategy is to avoid exposure to excess noises. Although 90 dB(SPL) is taken as a damage threshold if the noise exposure causes ringing in the ears, especially if the ringing lasts longer than the length of exposure, it may be that damage may be occurring even if the sound level is less than 90 dB(SPL).

There are a few situations where potential damage is more likely.

- The first is when listening to recorded music over headphones, as even small ones are capable of producing damaging sound levels.
- The second is when one is playing music, with either acoustic or electric instruments, as these are also capable of producing damaging sound levels, especially in small rooms with a 'live' acoustic, see Chapter 6.

In both cases the levels are under your control and so can be reduced. However, there is an effect called the acoustic reflex (see Section 2.1.2), which reduces the sensitivity of your hearing when loud sounds occur. This effect, combined with the effects of temporary threshold shifts, can result in a sound level increase spiral, where there is a tendency to increase the sound level 'to hear it better' which results in further dulling, etc. The only real solution is to avoid the loud sounds in the first place. However, if this situation does occur then a rest away from the excessive noise will allow some sensitivity to return.

There are sound sources over which one has no control, such as bands, discos, night clubs, and power tools. In these situations it is a good idea either to limit the noise dose or, better still, use some hearing protection. For example, one can keep a reasonable distance away from the speakers at a concert or disco. It takes a few days, or even weeks in the case of hearing acuity, to recover from a large noise dose so one should avoid going to a loud concert, or night club, every day of the week! The authors regularly use small 'in-ear' hearing protectors when they know they are going to be exposed to high sound levels, and many professional sound engineers also do the same. These have the advantage of being unobtrusive and reduce the sound level by a modest, but useful, amount (15–20 dB) while still allowing conversation to take place at the speech levels required to compete with the noise! These devices are also available with a 'flat' attenuation characteristic with frequency and so do not alter the sound balance too much, and cost less than a CD recording. For very loud sounds, such as power tools, then a more extreme form of hearing protection may be required, such as headphone style ear defenders.

Your hearing is essential, and irreplaceable, for both the enjoyment of music, for communicating, and socialising with other people. Now and in the future, it is worth taking care of.

2.6 Perception of sound source direction

How do we perceive the direction that a sound arrives from? The answer is that we make use of our two ears, but how?

Figure 2.21 The effect of the direction of a sound source with respect to the head.

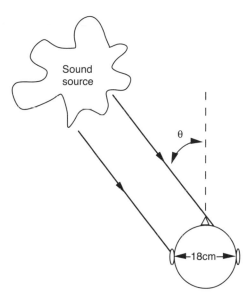

Because our two ears are separated by our head, this has an acoustic effect which is a function of the direction of the sound. There are two effects of the separation of our ears on the sound wave: firstly the sounds arrive at different times and secondly they have different intensities. These two effects are quite different so let us consider them in turn.

2.6.1 Interaural time difference (ITD)

Consider the model of the head, shown in Figure 2.21, which shows the ears relative to different sound directions in the horizontal plane. Because the ears are separated by about 18 cm there will be a time difference between the sound arriving at the ear nearest the source and the one further away. So when the sound is off to the left the left ear will receive the sound and when it is off to the right the right ear will hear it first. If the sound is directly in front, or behind, or anywhere on the median plane, the sound will arrive at both ears simultaneously. The time difference between the two ears will depend on the difference in the lengths that the two sounds have to travel. A simplistic view might just allow for the fact that the ears are separated by a distance d and calculate the effect of angle on the relative time difference by considering the extra length introduced purely due to the angle of incidence, as shown in Figure 2.22. This assumption will give the following equation for the time difference due to sound angle:

Figure 2.22 A simple model for the interaural time difference.

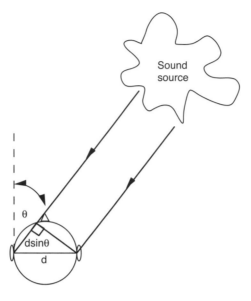

$$\Delta t = \frac{d \, \sin(\theta)}{c}$$

where Δt = the time difference between the ears (in s)
   d = the distance between the ears (in m)
   θ = the angle of arrival of the sound from the median (in radians)
   and c = the speed of sound (in ms^{-1})

Unfortunately this equation is wrong. It underestimates the delay between the ears because it ignores the fact that the sound must travel around the head in order to get to them. This adds an additional delay to the sound, as shown in Figure 2.23. This additional delay can be calculated, providing one assumes that the head is spherical, by recognising that the distance travelled around the head for a given angle of incidence is give by:

$$\Delta d = r\theta$$

where Δd = the extra path round the head at a given angle of incidence (in m)
   and r = half the distance between the ears (in m)

This equation can be used in conjunction with the extra pathlength due to the angle of incidence, which is now a function of r, as shown in Figure 2.24, to give a more accurate equation for the ITD as:

$$ITD = \frac{r(\theta + \sin(\theta))}{c} \tag{2.8}$$

Figure 2.23 The effect of the pathlength around the head on the interaural time difference.

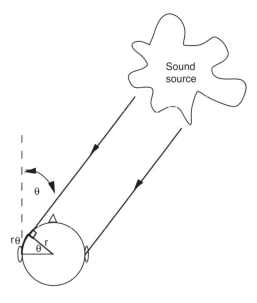

Figure 2.24 A better model for the interaural time difference.

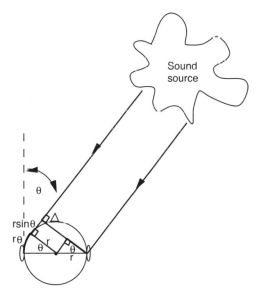

Using this equation we can find that the maximum ITD, which occurs at 90° or ($\pi/2$ radians) to be:

$$ITD_{max} = \frac{0.09 \text{ m} \times (\pi/2 + \sin(\pi/2))}{344 \text{ ms}^{-1}} = 6.73 \times 10^{-4} \text{ s } (673 \text{ μs})$$

Figure 2.25 The interaural time difference (ITD) as a function of angle

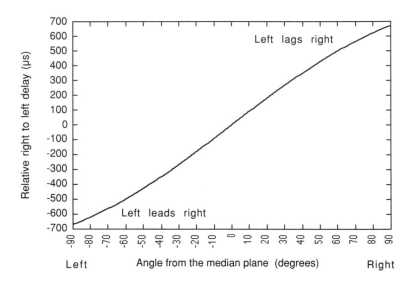

This is a very small delay but a variation from this to zero determines the direction of sounds at low frequencies. Figure 2.25 shows how this delay varies as a function of angle, where positive delay corresponds to a source at the right of the median plane and negative delay corresponds to a source on the left. Note that there is no difference in the delay between front and back positions at the same angle. This means that we must use different mechanisms and strategies to differentiate between front and back sounds. There is also a frequency limit to the way in which sound direction can be resolved by the ear in this way. This is due to the fact that the ear appears to use the phase shift in the wave caused by the interaural time difference to resolve the direction. That is, the ear measures the phase shift given by:

$$\Phi_{\text{ITD}} = 2\pi fr(\theta + \sin(\theta))$$

where Φ_{ITD} = the phase difference between the ears (in radians)
and f = the frequency (in Hz)

When this phase shift is greater than π radians (180°) there will be an unresolvable ambiguity in the direction because there are two possible angles—one to the left and one to the right—that could cause such a phase shift. This sets a maximum frequency, at a particular angle, for this method of sound localisation, which is given by:

$$f_{\text{max}}(\theta) = \frac{1}{2 \times 0.09 \text{ m} \times (\theta + \sin(\theta))}$$

which for an angle of 90° is:

$$f_{max} (\theta = \pi/2) = \frac{1}{2 \times 0.09 \text{ m} \times (\pi/2 + \sin(\pi/2))} = 743 \text{ Hz}$$

Thus for sounds at 90° the maximum frequency that can have its direction determined by phase is 743 Hz. However, the ambiguous frequency limit would be higher at smaller angles.

2.6.2 Interaural intensity difference (IID)

The other cue that is used to detect the direction of the sound is the differing levels of intensity that result at each ear due to the shading effect of the head. This effect is shown in Figure 2.26 which shows that the levels at each ear is equal when the sound source is on the median plane but that the level at one ear progressively reduces, and increases at the other, as the source moves away from the median plane. The level reduces in the ear that is furthest away from the source. The effect of the shading of the head is harder to calculate but experiments seem to indicate that as the intensity ratio between the two ears varies sinusoidally from 0 dB up to 20 dB, depending on frequency, with the sound direction angle, as shown in Figure 2.27. However, as we saw in Chapter 1, an object is not significant as

Figure 2.26 The effect of the head on the interaural intensity difference.

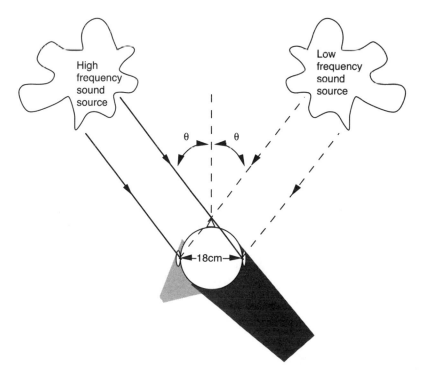

Figure 2.27 The interaural intensity difference (ITD) as a function of angle and frequency (data from Gulick 1971).

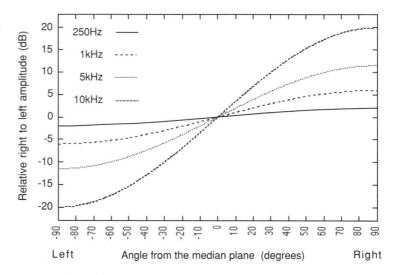

a scatterer or shader of sound until its size is about two thirds of a wavelength (⅔λ), although it will be starting to scatter an octave below that frequency. This means that there will be a minimum frequency below which the effect of intensity is less useful for localisation which will correspond to when the head is about one third of a wavelength in size (⅓λ). For a head the diameter of which is 18 cm, this corresponds to a minimum frequency of:

$$f_{\min(\theta = \pi/2)} = \frac{1}{3}\left(\frac{c}{d}\right) = \frac{1}{3} \times \left(\frac{344 \text{ ms}^{-1}}{0.18 \text{ m}}\right) = 637 \text{ Hz}$$

Thus the interaural intensity difference is a cue for direction at high frequencies whereas the interaural time difference is a cue for direction at low frequencies. Note that the cross-over between the two techniques starts at about 700 Hz and would be complete at about four times this frequency at 2.8 kHz. In between these two frequencies the ability of our ears to resolve direction is not as good as at other frequencies.

2.6.3 Pinnae and head movement effects

The above models of directional hearing do not explain how we can resolve front to back ambiguities or the elevation of the source. There are in fact two ways which are used by the human being to perform these tasks.

The first is to use the effect of our ears on the sounds we receive to resolve the angle and direction of the sound. This is due to the fact that sounds striking the pinnae are reflected into the ear canal by the complex set of ridges that exist on the ear. These pinnae reflections will be delayed, by a very small but significant amount, and so will form comb filter interference effects on the sound the ear receives. The delay that a sound wave experiences will be a function of its direction of arrival, in all three dimensions, and we can use these cues to help resolve the ambiguities in direction that are not resolved by the main directional hearing mechanism. The delays are very small and so these effects occur at high audio frequencies, typically above 5 kHz. The effect is also person specific, as we all have differently shaped ears and learn these cues as we grow up. Thus we get confused for a while when we change our acoustic head shape radically, by cutting very long hair short for example. We also find that if we hear sound recorded through other people's ears that we have a poorer ability to localise the sound, because the interference patterns are not the same as those for our ears.

The second, and powerful, means of resolving directional ambiguities is to move our heads. When we hear a sound that we wish to attend to, or resolve its direction, we move our head towards the sound and may even attempt to place it in front of us in the normal direction, where all the delays and intensities will be the same. The act of moving our head will change the direction of the sound arrival and this change of direction will depend on the sound source position relative to us. Thus a sound from the rear will move in different direction compared to a sound in front of or above the listener. This movement cue is one of the reasons that we perceive the sound from headphones as being 'in the head'. Because the sound source tracks our head movement it cannot be outside and hence must be in the head. There is also an effect due to the fact that the headphones also do not model the effect of the head. Experiments with headphone listening which correctly model the head and keep the source direction constant as the head moves give a much more convincing illusion.

2.6.4 ITD and IID trading

Because both intensity and delay cues are used for the perception of sound source direction one might expect the mechanisms to be in similar areas of the brain and linked together. If this were the case one might also reasonably expect that there was some overlap in the way the cues were interpreted such that intensity might be confused with delay and vice versa in the brain. This allows for the possibility that the effect of one cue, for example delay, could be cancelled out by the other, for example intensity. This effect does in fact happen and is known

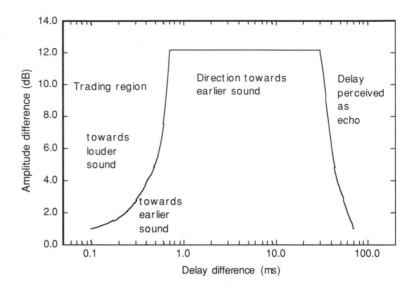

Figure 2.28 Delay versus intensity trading (data from Madsen, 1990).

as interaural time difference versus interaural intensity difference trading. In effect, within limits, an interaural time delay can be compensated for by an appropriate interaural intensity difference, as shown in Figure 2.28, which has several interesting features. Firstly, as expected time delay versus intensity trading is only effective over the range of delay times which correspond to the maximum interaural time delay of 673 µs, beyond this amount of delay small intensity differences will not alter the perceived direction of the image. Instead the sound will appear to come from the source which arrives first. This effect occurs between 673 µs and 30 ms. However, if the delayed sound's amplitude is more than 12 dB greater than the first arrival then we will perceive the direction of the sound to be towards the delayed sound. After 30 ms the delayed signal is perceived as an echo and so the listener will be able to differentiate between the delayed and undelayed sound. The implication of these results are two-fold; firstly, it should be possible to provide directional information purely through either only delay cues or only intensity cues. Secondly, once a sound is delayed by greater than about 700 µs the ear attends to the sound that arrives first almost irrespective of their relative levels, although clearly if the earlier arriving sound is significantly lower in amplitude, compared to the delayed sound, then the effect will disappear.

2.6.5 The Haas effect

The second of the ITD and IID trading effects is also known as the Haas, or precedence, effect, named after the experimenter who quantified this behaviour of our ears. The effect can be summarised as follows:

- The ear will attend to the direction of the sound that arrives first and will not attend to the reflections providing they arrive within 30 ms of the first sound.
- The reflections arriving before 30 ms are fused into the perception of the first arrival. However, if they arrive after 30 ms they will be perceived as echoes.

These results have important implications for studios, concert halls and sound reinforcement systems. In essence it is important to ensure that the first reflections arrive at the audience earlier than 30 ms to avoid them being perceived as echoes. In fact it seems that our preference is for a delay gap of less than 20 ms if the sound of the hall is to be classed as 'intimate'. In sound reinforcement systems the output of the speakers will often be delayed with respect to their acoustic sound but, because of this effect, we perceive the sound as coming from the acoustic source, unless the level of sound from the speakers is very high.

2.6.6 Stereophonic listening

Because of the way we perceive directional sound it is possible to fool the ear into perceiving a directional sound through just two loudspeakers or a pair of headphones in stereo listening. This can be achieved in basically three ways, two using loudspeakers and one using headphones. The first two ways are based on the concept of providing only one of the two major directional cues in the hearing system. That is using either intensity or delay cues and relying on the effect of the ear's time–intensity trading mechanisms to fill in the gaps. The two systems are as follows:

- *Delay stereo:* this system is shown in Figure 2.29 and consists of two omni-directional microphones spaced a reasonable distance apart and away from the performers. Because of the distance of the microphones a change in performer position does not alter the sound intensity much, but does alter the delay. So the two channels when presented over loudspeakers contain predominantly directional cues based on delay to the listener.
- *Intensity stereo:* this system is shown in Figure 2.30 and consists of two directional microphones placed together and pointing at the left and right extent of the performers' positions. Because the microphones are closely spaced, a

Figure 2.29 Delay stereo recording.

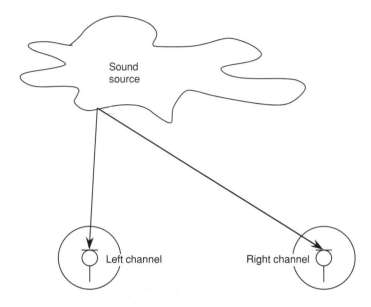

change in performer position does not alter the delay between the two sounds. However, because the microphones are directional the intensity received by the two microphones does vary. So the two channels when presented over loudspeakers contain predominantly directional cues based on intensity to the listener. Intensity stereo is the method that is mostly used in pop music production as the pan-pots on

Figure 2.30 Intensity stereo recording.

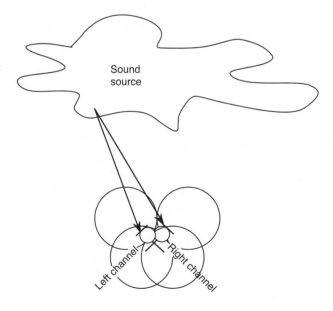

Figure 2.31 The effect of the 'pan-pots' in a mixing desk on intensity of the two channels.

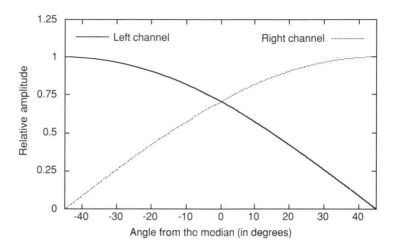

a mixing desk, which determine the position of a track in the stereo image, vary the relative intensities of the two channels, as shown in Figure 2.31.

These two methods differ primarily in the method used to record the original performance and are independent of the listening arrangement, so which method is used is determined by the producer or engineer on the recording. It is also possible to mix the two cues by using different types of microphone arrangement—for example slightly spaced directional microphones—and these can give stereo based on both cues. Unfortunately they also provide spurious cues, which confuse the ear, and getting the balance between the spurious and wanted cues, and so providing a good directional illusion is difficult.

• *Binaural stereo:* the third major way of providing a directional illusion is to use binaural stereo techniques. This system is shown in Figure 2.32 and consists of two omni-directional microphones placed on a head, real or more usually artificial and presenting the result over headphones. Because the distance of the microphones is identical to the ear spacing and they are placed on an object which shades the sound in same way a change in performer position provides both intensity and delay cues to the listener, the results can be very effective but they must be presented over headphones because any cross-ear-coupling of the two channels, as would happen with loudspeaker reproduction, will cause spurious cues and so destroy the illusion. Note that this effect happens in reverse when listening to loudspeaker stereo over headphones which is another reason that the sound is always 'in the head'.

Figure 2.32 Binaural stereo recording.

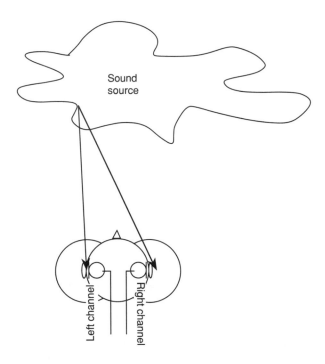

The main compromise in stereo sound reproduction is the presence of spurious direction cues in the listening environment because the loudspeakers and environment will all contribute cues about their position in the room, which have nothing to do with the original recording.

References

von Békésy (1960). *Experiments in Hearing.* New York: McGraw-Hill.

Fletcher, H. and Munson, W. (1933). Loudness, its measurement and calculation. *Journal of the Acoustical Society of America*, **5**, 82–108.

Gulick, L.W. (1971). *Hearing: Physiology and Psychophysics.* Oxford: Oxford University Press, pp. 188–189.

Madsen, E.R. (1990). In *The Science of Sound* (T.D. Rossing, ed.), 2nd edition, Addison Wesley, p. 500.

Moore, B.C.J. and Glasberg, B.P. (1983). Suggested formulae for calculating auditory-filter bandwidths and excitation patterns. *Journal of the Acoustical Society of America*, **74**, (3), pp 750–753.

Pickles, J.O. (1982). *An introduction to the physiology of hearing.* London: Academic Press.

Scharf, B. (1970). Critical bands. In *Foundations of Modern Auditory Theory*, Vol. 1 (J.V. Tobias, ed.), pp. 159–202. London: Academic Press.

3 Notes and harmony

3.1 Musical notes

Music of all cultures is based on the use of instruments (including the human voice) which produce notes of different pitches. The particular set of pitches used by any culture may be unique but the psychoacoustic basis on which pitch is perceived is basic to all human listeners. This chapter explores the acoustics of musical notes which are perceived as having a pitch and the psychoacoustics of pitch perception. It then considers the acoustics and psychoacoustics of different tuning systems that have been used in Western music.

The representation of musical pitch can be confusing because a number of different notation systems are in use. In this book the system which uses A4 to represent the A above middle C has been adopted. The number changes between the notes B and C, and capital letters are always used for the names of the notes. Thus middle C is C4, the B immediately below it is B3, etc. The bottom note on an 88-note piano keyboard is therefore A0 since it is the fourth A below middle C, and the top note on an 88-note piano keyboard is C8. This notation system is shown for reference against a keyboard in Figure 3.21.

3.1.1 Musical notes and their fundamental frequency

When we listen to a note played on a musical instrument and we perceive it as having a clear unambiguous musical pitch, this

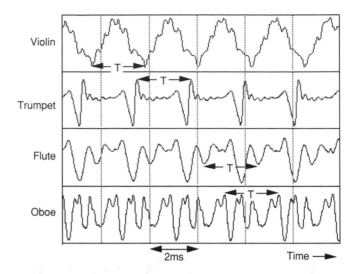

Figure 3.1 Acoustic pressure waveform of A4 (440 Hz) played on a violin, trumpet, flute and oboe.

Violin

Trumpet

Flute

Oboe

2ms   Time →

is because that instrument produces an acoustic pressure wave which repeats regularly. For example, consider the acoustic pressure waveforms recorded by means of a microphone and shown in Figure 3.1 for A4 played on four orchestral instruments: violin, trumpet, flute and oboe. Notice that in each case, the waveshape repeats regularly, or the waveform is 'periodic' (see Chapter 1). Each section that repeats is known as a 'cycle' and the time for which each cycle lasts is known as the 'fundamental period' or 'period' of the waveform. The number of cycles which occur in one second gives the fundamental frequency of the note in Hertz or Hz. The fundamental frequency is often notated as 'f_0', pronounced 'F zero' or 'F nought', a practice which will be used throughout the rest of this book. Thus f_0 of any waveform can be found from its period as:

$$(f_0 \text{ in Hz}) = \frac{1}{(\text{period in seconds})} \tag{3.1}$$

and the period from a known f_0 as:

$$(\text{period in seconds}) = \frac{1}{f_0 \text{ in Hz}} \tag{3.2}$$

Example 3.1 Find the period of the note G5, and the note an instrument is playing if its measured period is 5.41 ms.

Figure 3.21 gives the f_0 of G5 as 784.0 Hz, therefore its period from Equation 3.2 is:

$$\text{Period of G5 in seconds} = \frac{1}{784.0} = 1.276 \times 10^{-3} \text{ or } 1.276 \text{ ms}$$

110

The f_0 of a note whose measured period is 5.405 ms can be found using Equation 3.1 as:

$$f_0 \text{ in Hz} = \frac{1}{5.41 \times 10^{-3}} = 184.8 \text{ Hz}$$

The note whose f_0 is nearest to 184.8 Hz (from Figure 3.21) is F#3.

For the violin note shown in Figure 3.1, the f_0 equivalent to any cycle can be found by measuring the period of that cycle from the waveform plot from which the f_0 can be calculated. The period is measured from any point in one cycle to the point in the next (or last) cycle where it repeats, for example, a positive peak, a negative peak or a point where it crosses the zero amplitude line. The distance marked 'T_{violin}' in the figure shows where the period could be measured between negative peaks, and this measurement was made in the laboratory to give the period as 2.28 ms. Using Equation 3.1:

$$f_0 = \frac{1}{2.27 \text{ ms}} = \frac{1}{(2.27 \times 10^{-3}) \text{ s}} = 440.5 \text{ Hz}$$

This is close to 440 Hz, which is the tuning reference f_0 for A4 (see Figure 3.21). Variation in tuning accuracy, intonation or, for example vibrato if the note were not played on an open string, will mean that the f_0 measured for any particular individual cycle is not likely to be exactly equivalent to one of the reference f_0 values in Figure 3.21. An average f_0 measurement over a number of individual periods might be taken in practice.

3.1.2 Musical notes and their harmonics

Figure 3.1 also shows the acoustic pressure waveforms produced by other instruments when A4 is played. Whilst the periods and therefore the f_0 values of these notes is similar, their waveform shapes are very different. The perceived pitch of each of these notes will be A4 and the distinctive sound of each of these instruments is related to the differences in the detailed shape of their acoustic pressure waveforms, which is how listeners recognise the difference between, for example, a violin, a clarinet and an oboe. This is because it is acoustic pressure variations produced by a musical instrument that impinge on the listener's tympanic membrane, resulting in the pattern of vibration set up on the basilar membrane of that ear which is then analysed in terms of the frequency components of which they are comprised (see Chapter 2). If the pattern of vibration on the basilar membrane varies when comparing different sounds, for example, a violin and

Figure 3.2 Spectra of waveforms shown in Figure 3.1 for A4 (f_0 = 440 Hz) played on a violin, trumpet, flute and oboe.

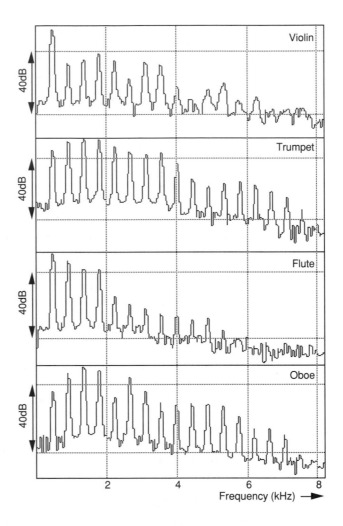

a clarinet, then the sounds are perceived as having a different 'timbre' (see Chapter 5) whether or not they have the same pitch.

Every instrument therefore has an underlying set of partials in its spectrum (see Chapter 1) from which we are able to recognise it from other instruments. These can be thought of as the frequency component 'recipe' underlying the particular sound of that instrument. Figure 3.1 shows the acoustic pressure waveform for different notes played on four orchestral instruments and Figure 3.2 shows the amplitude–frequency spectrum for each. Notice that the shape of the waveform for each of the notes is different and so is the recipe of frequency components. Each of these notes would be perceived as being the note A4 but as having different timbres. The frequency components of notes produced by any pitched instrument, such as a violin, oboe,

Table 3.1 The relationship between overtone series, harmonic series and fundamental frequency for the first ten components of a period waveform

Integer (N)	Overtone series $((N-1) \times f_0)$ when $N > 1$	Harmonic series $(N \times f_0)$	Component frequency (Hz)
1	fundamental frequency (f_0)	1st harmonic	$1\ f_0$
2	1st overtone	2nd harmonic	$2\ f_0$
3	2nd overtone	3rd harmonic	$3\ f_0$
4	3rd overtone	4th harmonic	$4\ f_0$
5	4th overtone	5th harmonic	$5\ f_0$
6	5th overtone	6th harmonic	$6\ f_0$
7	6th overtone	7th harmonic	$7\ f_0$
8	7th overtone	8th harmonic	$8\ f_0$
9	8th overtone	9th harmonic	$9\ f_0$
10	9th overtone	10th harmonic	$10\ f_0$

clarinet, trumpet, etc., are harmonics, or integer (1, 2, 3, 4, 5, etc.) multiples of f_0 (see Chapter 1). Thus the only possible frequency components for the acoustic pressure waveform of the violin note shown in Figure 3.1 whose f_0 is 440.5 Hz are: 440.5 Hz (1×440.5 Hz); 881.0 Hz (2×440.5 Hz); 1321.5 Hz (3×440.5 Hz); 1762 Hz (4×440.5 Hz); 2202.5 Hz (5×440.5 Hz), etc. Figure 3.2 shows that these are the only frequencies at which peaks appear in each spectrum (see Chapter 1). These harmonics are generally referred to by their 'harmonic number', which is the integer by which f_0 is multiplied to calculate the frequency of the particular component of interest.

An earlier term still used by many authors for referring to the components of a periodic waveform is 'overtones'. The first overtone refers to the first frequency component that is 'over' or above f_0, which is the second harmonic. The second overtone is the third harmonic and so on. Table 3.1 summarises the relationship between f_0, overtones and harmonics for integer multipliers from one to ten.

Example 3.2 Find the fourth harmonic of a note whose f_0 is 101 Hz, and the sixth overtone of a note whose f_0 is 120 Hz.

The fourth harmonic has a frequency which is ($4\ f_0$), which is (4×101) Hz = 404 Hz.
The sixth overtone has a frequency which is ($7\ f_0$), which is (7×120) Hz = 840 Hz.

There is no theoretical upper limit to the number of harmonics which could be present in the output from any particular instrument, although for many instruments there are acoustic limits imposed by the structure of the instrument itself. An upper limit can be set though, in terms of the number of harmonics which could be present based on the upper frequency limit of the hearing system, for which a practical limit might be 16 000 Hz (see Chapter 2). Thus an instrument playing the A above middle C, which has an f_0 of 440 Hz, could theoretically contain 36 (=16 000/440) harmonics within the human hearing range. If this instrument played a note an octave higher, f_0 is doubled to 880 Hz, and the output could now theoretically contain 18 (=16 000/880) harmonics. This is an increasingly important consideration since although there is often an upper frequency limit to an acoustic instrument which is well within the practical upper frequency range of human hearing, it is quite possible with electronic synthesisers to produce sounds with harmonics which extend beyond this upper frequency limit.

3.1.3 Musical intervals between harmonics

Acoustically, a note perceived to have a distinct pitch contains frequency components that are integer multiples of f_0 usually known as harmonics. Each harmonic is a sine wave and since the hearing system analyses sounds in terms of their frequency components it turns out to be highly instructive in terms of understanding how to analyse and synthesise periodic sounds, as well as being central to the development of Western musical harmony to consider the musical relationship between the individual harmonics themselves. The frequency ratios of the harmonic series are known (see Table 3.1) and their equivalent musical intervals, frequency ratios and staff notation in the key of C are shown in Figure 3.3 for the first ten harmonics. The musical intervals (apart from the octave) are only approximated

Figure 3.3 Frequency ratios and common musical intervals between the first ten harmonics of the natural harmonic series of C3 against a musical stave and keyboard.

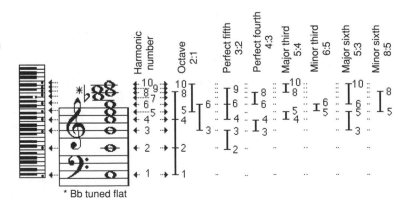

* Bb tuned flat

on a modern keyboard due to the tuning system used, as discussed in Section 3.3.

The musical intervals of adjacent harmonics in the natural harmonic series starting with the fundamental or first harmonic, illustrated on a musical stave and as notes on a keyboard in Figure 3.3, are: octave (2:1), perfect fifth (3:2), perfect fourth (4:3), major third (5:4), minor third (6:5), flat minor third (7:6), sharp major second (8:7), a major whole tone (9:8), and a minor whole tone (10:9). The frequency ratios for intervals between non-adjacent harmonics in the series can also be inferred from the figure. For example, the musical interval between the fourth harmonic and the fundamental is two octaves and the frequency ratio is 4:1, equivalent to a doubling for each octave. Similarly the frequency ratio for three octaves is 8:1, and for a twelfth (octave and a fifth) is 3:1.

Intervals for other commonly used musical intervals can be found from these (musical intervals which occur within an octave are illustrated in Figure 3.15). To demonstrate this for a known result, the frequency ratio for a perfect fourth (4:3) can be found from that for a perfect fifth (3:2) since together they make one octave (2:1): C to G (perfect fifth) and G to C (perfect fourth). The perfect fifth has a frequency ratio 3:2 and the octave a ratio of 2:1. Bearing in mind that musical intervals are ratios in terms of their frequency relationships and that any mathematical manipulation must therefore be carried out by means of division and multiplication, the ratio for a perfect fourth is that for an octave divided by that for a perfect fifth, or up one octave and down a fifth:

$$\text{frequency ratio for a perfect fourth} = \frac{2}{1} \div \frac{3}{2} = \frac{2}{1} \times \frac{2}{3} = \frac{4}{3}$$

Two other common intervals are the major sixth and minor sixth and their frequency ratios can be found from those for the minor third and major third respectively since in each case, they combine to make one octave.

Example 3.3 Find the frequency ratio for a major and a minor sixth given the frequency ratios for an octave (2:1), a minor third (6:5) and a major third (5:4).

A major sixth and a minor third together span one octave. Therefore:

$$\text{Frequency ratio for a major sixth} = \frac{2}{1} \div \frac{6}{5} = \frac{2}{1} \times \frac{5}{6} = \frac{10}{6} = \frac{5}{3}$$

A minor sixth and a major third together span one octave. Therefore:

$$\text{Frequency ratio for a minor sixth} = \frac{2}{1} \div \frac{5}{4} = \frac{2}{1} \times \frac{4}{5} = \frac{8}{5}$$

These ratios can also be inferred from knowledge of the musical intervals and the harmonic series. Figure 3.3 shows that the major sixth is the interval between the fifth and third harmonics, in this example these are G4 and E5; therefore their frequency ratio is 5:3. Similarly the interval of a minor sixth is the interval between the fifth and eighth harmonics, in this case E5 and C6; therefore the frequency ratio for the minor sixth is 8:5. Knowledge of the notes of the harmonic series is both musically and acoustically useful and is something that all brass players and organists who understand mutation stops (see Section 5.4) are particularly aware of.

Figure 3.4 shows the positions of the first ten harmonics of A3 (f_0 = 220.0 Hz), plotted on a linear and a logarithmic axis. Notice that the distance between the harmonics is equal on the linear plot and that it becomes progressively closer together as frequency increases on the logarithmic axis. Whilst the logarithmic plot might appear more complex than the linear plot at first sight in terms of the distribution of the harmonics themselves, particularly given that nature often appears to make use of the

Figure 3.4 The positions of the first ten harmonics of A3 (f_0 = 220 Hz), E4 (f_0 = 330 Hz), and A4 (f_0 = 440 Hz) on linear (upper) and logarithmic (lower) axes.

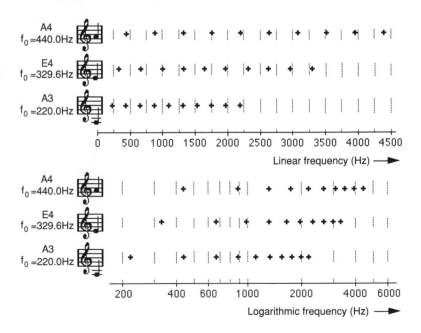

Figure 3.5 Octaves, perfect fifths, perfect fourths, major sixths and minor thirds plotted on a logarithmic scale relative to A1 (f_0 = 55 Hz).

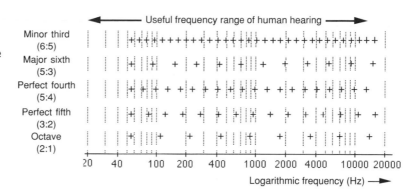

most efficient process, notice that when different notes are plotted, in this case, E4 (f_0 = 329.6 Hz) and A4 (f_0 = 440.0 Hz), the patterning of the harmonics remains constant on the logarithmic scale but they are spaced differently on the linear scale. This is an important aspect of timbre perception which will be explored further in Chapter 5.

Bearing in mind that the hearing system carries out a frequency analysis due to the place analysis which is based on a logarithmic distribution of position with frequency on the basilar membrane, the logarithmic plot most closely represents the perceptual weighting given to the harmonics of a note played on a pitched instrument.

The use of a logarithmic representation of frequency in perception has the effect of giving equal weight to the frequencies of components analysed by the hearing system that are in the same ratio. Figure 3.5 shows a number of musical intervals plotted on a logarithmic scale and in each case they continue to around the upper useful frequency limit of the hearing system. In this case they are all related to A1 (f_0 = 55 Hz) for convenience. Such a plot could be produced relative to an f_0 value for any note and it is important to notice that the intervals themselves would remain a constant distance on a given logarithmic scale. This can be readily verified with a ruler, for example by measuring the distance equivalent to an octave from 100 Hz (i.e. between 100 Hz and 200 Hz, 200 Hz and 400 Hz, 400 Hz and 800 Hz, etc.) on the x axis of Figure 3.5 and comparing this with the distance between any of the points on the octave plot. The distance anywhere on a given logarithmic axis that is equivalent to a particular ratio such as 2:1, 3:2, 4:3, etc., will be the same no matter where on the axis it is measured. A musical interval ruler could be made which is calibrated in musical intervals to enable the frequencies of notes separated by particular intervals to be

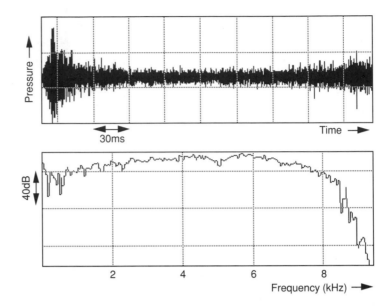

Figure 3.6 Acoustic pressure waveform (upper) and spectrum (lower) for a snare drum being brushed.

readily found on a logarithmic axis. Such a calibration must, however, be carried out with respect to the length of the ratios of interest: octave (2:1), perfect fifth (3:2), major sixth (5:3), etc. If the distance equivalent to a perfect fifth is added to the distance equivalent to a perfect fourth, the distance for one octave will be obtained since a fifth plus a fourth equals one octave. Similarly, if the distance equivalent to a major sixth is added to that for a minor third, the distance for one octave will again be obtained since a major sixth plus a minor third equals one octave (see Example 3.3).

A doubling (or halving) of a value anywhere on a logarithmic frequency scale is equivalent perceptually to a raising (or lowering) by a musical interval of one octave, and multiplying by ½ (or by ⅔) is equivalent perceptually to a raising (or lowering) by a musical interval of a perfect fifth, and so on. We perceive a given musical interval (octave, perfect fifth, perfect fourth, major third, etc.) as being similar no matter where in the frequency range it occurs. For example, a two-note chord a major sixth apart whether played on two double basses or two flutes give a similar perception of the musical interval. In this way, the logarithmic nature of the place analysis mechanism provides a basis for understanding the nature of our perception of musical intervals and of musical pitch.

By way of contrast and to complete the story, sounds which have no definite musical pitch (but a pitch, nevertheless—see below) associated with them, such as the 'ss' in *sea* (Figure 3.9) have an acoustic pressure waveform that does not repeat regularly, and is often random in its variation with time and is therefore not periodic. Such a waveform is referred to as being 'aperiodic' (see Chapter 1). The spectrum of such sounds contains frequency components that are not related as integer multiples of some frequency and there are no harmonic components, and it will often contain all frequencies, in which case it is known as a 'continuous' spectrum. An example of an acoustic pressure waveform and spectrum for a non-periodic sound is illustrated in Figure 3.6 for a snare drum being brushed.

3.2 Hearing pitch

The perception of pitch is basic to the hearing of tonal music. Familiarity with current theories of pitch perception as well as other aspects of psychoacoustics enable a well founded understanding of musically important matters such as tuning, intonation, perfect pitch, vibrato, electronic synthesis of new sounds, and pitch paradoxes (see Chapter 5).

Pitch relates to the perceived position of a sound on a scale from low to high and its formal definition by the American National Standards Institute (1960) is couched in these terms as: 'pitch is that attribute of auditory sensation in terms of which sounds may be ordered on a scale extending from low to high'. The measurement of pitch is therefore 'subjective' because it requires a human listener (the 'subject') to make a perceptual judgement. This is in contrast to the measurement in the laboratory of, for example, the fundamental frequency (f_0) of a note, which is an 'objective' measurement.

In general, sounds which have a periodic acoustic pressure variation with time are perceived as having a pitch associated with them, and sounds whose acoustic pressure waveform is non-periodic are perceived as having no pitch. The relationship between the waveforms and spectra of pitched and non-pitched sounds is summarised in Table 3.2 and examples of each have been discussed

Table 3.2 The nature of the waveforms and spectra for pitched and non-pitched sounds

	Pitched	Non-pitched
Waveform (time domain)	Periodic *regular repetitions*	Non-periodic *no regular repetitions*
Spectrum (frequency domain)	Line *harmonic components*	Continuous *no harmonic components*

in relation to Figures 3.2 and 3.6. The terms 'time domain' and 'frequency domain' are widely used when considering time (waveform) and frequency (spectral) representations of signals.

The pitch of a note varies as its f_0 is changed, the greater the f_0 the higher the pitch and vice versa. Although the measurement of pitch and f_0 are subjective and objective and measured on a scale of high/low and Hz respectively, a measurement of pitch can be given in Hz. This is achieved by asking a listener to compare the sound of interest by switching between it and a sine wave with a variable frequency. The listener would adjust the frequency of the sine wave until the pitch of the two sounds are perceived as being equal, at which point the pitch of the sound of interest is equal to the frequency of the sine wave in Hz.

Two basic theories of pitch perception have been proposed to explain how the human hearing system is able to locate and track changes in the f_0 of an input sound: the 'place' theory and the 'temporal' theory. These are described below along with their limitations in terms of explaining observed pitch perception effects.

3.2.1 Place theory of pitch perception

The place theory of pitch perception relates directly to the frequency analysis carried out by the basilar membrane in which different frequency components of the input sound stimulate different positions, or places, on the membrane. Neural firing of the hair cells occurs at each of these places, indicating to higher centres of neural processing and the brain which frequency components are present in the input sound. For sounds in which all the harmonics are present, the following are possibilities for finding the value of f_0 based on a place analysis of the components of the input sound and allowing for the possibility of some 'higher processing' of the component frequencies at higher centres of neural processing and/or the brain.

- *Method 1:* locate the f_0 component itself.
- *Method 2:* find the minimum frequency difference between adjacent harmonics. The frequency difference between the $(n + 1)$th and the (n)th harmonic, which are adjacent by definition if all harmonics are present, is:

$$((n + 1)f_0) - (n\, f_0) = (n\, f_0) + (1\, f_0) - (nf_0) = f_0$$

where $n = 1, 2, 3, 4, \ldots$

- *Method 3:* find the highest common factor (the highest number that will divide into all the frequencies present giving an integer result) of the components present. Table 3.3 illustrates

Table 3.3 Processing method to find the highest common factor of the frequencies of the first ten harmonics of a sound whose f_0 = 100 Hz (calculations to four significant figures)

Place analysis $n \times f_0$ (÷1) (Hz)	Higher processing								
	÷2 (Hz)	÷3 (Hz)	÷4 (Hz)	÷5 (Hz)	÷6 (Hz)	÷7 (Hz)	÷8 (Hz)	÷9 (Hz)	÷10 (Hz)
100	50.00	33.33	25.00	20.00	16.67	14.29	12.50	11.11	10.00
200	100.0	66.67	50.00	40.00	33.33	28.57	25.00	22.22	20.00
300	150.0	100.0	75.00	60.00	50.00	42.86	37.30	33.33	30.00
400	200.0	133.3	100.0	80.00	66.67	57.14	50.00	44.44	40.00
500	250.0	166.7	125.0	100.0	83.33	71.43	62.50	55.56	50.00
600	300.0	200.0	150.0	120.0	100.0	85.71	75.00	66.67	60.00
700	350.0	233.3	175.0	140.0	116.7	100.0	87.50	77.78	70.00
800	400.0	266.7	200.0	160.0	133.3	114.3	100.0	88.89	80.00
900	450.0	300.0	225.0	180.0	150.0	128.6	112.5	100.0	90.00
1000	500.0	333.3	250.0	200.0	166.7	142.9	125.0	111.1	100.0

this for a sound consisting of the first ten harmonics whose f_0 is 100 Hz, by dividing each frequency by integers, in this case up to 10, and looking for the largest number in the results which exists for every frequency. The frequencies of the harmonics are given in the left-hand column (the result of a place analysis), and each of the other columns show the result of dividing the frequency of each component by integers (m = 2 to 10). The highest common factor is the highest value appearing in all rows of the table, including the frequencies of the components themselves (f_0 ÷ 1) or (m = 1), is 100 Hz which would be perceived as the pitch. In addition, it is of interest to notice that every value which appears in the row relating to the f_0, in this case 100 Hz, will appear in each of the other rows if the table were extended far enough to the right. This is the case because by definition, 100 divides into each harmonic frequency to give an integer result (n) and all values appearing in the 100 Hz row are found by integer (m) division of 100 Hz; therefore all values in the 100 Hz row can be gained by division of harmonic frequencies by ($m \times n$) which must itself be an integer. These are f_0 values (50 Hz, 33 Hz, 25 Hz, 20 Hz, etc.) whose harmonic series also contain all the given components, and they are known as 'sub-harmonics'. This is why it is the *highest* common factor which is used.

One of the earliest versions of the place theory suggests that the pitch of a sound corresponds to the place stimulated by the lowest frequency component in the sound which is f_0 (method 1 above). The assumption underlying this is that f_0 is always present in sounds and the theory was encapsulated by Ohm in his second or 'acoustical' law (his first law being basic to electrical work: voltage = current × resistance): 'a pitch corresponding to a certain

Figure 3.7 An idealised spectrum for a sound with odd harmonics only to show the spacing between adjacent harmonics when the fundamental frequency component (shown dashed) is present or absent.

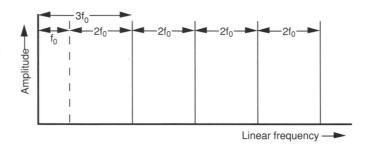

frequency can only be heard if the acoustic wave contains power at that frequency'.

This theory came under close scrutiny when it became possible to carry out experiments in which sounds could be synthesised with known spectra. Schouten (1940) demonstrated that the pitch of a pulse wave remained the same when the fundamental component was removed, thus demonstrating: (i) that f_0 did not have to be present for pitch perception, and (ii) that the lowest component present is not the basis for pitch perception because the pitch does not jump up by one octave (since the second harmonic is now the lowest component after f_0 has been removed). This experiment has become known as: 'the phenomenon of the missing fundamental', and suggests that Method 1 cannot account for human pitch perception.

Method 2 seems to provide an attractive possibility since the place theory gives the positions of the harmonics, whether or not f_0 is present, and it should provide a basis for pitch perception providing some adjacent harmonics are present. For most musical sounds, adjacent harmonics are indeed present. However, researchers are always looking for ways of testing psychoacoustic theories, in this case pitch perception, by creating sounds for which the perceived pitch cannot be explained by current theories. Such sounds are often generated electronically to provide accurate control over their frequency components and temporal development. Figure 3.7 shows an idealised spectrum of a sound which contains just odd harmonics ($1 f_0$, $3 f_0$, $5 f_0$, ...) and shows that measurement of the frequency distance between adjacent harmonics would give f_0, $2 f_0$, $2 f_0$, $2 f_0$, etc. The minimum spacing between the harmonics is f_0 which gives a possible basis for pitch perception. However, if the f_0 component were removed (imagine removing the dotted f_0 component in Figure 3.7), the perceived pitch would not change. Now however, the spacings between adjacent harmonics is $3 f_0$, $2 f_0$, $2 f_0$, $2 f_0$, etc. and the minimum spacing is $2 f_0$ but the pitch does not jump up by an octave.

Table 3.4 Illustration of how finding the highest common factor of the frequencies of the three components: 1040 Hz, 1240 Hz and 1440 Hz gives a basis for explaining a perceived pitch of approximately 207 Hz (calculations to four significant figures)

Component frequency (Hz)	÷2 (Hz)	÷3 (Hz)	÷4 (Hz)	÷5 (Hz)	÷6 (Hz)	÷7 (Hz)	÷8 (Hz)	÷9 (Hz)	÷10 (Hz)
1040	520.0	346.7	260.0	208.0	173.3	148.6	130.0	115.6	104.0
1240	620.0	413.3	310.0	248.0	206.7	177.1	155.0	137.8	124.0
1440	720.0	480.0	360.0	288.0	240.0	205.7	180.0	160.0	144.0

The third method will give an appropriate f_0 for: (i) sounds with missing f_0 components (see Table 3.3 and ignore the f_0 row), (ii) sounds with odd harmonic components only (see Table 3.3 and ignore the rows for the even harmonics), and (iii) sounds with odd harmonic components only with a missing f_0 component (see Table 3.3 and ignore the rows for f_0 and the even harmonics). In each case, the highest common factor of the components is f_0. This method also provides a basis for understanding how a pitch is perceived for non-harmonic sounds, such as bells or chime bars, whose components are not exact harmonics (integer multipliers) of the resulting f_0.

As an example of such a non-harmonic sound, Schouten in one of his later experiments produced sounds whose component frequencies were 1040 Hz, 1240 Hz, and 1440 Hz and found that the perceived pitch was approximately 207 Hz. The f_0 for these components, based on the minimum spacing between the components (Method 2), is 200 Hz. Table 3.4 shows the result of applying Method 3 (searching for the highest common factor of these three components) up to an integer divisor of ten. Schouten's proposal can be interpreted in terms of this table by looking for the closest set of values in the table which would be consistent with the three components being true harmonics and taking their average to give an estimate of f_0. In this case, taking 1040 Hz as the fifth 'harmonic', 1240 Hz as the sixth 'harmonic' and 1440 Hz as the seventh 'harmonic' gives 208 Hz, 207 Hz and 206 Hz respectively. The average of these values is 207 Hz, and Schouten referred to the pitch perceived in such a situation as the 'residue pitch' or 'pitch of the residue'. It is also sometimes referred to as 'virtual pitch'.

By way of a coda to this discussion, it is interesting to note that these components 1040 Hz, 1240 Hz and 1440 Hz do, in fact, have a true f_0 of 40 Hz of which they are the 26th, 31st and 36th harmonics, which would appear if the table were continued well over to the right. However, the auditory system appears to find an f_0 for which the components present are adjacent harmonics.

Figure 3.8 Just noticeable difference for pitch perception and the equivalent rectangular bandwidth.

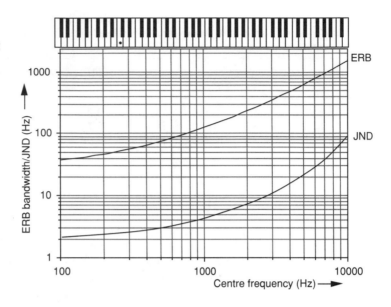

3.2.2 Problems with the place theory

The place theory provides a basis for understanding how f_0 could be found from a frequency analysis of components. However, there are a number of problems with the place theory because it does not explain:

* the fine degree of accuracy observed in human pitch perception,
* pitch perception of sounds whose frequency components are not resolved by the place mechanism,
* the pitch perceived for some sounds which have continuous (non-harmonic) spectra, or
* pitch perception for sounds with an f_0 less than 50 Hz.

Each will be considered in turn.

Psychoacoustically, the ability to discriminate between sounds that are nearly the same except for a change in one aspect (f_0, intensity, duration, etc.) is measured as a 'difference limen' (DL), or 'just noticeable difference' (JND). JND is preferred in this book. The JND for human pitch perception is shown graphically in Figure 3.8 along with the critical bandwidth curve. This JND graph is based on an experiment by Zwicker *et al.* (1957) in which sinusoidal stimuli were used (fixed waveshape) and the sound intensity level and sound

Table 3.5 Illustration of resolution of place mechanism for an input consisting of the first ten harmonics and an f_0 of 110 Hz (calculations to four significant figures). Key: CB = critical bandwidth, CF = centre frequency

Harmonic frequency (Hz)	CB of local filter (Hz)	CF of mid harmonic filter (Hz)	CB of mid harmonic filter (Hz)	Resolved?
110	38.87	165.0	44.10	Yes
220	49.37	275.0	54.67	Yes
330	60.02	385.0	65.40	Yes
440	70.82	495.0	76.28	Yes
550	81.77	605.0	87.30	Yes
660	92.87	715.0	98.48	Yes
770	104.1	825.0	109.8	Yes
880	115.5	935.0	121.3	No
990	127.1	1045	132.9	No
1100	138.8	1155	144.7	No

duration remained constant. It turns out that the JND is approximately one thirtieth of the critical bandwidth across the hearing range. Musically, this is equivalent to approximately one twelfth of a semitone. Thus the JND in pitch is much smaller than the resolution of the analysis filters (critical bandwidth).

The place mechanism will resolve a given harmonic of an input sound provided that the critical bandwidth of the filter concerned is sufficiently narrow to exclude adjacent harmonics. It turns out that, no matter what the f_0 of the sound is, only the first 5 to 7 harmonics are resolved by the place analysis mechanism. This can be illustrated for an example as follows with reference to Table 3.5.

Consider a sound consisting of all harmonics (f_0, 2 f_0, 3 f_0, 4 f_0, 5 f_0, etc.) whose f_0 is 110 Hz. The frequencies of the first ten harmonics are given in the left hand column of the table. The next column shows the critical bandwidth of a filter centred on each of these harmonics by calculation using Equation 2.8. The critical bandwidth increases with filter centre frequency (e.g. see Figure 3.8), and the frequency analysis action of the basilar membrane is equivalent to a bank of filters. Harmonics will cease to be resolved by the place mechanism when the critical bandwidth exceeds the frequency spacing between adjacent harmonics, which is f_0 when all adjacent harmonics are present. In the table, it can be seen that the critical bandwidth is greater than f_0 for the filter centred at 880 Hz (the eighth harmonic), but this filter will resolve the eighth harmonic since it is centred over it and its bandwidth extends half above and half below 880 Hz. In order to establish when harmonics are not resolved, consider the filters centred mid way between adjacent harmonic positions (their centre frequencies and critical bandwidths are shown in the table). The filter centred between the eighth and ninth harmonic has a critical bandwidth of 121 Hz which exceeds f_0 (110 Hz) in this example and therefore the eighth and ninth harmonics will not be resolved by this filter, and the filter between the seventh and eighth harmonic has a critical bandwidth of 110 Hz and the seventh and eighth harmonics would not be resolved either, especially as the frequency range 'seen' by a band-pass filter extends beyond its nominal bandwidth (see Chapter 1). Due to the continuous nature of the travelling wave along the basilar membrane, no harmonics will in this example be resolved above the seventh (possibly above the sixth) since there will be areas on the membrane responding to at least adjacent pairs of harmonics everywhere above the place where the seventh (or possibly the sixth) harmonic stimulates it. Appendix 1 shows a method for finding the filter centre frequency whose critical bandwidth is equal to a given f_0 by solving Equation 2.8 mathematically.

Example 3.4 Confirm the result illustrated in table 3.5 that the 7th and 8th harmonics of an f_0 of 110 Hz will only just be resolved.

Find the positive centre frequency (f_c) for which the critical bandwidth equals 110.0 Hz by substituting 110 Hz for Y (= ERB = critical bandwidth) in the equation AP1.1 in appendix 1:

$$(f_c)_{1,2} = \frac{-[93.39*10^{-3}] \pm \sqrt{[93.39*10^{-3}]^2 - \{[24.92*10^{-6}][28.52 - 110]\}}}{[12.46*10^{-6}]}$$

Taking the positive result to three significant figures:

$$(f_c)_1 = 827 \text{ Hz}$$

This is the centre frequency of the filter whose critical bandwidth is 110.0 Hz and it is very close to 825 Hz which is the centre frequency of the filter mid way between the 7th and 8th harmonics which will therefore only just be resolved.

Observation of the relationship between the critical bandwidth and centre frequency plotted in Figure 3.8 allows the general conclusion that no harmonic above about the fifth to seventh is resolved for any f_0 to be approximately validated as follows. The centre frequency for which the critical band exceeds the f_0 of the sound of interest is found from the graph and no harmonic above this centre frequency will be resolved. To find the centre frequency, plot a horizontal line for the f_0 of interest on the y axis, and find the frequency on the x axis where the line intersects the critical band curve. Only harmonics below this frequency will be resolved and those above will not. It is worth trying this exercise for a few f_0 values to reinforce the general conclusion about resolution of harmonics, since this is vital to understanding of other aspects of psychoacoustics as well as pitch perception.

There are sounds which have non-harmonic spectra for which a pitch is perceived; these are exceptions to the second part of the general statement given earlier that: 'sounds whose acoustic pressure waveform is non-periodic are perceived as having no pitch'. For example, listen to examples of the 'ss' in *sea* and the 'sh' in *shell* (produce these yourself or ask someone else to) in terms of which one has the higher pitch. Most listeners will readily respond that 'ss' has a higher pitch than 'sh'. The spectrum of both sounds is continuous and an example for each is shown in Figure 3.9. Notice that the energy is biased more towards lower frequencies for the 'sh' with a peak around 2.5 kHz, compared to the 'ss' where the energy has a peak at about 5 kHz. This 'centre of gravity' of the spectral energy of a sound is thought to convey a sense of higher or lower pitch for such sounds which are noise based, but the pitch sensation is far weaker than for that perceived for periodic sounds. This pitch phenomenon is, however, important in music when considering the perception of the non-periodic sounds produced, for example by some groups of instruments in the percussion section, but the majority of instruments on which notes are played in musical performance produce periodic acoustic pressure variations. Pitch sensations arising associated with non-periodic acoustic waveforms cannot be described by the place theory since there are no frequency components on which to base the place analysis since the spectra of such sounds is continuous.

The final identified problem is that the pitch perceived for sounds with components only below 50 Hz cannot be explained by the place theory, because the pattern of vibration on the basilar membrane does not appear to change in that region.

127

Figure 3.9 Waveforms and spectra for 'ss' as in *sea* and 'sh' as in *shoe*.

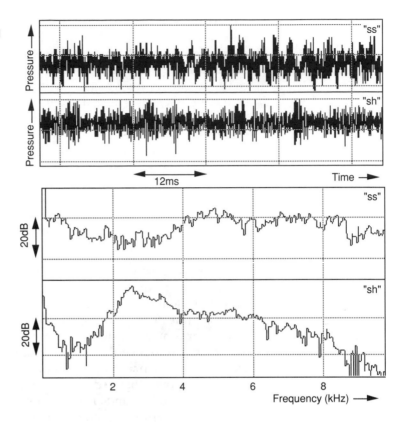

Sounds of this type are rather unusual, but not impossible to create by electronic synthesis. Since the typical lowest audible frequency for humans is 20 Hz, a sound with an f_0 of 20 Hz would have harmonics at 40 Hz, 60 Hz etc., and only the first two fall within this region which no change is observed in basilar membrane response. Harmonics falling above 50 Hz will be analysed by the place mechanism in the usual manner. Sinusoids in the 20 Hz to 50 Hz range are perceived as having different pitches and the place mechanism cannot explain this.

These are some of the key problems which the place mechanism cannot explain, and attention will now be drawn to the temporal theory of pitch perception which was developed to explain some of these problems with the place theory.

3.2.3 Temporal theory of pitch perception

The temporal theory of pitch perception is based on the fact that the waveform of a sound with a strong musical pitch repeats or is periodic (see Table 3.2). An example is shown in Figure 3.1 for

Figure 3.10 Output from a transputer-based model of human hearing to illustrate the nature of basilar membrane vibration at different places along its length for C4 played on a violin.

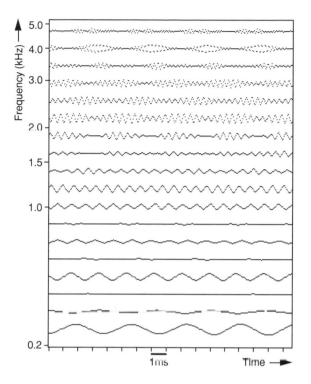

A4 played on four instruments. The f_0 for a periodic sound can be found from a measurement of the period of a cycle of the waveform using Equation 3.1.

The temporal theory of pitch perception relies on the timing of neural firings generated in the organ of Corti (see Figure 2.3) which occur in response to vibrations of the basilar membrane. The place theory is based on the fact that the basilar membrane is stimulated at different places along its length according to the frequency components in the input sound. The key to the temporal theory is the detailed nature of the actual waveform exciting the different places along the length of the basilar membrane. This can be modelled using a bank of electronic band-pass filters whose centre frequencies and bandwidths vary according to the critical bandwidth of the human hearing system as illustrated, for example in Figure 3.8.

Figure 3.10 shows the output waveforms from such a bank of electronic filters, implemented using transputers by Howard *et al.* (1995), with critical bandwidths based on the ERB equation (Equation 2.6) for C4 played on a violin. The nominal f_0 for C4 is 261.6 Hz (see Figure 3.21). The output waveform from the filter with a centre frequency just above 200 Hz, the lowest centre frequency represented in the figure, is a sine wave at f_0. This is because the f_0 component is resolved by the analysing filter, and an individual harmonic of a complex periodic waveform is a sine wave (see Chapter 1). The place theory suggests (see calculation associated with Table 3.5) that the first five to seven harmonics will be resolved by the basilar membrane. It can be seen in the example note shown in Figure 3.10 that the second (around 520 Hz), third (around 780 Hz), fourth (around 1040 Hz) and fifth (around 1300 Hz) harmonics are resolved and their waveforms are sinusoidal. Some amplitude variation is apparent on these sine waves, particularly on the fourth and fifth, indicating the dynamic nature of the acoustic pressure output from a musical instrument. The sixth harmonic (around 1560 Hz) has greater amplitude variation, but the individual cycles are clear.

Output waveforms for filter centre frequencies above the sixth harmonic in this example are not sinusoidal because these harmonics are not resolved individually, demonstrating that the frequency range of a band-pass filter extends beyond its nominal bandwidth (see Chapter 1). At least two harmonics are combined in the outputs from filters which are not sinusoidal in Figure 3.10. When two components close in frequency are combined, they produce a 'beat' waveform whose amplitude rises and falls regularly if the components are harmonics of some fundamental. The period of the beat is equal to the difference between the frequencies of the two components. Therefore if the components are adjacent harmonics, then the beat frequency is equal to their f_0 and the period of the beat waveform is $(1/f_0)$. This can be observed in the figure by comparing the beat period for filter outputs above 1.5 kHz with the period of the output sinewave at f_0. Thus the period of output waveforms for filters with centre frequencies higher than the sixth harmonic will be at $(1/f_0)$ for an input consisting of adjacent harmonics.

The periods of all the output waveforms which stimulate the neural firing in the organ of Corti forms the basis of the temporal theory of pitch perception. There are nerve fibres available to fire at all places along the basilar membrane, and they do so in such a manner that a given nerve fibre may only fire at one phase or instant in each cycle of the stimulating waveform, a process known as 'phase-locking'. Although the nerve firing is phase locked to one instant in each cycle of the stimulating waveform, it has been observed that no single nerve fibre is able to fire continuously at frequencies above approximately 300 Hz. It turns out that the nerve does not necessarily fire in every cycle and that the cycle in which it fires tends to be random, which according to Pickles (1982) may be 'perhaps as little as once every hundred cycles on average'. However due to phase locking, the time between firings for any particular nerve will always be an integer (1, 2, 3, 4, ...) multiple of periods of the stimulating waveform and there are a number of nerves involved at each place. A 'volley firing' principle has also been suggested by Wever (1949) in which groups of nerves work together, each firing in different cycles to enable frequencies higher than 300 Hz to be coded. A full discussion of this area is beyond the scope of this book, and the interested reader is encouraged to consult, for example Pickles (1982); Moore (1982, 1986); Roederer (1975). What follows relies on the principle of phase locking.

The minimum time between firings (1 period of the stimulating waveform) at different places along the basilar membrane can be inferred from Figure 3.10 for the violin playing C4, since it will be equivalent to the period of the output waveform from the analysis filter. For places which respond to frequencies below about the sixth harmonic, the minimum time between firings is at the period of the harmonic itself, and for places above, the minimum time between firings is the period of the input waveform itself (i.e. $1/f_0$).

The possible instants of nerve firing are illustrated in Figure 3.11. This figure enables the benefit to be illustrated that results from the fact that nerves fire phase locked to the stimulating waveform but not necessarily during *every* cycle. The figure shows an idealised unrolled basilar membrane with the places corresponding to where maximum stimulation would occur for input components at multiples of f_0 up to the sixteenth harmonic, for *any* f_0 of input sound. The assumption on which the figure is based is that harmonics up to and including the seventh are analysed separately. The main part of the figure shows the possible instants where nerves could fire based on phase locking and the fact that nerves may not fire every cycle, and the lengths of the vertical lines illustrate the proportion of firings which might occur at that position, on the basis that more firings are likely with lower times between them. These approximate to the idea of a histogram of firings being built up, sometimes referred to as an 'inter-spike interval' histogram, where a 'spike' is a single nerve firing.

Thus at the place on the basilar membrane stimulated by the f_0 component, possible times between nerve firing are: $(1/f_0)$, $(2/f_0)$ and $(3/f_0)$ in this figure as shown, with less firings at the higher intervals. For the place stimulated by the second harmonic, possible firing times are: $[1/(2\ f_0)]$, $[2/(2\ f_0)]$ or $(1/f_0)$, $[3/(2\ f_0)]$,

Figure 3.11 The possible instants for nerve firing across the places on the basilar membrane for the first 16 harmonics of an input sound.

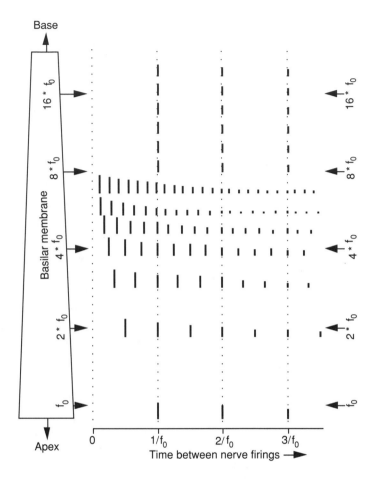

$[4/(2 f_0)]$ or $(2/f_0)$ and so on. This is the case for each place stimulated by a harmonic of f_0 up to the seventh. For place corresponding to higher frequencies than $(7\ f_0)$, the stimulating waveform is beat like and its fundamental period is $(1/f_0)$, and therefore the possible firing times are: $(1/f_0)$, $(2/f_0)$ and $(3/f_0)$ in this figure as shown. Visually it can be seen in the figure that if the entries in all these inter-spike interval histograms were added together vertically (i.e. for each firing time interval), then the maximum entry would occur for the period of f_0. This is reinforced when it is remembered that all places higher than those shown in the figure would exhibit outputs similar to those shown above the eighth harmonic. Notice how all the places where harmonics are resolved have an entry in their histograms at the fundamental period as a direct result of the fact that nerve may not fire in every cycle. This is the basis on which the temporal theory of pitch perception is thought to function.

3.2.4 Problems with the temporal theory

The temporal theory gives a basis for understanding how the fundamental period could be found from an analysis of the nerve firing times from all places across the basilar membrane. However, not all observed pitch perception abilities can be explained by the temporal theory alone, the most important being the pitch perceived for sounds whose f_0 is greater than 5 kHz. This cannot be explained by the temporal theory because phase locking breaks down above 5 kHz. Any ability to perceive the pitches of sounds with f_0 greater than 5 kHz is therefore thought to be due to the place theory alone. Given that the upper frequency limit of human hearing is at best 20 kHz for youngsters, with a more practical upper limit being 16 kHz for those over 20 years of age, a sound with an f_0 greater than 5 kHz is only going to provide the hearing system with two harmonics (f_0 and $2 f_0$) for analysis. In practice it has been established that human pitch perception for sounds whose f_0 is greater than 5 kIIz is rather poor with many musicians finding it difficult to judge accurately musical intervals in this frequency range. Moore (1982) notes that this ties in well with f_0 for the upper note of the piccolo being approximately 4.5 kHz. On large organs, some stops can have pipes whose f_0 exceeds 8 kHz, but these are provided to be used in conjunction with other stops (see Section 5.4).

3.2.5 Contemporary theory of pitch perception

Psychoacoustic research has tended historically to consider human pitch perception with reference to the place or the temporal theory and it is clear that neither theory alone can account for all observed pitch perception abilities. In reality, place analysis occurs giving rise to nerve firings from each place on the basilar membrane that is stimulated. Thus nerve centres and the parts of the brain concerned with auditory processing are provided not only with an indication of the place where basilar membrane stimulation occurs (frequency analysis) but also with information about the nature of that stimulation (temporal analysis). Therefore neither theory is likely to explain human pitch perception completely, since the output from either the place or temporal analysis makes use of the other in communicating itself on the auditory nerve.

Figure 3.12 shows a model for pitch perception of complex tones based on that of Moore (1982) which encapsulates the benefits described for both theories. The acoustic pressure wave is modified by the frequency response of the outer and middle ears

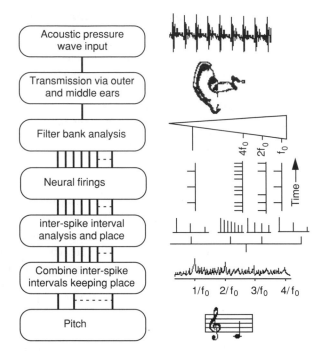

Figure 3.12 A model for human pitch perception based on Moore (1982).

(see Chapter 2), and analysed by the place mechanism which is equivalent to a filter bank analysis. Neural firings occur stimulated by the detailed vibration of the membrane at places equivalent to frequency components of the input sound based on phase locking but not always once per cycle, the latter is illustrated on the right-hand side of the figure. The fact that firing is occurring from particular places provides the basis for the place theory of pitch perception. The intervals between neural firings (spikes) are analysed and the results are combined to allow common intervals to be found which will tend to be at the fundamental period and its multiples, but predominantly at $(1/f_0)$. This is the basis of the temporal theory of pitch perception. The pitch of the sound is based on the results.

3.2.6 Secondary aspects of pitch perception

The perceived pitch of a sound is primarily affected by changes in f_0, which is why the pitch of a note is usually directly related to its f_0, for example, by stating that A4 has an f_0 of 440 Hz as a standard pitch reference. The estimation of f_0 forms the basis of both the place and temporal theories of pitch perception. A change in pitch of a particular musical interval manifests itself if the f_0 values of the notes concerned are in the appropriate frequency ratio to give the primary acoustic (objective) basis for

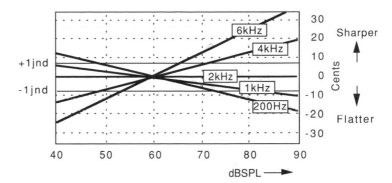

Figure 3.13 The pitch shifts perceived when the intensity of a sine wave with a constant fundamental frequency is varied (after Rossing, 1989).

the perceived (subjective) pitch of the notes and hence the musical interval. Changes in pitch are also, however, perceived by modifying the intensity or duration of a sound while keeping f_0 constant. These are by far secondary pitch change effects compared to the result of varying f_0, and they are often very subtle.

These secondary pitch effects are summarised as follows. If the intensity of a sine wave is varied between 40 dBSPL and 90 dBSPL while keeping its f_0 constant, a change in pitch is perceived for all f_0 values other than those around 1–2 kHz. For f_0 values greater than 2 kHz the pitch becomes sharper as the intensity is raised, and for f_0 values below 1 kHz the pitch becomes flatter as the intensity is raised. This effect is illustrated in Figure 3.13, and the JND for pitch is shown with reference to the pitch at 60 dBSPL to enable the frequencies and intensities of sine waves for which the effect might be perceived to be inferred. This effect is for sine waves which are rarely encountered in music, although electronic synthesisers have made them widely available. With complex tones the effect is less well defined; Rossing (1989) suggests around 17 cents (0.17 of a semitone) for an intensity change between 65 dBSPL and 95 dBSPL. Rossing give two suggestions as to where this effect could have musical consequences: (i) he cites Parkin (1974) to note that this pitch shift phenomenon is apparent when listening in a highly reverberant building to the release of a final loud organ chord which appears to sharpen as the sound level diminishes, and (ii) he suggests that the pitch shift observed for sounds with varying rates of waveform amplitude change, while f_0 is kept constant, should be 'taken into account when dealing with percussion instruments'.

The effect that the duration of a sound has on the perception of the pitch of a note is not a simple one, but it is summarised graphically in Figure 3.14 in terms of the minimum number of

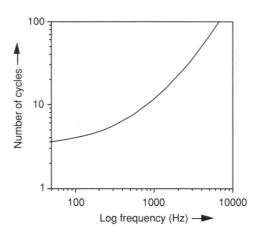

Figure 3.14 The effect of duration on pitch in terms of the number of cycles needed for a definite distinct pitch to be perceived for a given fundamental frequency (data from Rossing, 1989).

cycles required at a given f_0 for a definite distinct pitch to be perceived. Shorter sounds may be perceived as being pitched rather than non-pitched, but the accuracy with which listeners can make such a judgement worsens as the duration of the sound drops below that shown in the figure.

By way of a coda to this section on the perception of pitch, a phenomenon known as 'repetition pitch' is briefly introduced, particularly now that electronic synthesis and studio techniques make it relatively straightforward to reproduce. Repetition pitch is perceived (by most but not all listeners) if a non-periodic noise-based signal, for example the sound of a waterfall, the consonants in *see*, *shoe*, *fee*, or a noise generator, is added to a delayed version of itself and played to listeners. When the delay is altered a change in pitch is perceived. The pitch is equivalent to a sound whose f_0 is equal to (1/delay), and the effect works for delays between approximately 1 ms and 10 ms depending on the listener, giving an equivalent f_0 range for the effect of 100 Hz to 1000 Hz. With modern electronic equipment it is quite possible to play tunes using this effect!

3.3 Hearing notes

The music of different cultures can vary considerably in many aspects including, for example, pitch, rhythm, instrumentation, available dynamic range, and the basic melodic and harmonic usage in the music. Musical taste is always evolving with time; what one composer is experimenting with may well become part of the established tradition a number of years later. The perception of chords and the development of different tuning systems is discussed in this section from a psychoacoustic perspective to

Figure 3.15 All two note musical intervals occurring up to an octave related to C4.

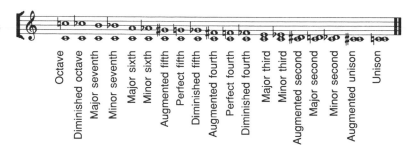

Octave · Diminished octave · Major seventh · Minor seventh · Major sixth · Minor sixth · Augmented fifth · Perfect fifth · Diminished fifth · Augmented fourth · Perfect fourth · Diminished fourth · Major third · Minor third · Augmented second · Major second · Minor second · Augmented unison · Unison

complement the acoustic discussion earlier in this chapter in consideration of the development of melody and harmony in Western music.

3.3.1 Harmonics and the development of Western harmony

Hearing harmony is basic to music appreciation, and in its basic form harmony is sustained by means of chords. A chord consists of at least two notes sounding together and it can be described in terms of the musical intervals between the individual notes which make it up.

A basis for understanding the psychoacoustics of a chord is given by considering the perception of any two notes sounding together. The full set of commonly considered two note intervals and their names are shown in Figure 3.15 relative to middle C. Each of the augmented and diminished intervals sound the same as another interval shown if played on a modern keyboard, for example the augmented unison and minor second, the augmented fourth and diminished fifth, the augmented fifth and minor sixth, and the major seventh and diminished octave, but they are notated differently on the stave and depending on the tuning system in use, these 'enharmonics' would sound different also.

The development of harmony in Western music can be viewed in terms of the decreasing musical interval size between adjacent members of the natural harmonic series as the harmonic number is increased. Figure 3.3 shows the musical intervals between the first ten harmonics of the natural harmonic series. The musical interval between adjacent harmonics must reduce as the harmonic number is increased since it is determined in terms of the f_0 of the notes concerned by the ratio of the harmonic numbers themselves (e.g. 2:1 > 3:2 > 4:3 > 5:4 > 6:5, etc.).

The earliest polyphonic Western music, known as 'organum', made use of the octave, the perfect fifth and its inversion, the perfect fourth. These are the intervals between the 1st and 2nd, the 2nd and 3rd, and the 3rd and 4th members of the natural harmonic series respectively (see Figure 3.3). Later, the major and minor third began to be accepted, the intervals between the 4th and 5th, and the 5th and 6th natural harmonics, with their inversions, the minor and major sixth respectively which are the intervals between the 5th and 8th, and the 3rd and 5th harmonics respectively. The major triad, consisting of a major third and a minor third, and the minor triad, a minor third and a major third, became the building block of Western tonal harmony. The interval of the minor seventh started to be incorporated, and its inversion the major second, the intervals between the 4th and 7th harmonic and the 7th and 8th harmonics respectively. Twentieth century composers have explored music composed using whole tones, the intervals between the 8th and 9th, and between the 9th and 10th harmonics, semitones, harmonics above the 11th are spaced by intervals close to semitones and microtones (intervals of less than a semitone), harmonics above the 16th are spaced by microtones.

3.3.2 Consonance and dissonance

The development of Western harmony follows a pattern where the intervals central to musical development have been gradually ascending the natural harmonic series. These changes have occurred partly as a function of increasing acceptance of intervals which are deemed to be musically 'consonant', or pleasing to listen to, as opposed to 'dissonant', or unpleasant to the listener. The psychoacoustic basis behind consonance and dissonance relates to critical bandwidth, which provides a means for determining the degree of consonance (or dissonance) of musical intervals.

Figure 3.16 The perceived consonance and dissonance of two pure tones (after Plomp and Levelt, 1965. Reproduced with permission).

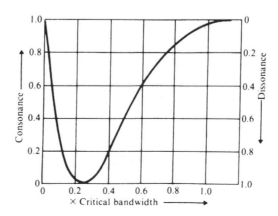

Figure 2.6 illustrates the perceived effect of two sine waves heard together when the difference between their frequencies was increased from 0 to above one critical bandwidth. Listeners perceive a change from 'rough' to 'smooth' when the frequency difference crosses the critical bandwidth. In addition, a change occurred between 'rough fused' to 'rough separate' as the frequency difference is increased within the critical bandwidth. Figure 3.16 shows the result of an experiment by Plomp and Levelt (1965) to determine to what extent two sine waves played together sound consonant or dissonant as their frequency difference is altered. Listeners with no musical training were asked to indicate the consonance or pleasantness of two sine waves played together. (Musicians were not used in the experiment since they would have preconceived ideas about musical intervals which are consonant.) The result is the continuous pattern of response shown in the figure, with no particular musical interval being prominent in its degree of perceived consonance. Intervals greater than a minor third were judged to be consonant for all frequency ratios. The following can be concluded:

- when the frequencies are equal (unison) the tones are judged to be 'perfectly consonant'
- when their frequency difference is greater than one critical bandwidth, they are judged consonant
- for frequency differences of between 5% and 50% of the critical bandwidth the interval is dissonant
- maximum dissonance occurs when the frequency difference is a quarter of a critical bandwidth.

Few musical instruments ever produce a sinusoidal acoustic waveform, and the results relating consonance and dissonance to pure tones can be extended to the perception of musical intervals heard when instruments which produce complex periodic tones play together. For each note of the chord, each harmonic that would be resolved by the hearing system if the note were played alone, that is all harmonics up to about the seventh, contributes to the overall perception of consonance or dissonance depending on its frequency proximity to a harmonic of another note in the chord. This contribution can be assessed based on the conclusions from Figure 3.16. The overall consonance (dissonance) of a chord is based on the total consonance (dissonance) contribution from each of these harmonics.

3.3.3 Hearing musical intervals

Musical intervals can be ordered by decreasing consonance on this psychoacoustic basis. To determine the degree of consonance

Table 3.6 The degree of consonance and dissonance of a two-note chord in which all harmonics are present for both notes a perfect fifth apart, the f_0 for the lower note being 220 Hz

			Perfect fifth (3:2); f_0 of lower note = 220 Hz			
First seven harmonics of lower note (Hz)	Harmonics of upper note (Hz)	Frequency difference (Hz)	Mid-frequency (Hz)	Mid-frequency critical band-width (Hz)	Mid-frequency half critical bandwidth (Hz)	Consonant consonant dissonant Dissonant (C, c, d, D)
220						
440	330	110	385	65	32.5	c
660	660	0	Unison	–	–	C
880	990	110	1045	133	66.5	d
1100						
1320	1320	0	Unison	–	–	C
1540	1650	110	1595	193.3	96.7	d

of a musical interval consisting of two complex tones, each with all harmonics present, the frequencies up to the frequency of the seventh harmonic of the lower notes are found, then the critical bandwidth at each frequency mid-way between harmonics of each note that are closest in frequency is found to establish whether or not they are within 5% to 50% of a critical bandwidth and therefore adding a dissonance contribution to the overall perception when the two notes are played together. If the harmonic of the upper note is mid-way between harmonics of the lower note, the test is carried out with the higher frequency pair since the critical bandwidth will be larger and the positions of table entries indicates this. (This exercise is similar to that carried out using the entries in Table 3.5.)

For example, Table 3.6 shows this calculation for two notes whose f_0 values are a perfect fifth apart (f_0 frequency ratio is 3:2), the lower note having an f_0 of 220 Hz. The frequency difference between each harmonic of each note and its closest neighbour harmonic in the other note is calculated (the higher of the two is used in the case of a tied distance), to give the entries in column 3, the frequency mid-way between these harmonic pairs is found (column 4), the critical bandwidth for these mid-frequencies is calculated (column 5). The contribution to dissonance of each of the harmonic pairs is given in the right-hand column as follows:

(i) if they are in unison (equal frequencies) they are 'perfectly consonant', shown as 'C' (note that their frequency difference is less than 5% of the critical bandwidth).

(ii) if their frequency difference is greater than the critical bandwidth of the frequency mid-way between them (i.e. the entry in column 3 is greater than that in column 5) they are 'consonant', shown as 'c'.

(iii) if their frequency difference is less than half the critical bandwidth of the frequency mid-way between them (i.e. the entry in column 3 is less than that in column 6) they are 'highly dissonant', shown as 'D'.

(iv) if their frequency difference is less than the critical bandwidth of the frequency mid-way between them but greater than half that critical bandwidth (i.e. the entry in column 3 is less than that in column 5 and greater than that in column 6) they are 'dissonant', shown as 'd'.

The contribution to dissonance depends on where the musical interval occurs between adjacent harmonics in the natural harmonic series. The higher up the series it occurs, the greater the dissonant contribution made by harmonics of the two notes concerned. The case of a two-note unison is trivial in that all harmonics are in unison with each other and all contribute as 'C'. For the octave, all harmonics of the upper note are in unison with harmonics of the lower note contributions as 'C'. Tables 3.6 to 3.10 show the contribution to dissonance and consonance for the intervals perfect fifth (3:2), perfect fourth (4:3), major third (5:4), minor third (6:5) and major whole tone (9:8) respectively. The dissonance of the chord in each case is related to the entries in the final column which indicate increased dissonance in the order C, c, d and D, and it can be seen that the dissonance increases as the harmonic number increases and the musical interval decreases.

Table 3.7 The degree of consonance and dissonance of a two-note chord in which all harmonics are present for both notes a perfect fourth apart, the f_0 for the lower note being 220 Hz

First seven harmonics of lower note (Hz)	Harmonics of upper note (Hz)	Frequency difference (Hz)	Mid-frequency (Hz)	Mid-frequency critical bandwidth (Hz)	Mid-frequency half critical bandwidth (Hz)	Consonant consonant dissonant Dissonant (C, c, d, D)
220	293	73.0	330	60.0	30.0	c
440						
660	586	73.3	623	89.2	44.6	d
880	879	Unison	–	–	–	C
1100	1172	73.3	1170	147	73.4	D
1320						
1540	1465	73.3	1500	183	91.5	D

Perfect fourth (4:3); f_0 of lower note = 220 Hz

Table 3.8 The degree of consonance and dissonance of a two-note chord in which all harmonics are present for both notes a major third apart, the f_0 for the lower note being 220 Hz

			Major third (5:4); f_0 of lower note = 220 Hz			
First seven harmonics of lower note (Hz)	Harmonics of upper note (Hz)	Frequency difference (Hz)	Mid-frequency (Hz)	Mid-frequency critical band-width (Hz)	Mid-frequency half-critical bandwidth (Hz)	Consonant consonant dissonant Dissonant (C, c, d, D)
220	275	55	248	52.0	26.0	c
440						
660	550	110	605	87.3	43.7	c
880	825	55	853	113	56.4	D
1100	1100	Unison	–	–	–	C
1320	1375	55	1350	166	82.9	D
1540	1650	110	1600	193	96.7	d

Table 3.9 The degree of consonance and dissonance of a two-note chord in which all harmonics are present for both notes a minor third apart, the f_0 for the lower note being 220 Hz

			Minor third (6:5); f_0 of lower note = 220 Hz			
First seven harmonics of lower note (Hz)	Harmonics of upper note (Hz)	Frequency difference (Hz)	Mid-frequency (Hz)	Mid-frequency critical band-width (Hz)	Mid-frequency half-critical bandwidth (Hz)	Consonant consonant dissonant Dissonant (C, c, d, D)
220	264	44	242	51.5	25.8	d
440	528	88	484	75.2	37.6	d
660						
880	792	82	833	111	55.3	d
1100	1056	44	1080	136	68.2	D
1320	1320	Unison	–	–	–	C
1540	1584	44	1560	190	94.8	D

The harmonics which are in unison with each other can be predicted from the harmonic number. For example, in the case of the perfect fourth the fourth harmonic of the lower note is in unison with the third of the upper note because their f_0 values are in the ratio (4:3). For the major whole tone (9:8), the unison will occur between harmonics (the eighth of the upper note and the ninth of the lower) which are not resolved by the auditory system for each individual note.

As a final point, the degree of dissonance of a given musical interval will vary depending on the f_0 value of the lower note, due to the nature of the critical bandwidth with centre frequency

Table 3.10 The degree of consonance and dissonance of a two-note chord in which all harmonics are present for both notes a major whole tone apart, the f_0 for the lower note being 220 Hz

			Major whole tone (9:8); f_0 of lower note = 220 Hz			
First seven harmonics of lower note (Hz)	Harmonics of upper note (Hz)	Frequency difference (Hz)	Mid-frequency (Hz)	Mid-frequency critical bandwidth (Hz)	Mid-frequency half-critical bandwidth (Hz)	Consonant consonant dissonant Dissonant (C, c, d, D)
220	247.5	27.5	234	50.7	25.4	d
440	495.0	55.0	477	74.4	37.2	d
660	742.5	82.5	701	97.1	48.6	d
880						
1100	990.0	110	1050	133	66.5	d
1320	1237.5	82.5	1280	158	79.1	d
1540	1485.0	55.0	1510	184	92.0	D

Table 3.11 The degree of consonance and dissonance of a two-note chord in which all harmonics are present for both notes a major third apart, the f_0 for the lower note being 110 Hz

			Major third (5:4); f_0 of lower note = 110 Hz			
First seven harmonics of lower note (Hz)	Harmonics of upper note (Hz)	Frequency difference (Hz)	Mid-frequency (Hz)	Mid-frequency critical bandwidth (Hz)	Mid-frequency half critical bandwidth (Hz)	Consonant consonant dissonant Dissonant (C, c, d, D)
110	137.5	27.5	124	40.2	20.1	d
220						
330	275.0	55.0	303	57.3	28.7	d
440	412.5	27.5	426	69.5	34.8	D
550	550.0	Unison	–	–	–	C
660	687.5	27.5	674	94.3	47.2	D
770	825.0	55.0	798	107	53.5	d

(e.g. see Figure 3.8). Tables 3.11 and 3.12 illustrate this effect for the major third where the f_0 of the lower note is one octave and two octaves below that used in Table 3.8 at 110 Hz and 55 Hz respectively. The number of 'D' entries increases in each case as the f_0 values of the two notes are lowered.

This increase in dissonance of any given interval, excluding the unison and octave which are equally consonant at any pitch on this basis, manifests itself in terms of preferred chord spacings in classical harmony. As a rule when writing four-part harmony such as SATB (soprano, alto, tenor, bass) hymns, the bass and tenor parts are usually no closer together than a fourth except

Table 3.12 The degree of consonance and dissonance of a two-note chord in which all harmonics are present for both notes a major third apart, the f_0 for the lower note being 55.0 Hz

			Major third (5:4); f_0 of lower note = 55 Hz			
First seven harmonics of lower note (Hz)	Harmonics of upper note (Hz)	Frequency difference (Hz)	Mid-frequency (Hz)	Mid-frequency critical band-width (Hz)	Mid-frequency half-critical bandwidth (Hz)	Consonant consonant dissonant Dissonant (C, c, d, D)
55.0	68.75	13.8	61.9	34.3	17.2	D
110						
165	137.5	27.5	151	42.8	21.4	d
220	206.3	13.8	213	48.7	24.4	D
275	275.0	Unison	–	–	–	C
330	343.8	13.8	337	60.7	30.4	D
385	412.5	27.5	399	66.8	33.4	D

Figure 3.17 Different spacings of the chord of C major. Play each and listen to the degree of 'muddiness' or 'harshness' each produces (see text).

when they are above the bass staff, because the result would otherwise sound 'muddy' or 'harsh'. Figure 3.17 shows a chord of C major in a variety of four-part spacings and inversions which illustrate this effect when the chords are played, preferably on an instrument producing a continuous steady sound for each note such as a pipe organ, instrumental group or suitable synthesiser sound. To realise the importance of this point, it is most essential to *listen* to the effect. The psychoacoustics of music is after all, about how music is perceived, not what it looks like on paper!

3.4 Tuning systems

Musical scales are basic to most Western music. Modern keyboard instruments have 12 notes per octave with a musical interval of one semitone between adjacent notes. All common Western scales incorporate octaves whose frequency ratios are (2:1). Therefore it is only necessary to consider notes in a scale over a range of one octave, since the frequencies of notes in other octaves can be found from them. Early scales were based on one or more of the musical intervals found between members of the natural harmonic series (e.g. see Figure 3.3).

3.4.1 Pythagorean tuning

The Pythagorean scale is built up from the perfect fifth. Starting for example from the note C and going up in 12 steps of a perfect fifth produces the 'circle of fifths': C, G, D, A, E, B, F#, C#, G#, D#, A#, E#, c. The final note after 12 steps around the circle of fifths, shown as c has a frequency ratio to the starting note, C, of the frequency ratio of the perfect fifth (3:2) multiplied by itself 12 times, or:

$$\frac{c}{C} = \left(\frac{3}{2}\right)^{12} = 129.746$$

An interval of twelve fifths is equivalent to 7 octaves, and the frequency ratio for the note (c′) which is 7 octaves above C is:

$$\frac{c'}{C} = 2^7 = 128.0$$

Thus twelve perfect fifths (C to c) is therefore slightly sharp compared with 7 octaves (C to c′) by the so-called 'Pythagorean comma' which has a frequency ratio:

$$\frac{c}{c'} - \frac{129.746}{128.0} = 1.01364$$

If the circle of fifths were established by descending by perfect fifths instead of ascending, the resulting note 12 fifths below the starting notes would be flatter than 7 octaves by 1.0136433, and *every* note of the descending circle would be slightly different to the members of the ascending circle. Figure 3.18 shows this effect

Figure 3.18 The Pythagorean scale is based on the circle of fifths formed either by ascending by 12 perfect fifths (outer) or descending by 12 perfect fifths (inner).

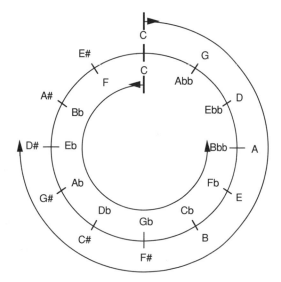

and the manner in which the notes can be notated. For example, notes such as D# and Eb, A# and Bb, Bbb and A are not the same and are known as 'enharmonics', giving rise to the pairs of intervals such as major third and diminished fourth, and major seventh and diminished octave shown in Figure 3.15. The Pythagorean scale can be built up on the starting note C by making F and G an exact perfect fourth and perfect fifth respectively (maintained a perfect relationship for the sub-dominant and dominant respectively):

$$\frac{F}{C} = \frac{4}{3}$$

$$\frac{G}{C} = \frac{3}{2}$$

The frequency ratios for the other notes of the scale are found by ascending in perfect fifths from G and when necessary, bringing the result down to be within an octave of the starting note. The resulting frequency ratios relative to the starting note C are:

$$\frac{D}{C} = \frac{3}{2} \times \frac{G}{C} \times \frac{1}{2} = \frac{3}{2} \times \frac{3}{2} \times \frac{1}{2} = \frac{9}{8}$$

$$\frac{A}{C} = \frac{3}{2} \times \frac{D}{C} = \frac{3}{2} \times \frac{9}{8} = \frac{27}{16}$$

$$\frac{E}{C} = \frac{3}{2} \times \frac{A}{C} \times \frac{1}{2} = \frac{3}{2} \times \frac{27}{16} \times \frac{1}{2} = \frac{81}{64}$$

$$\frac{B}{C} = \frac{3}{2} \times \frac{E}{C} = \frac{3}{2} \times \frac{81}{64} = \frac{243}{128}$$

The frequency ratios of the members of the Pythagorean major scale are shown in Figure 3.19 relative to C for convenience. The frequency ratios between adjacent notes can be calculated by dividing the frequency ratios of the upper note of the pair to C by that of the lower. For example:

$$\text{Frequency ratio between A and B} = \frac{243}{128} \div \frac{27}{16} = \frac{243}{128} \times \frac{16}{27} = \frac{9}{8}$$

$$\text{Frequency ratio between E and F} = \frac{4}{3} \div \frac{81}{64} = \frac{4}{3} \times \frac{64}{81} = \frac{256}{243}$$

Figure 3.18 shows the frequency ratios between adjacent notes of the Pythagorean major scale. A major scale consists of the following intervals: tone, tone, semitone, tone, tone, tone, semitone, and it can be seen that:

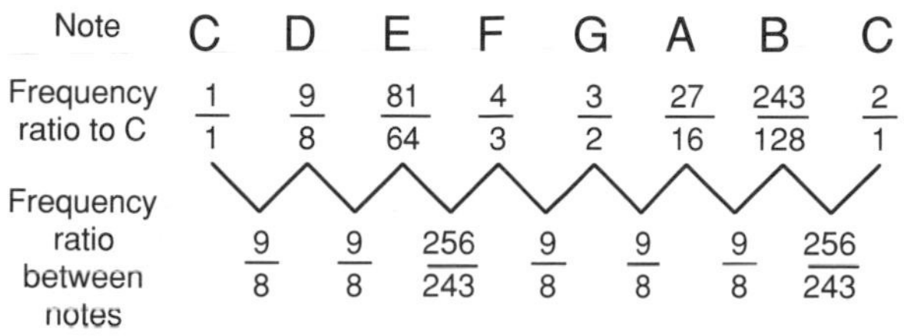

Figure 3.19 Frequency ratios between the notes of a C major Pythagorean scale and the tonic (C).

Note	C	D	E	F	G	A	B	C
Frequency ratio to C	$\frac{1}{1}$	$\frac{9}{8}$	$\frac{81}{64}$	$\frac{4}{3}$	$\frac{3}{2}$	$\frac{27}{16}$	$\frac{243}{128}$	$\frac{2}{1}$
Frequency ratio between notes		$\frac{9}{8}$	$\frac{9}{8}$	$\frac{256}{243}$	$\frac{9}{8}$	$\frac{9}{8}$	$\frac{9}{8}$	$\frac{256}{243}$

$$\text{Frequency ratio of the Pythagorean semitone} = \frac{256}{243}$$

$$\text{Frequency ratio of the Pythagorean tone} = \frac{9}{8}$$

3.4.2 Just tuning

Another important scale is the 'just diatonic' scale which is made by keeping the intervals which make up the major triads pure: the octave (2:1), the perfect fifth (3:2) and the major third (5:4) for triads on the tonic, dominant and sub-dominant. The dominant and sub-dominant keynotes are a perfect fifth above and below the key note respectively. This produces all the notes of the major scale (any of which can be harmonised using one of these three chords). Taking the note C being used as a starting reference for convenience, the major scale is built as follows. The notes E and G are a major third (5:4) and a perfect fifth (3:2) respectively above the tonic, C:

$$\frac{E}{C} = \frac{5}{4}$$

$$\frac{G}{C} = \frac{3}{2}$$

The frequency ratios of B and D are a major third (5:4) and a perfect fifth (3:2) respectively above the dominant, G and they are related to C as:

$$\frac{B}{C} = \frac{5}{4} \times \frac{G}{C} = \frac{5}{4} \times \frac{3}{2} = \frac{15}{8}$$

$$\frac{D}{C} = \frac{3}{2} \times \frac{G}{C} \times \frac{1}{2} = \frac{3}{2} \times \frac{3}{2} \times \frac{1}{2} = \frac{9}{8}$$

(The result for the D is brought down one octave to keep it within an octave of the C.)

Figure 3.20 Frequency ratios between the notes of a C major just diatonic scale and the tonic (C).

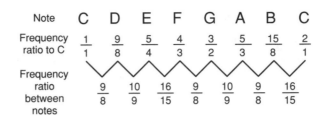

The frequency ratios of A and C are a major third (5:4) and a perfect fifth (3:2) respectively above the sub-dominant, F (the F is therefore a perfect fourth (4:3) above the C (perfect fourth plus a perfect fifth is an octave):

$$\frac{F}{C} = \frac{4}{3}$$

$$\frac{A}{C} = \frac{5}{4} \times \frac{F}{C} = \frac{5}{4} \times \frac{4}{3} = \frac{20}{12} = \frac{5}{3}$$

The frequency ratios of the members of the just diatonic major scale are shown in Figure 3.20 relative to C for convenience, along with the frequency ratios between adjacent notes (calculated by dividing the frequency ratio of the upper note of each pair to C by that of the lower). The figure shows that the just diatonic major scale (tone, tone, semitone, tone, tone, tone, semitone) has equal semitone intervals, but two different tone intervals, the larger of which is known as a 'major whole tone' and the smaller as a 'minor whole tone':

$$\text{Frequency ratio of the just diatonic semitone} = \frac{16}{15}$$

$$\text{and frequency ratio of the just diatonic major whole tone} = \frac{9}{8}$$

$$\text{and frequency ratio of the just diatonic minor whole tone} = \frac{10}{9}$$

The two whole tone and the semitone intervals appear as members of the musical intervals between adjacent members of the natural harmonic series (see Figure 3.3) which means that the notes of the scale are as consonant with each other as possible for both melodic and harmonic musical phrases. However, the presence of two whole tone intervals means that this scale can only be used in one key since each key requires its own tuning. This means, for example, that the interval between D and A is:

$$\frac{A}{D} = \frac{5}{3} \div \frac{9}{8} = \frac{5}{3} \times \frac{8}{9} = \frac{40}{27}$$

which is a musically flatter fifth than the perfect fifth (3:2).

In order to tune a musical instrument for practical purposes to enable it to be played in a number of different keys, the Pythagorean comma has to be distributed amongst some of the fifths in the circle of fifths such that the note reached after twelve fifths is exactly seven octaves above the starting note (see Figure 3.18). This can be achieved by flattening some of the fifths, possibly by different amounts, while leaving some perfect, or flattening all of the fifths by varying amounts, or even by additionally sharpening some and flattening others to compensate. There are therefore an infinite varieties of possibilities, but none will result in just tuning in all keys. Many tuning systems were experimented with which provided tuning of thirds and fifths which were close to just in some keys at the expense of other keys whose tuning could end up being so out of tune as to be unusable musically. Padgham (1986) gives a fuller discussion of tuning systems. A number of keyboard instruments have been experimented with which had split black notes (in either direction) to provide access to their enharmonics, giving C# and Db, D# and Eb, F# and Gb, G# and Ab, and A# and Bb—for example, the McClure pipe organ in the Faculty of Music of the University of Edinburgh discussed by Padgham (1986)—but these have never become popular with keyboard players.

3.4.3 Equal tempered tuning

The spreading of the Pythagorean comma unequally amongst the fifths in the circle results in an 'unequal temperament'. Another possibility is to spread it evenly to give 'equal temperament' which makes modulation to all keys possible where each one is equally out of tune with the just scale. This is the tuning system commonly found on today's keyboard instruments. All semitones are equal to one twelfth of an octave. Therefore the frequency ratio (r) for an equal tempered semitone is therefore a number which when multiplied by itself 12 times is equal to 2, or:

$$r = \sqrt[12]{2} = 1.0595$$

The equal tempered semitone is subdivided into 'cents', where one cent is one hundredth of an equal tempered semitone. The frequency ratio for one cent (*c*) is therefore:

$$c = \sqrt[100]{r} = \sqrt[100]{1.0595} = 1.000578$$

Figure 3.21 Fundamental frequency values to four significant figures for eight octaves of notes, four either side of middle C, tuned in equal temperament with a tuning reference of A4 = 440 Hz. (Middle C is marked with a black spot.)

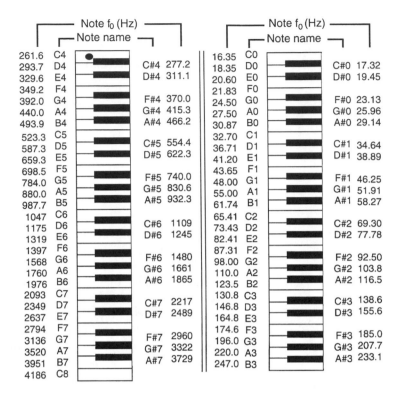

Cents are widely used in discussions of pitch intervals and the results of psychoacoustic experiments involving pitch. Appendix 2 gives an equation for converting frequency ratios to cents and vice versa.

Music can be played in all keys when equal tempered tuning is used as all semitones and tones have identical frequency ratios. However, no interval is in tune in relation to the intervals between adjacent members of the natural harmonic series (see Figure 3.3), therefore none is perfectly consonant. However, intervals of the equal tempered scale can still be considered in terms of their consonance and dissonance, because although harmonics of pairs of notes that are in unison for pure intervals (e.g. see Tables 3.6 to 3.12) are not identical in equal temperament, the difference is within the 5% critical bandwidth criterion for consonance. Beats (see Figure 2.6) will exist between some harmonics in equal tempered chords which are not present in their pure counterparts.

Figure 3.21 shows the f_0 values and the note naming convention used in this book for eight octaves, four either side of middle C, tuned in equal temperament with a tuning reference of 440 Hz for A4; the A above middle C. The equal tempered system is found on modern keyboard instruments, but there is increasing interest amongst performing musicians and listeners alike in the use of unequal temperament. This may involve the use of original instruments or electronic synthesisers which incorporate various tuning systems. Padgham (1986) lists approximately 100 pipe organs in Britain which are tuned to an unequal temperament in addition to the McClure organ.

References

Howard, D.M., Hirson, A., Brookes, T. and Tyrrell, A.M. (1995). Spectrography of disputed speech samples by peripheral human hearing modelling. *Forensic Linguistics*, **2**, (1), 28–38.

Moore, B.C.J. (1982). *An introduction to the Psychology of Hearing*. London: Academic Press.

Moore, B.C.J. (1986). *Frequency Selectivity in Hearing*. London: Academic Press.

Padgham, C.A. (1986). *The Well-tempered Organ*. Oxford: Positif Press.

Parkin, P.H. (1974). Pitch change during reverberant decay. *Journal of Sound and Vibration*, **32**, 530.

Pickles, J.O. (1982). *An Introduction to the Physiology of Hearing*. London: Academic Press.

Plomp, R. and Levelt, W.J.M. (1965). Tonal consonance and critical bandwidth. *Journal of the Acoustical Society of America*, **38**, 548.

Roederer, J.G. (1975). *Introduction to the Physics and Psychophysics of Music*. New York: Springer-Verlag.

Rossing, T.D. (1989). *The Science of Sound*. New York: Addison Wesley.

Schouten, J.F. (1940). The perception of pitch. *Philips Technical Review*, **5**, 286.

Wever, E.G. (1949). *Theory of Hearing*. New York: Wiley.

Zwicker, E., Flottorp, G. and Stevens, S.S. (1957). Critical bandwidth in loudness summation. *Journal of the Acoustical Society of America*, **29**, 548.

4 Acoustic model for musical instruments

4.1 A 'black box' model of musical instruments

In this chapter a simple model is developed which allows the acoustics of all musical instruments to be discussed and, it is hoped, readily understood. The model is used to explain the acoustics of stringed, wind and percussion instruments as well as the singing voice. Any acoustic instrument has two main components:

- a sound source, and
- sound modifiers.

For the purposes of our simple model, the sound source is known as the 'input' and the sound modifiers are known as the 'system'. The result of the input passing through the system is known as the 'output'. Figure 4.1 shows the complete input/system/output model.

This model provides a framework within which the acoustics of musical instruments can be usefully discussed, reviewed and understood. Notice that the 'output' relates to the actual output

Figure 4.1 An input/system/output model for describing the acoustics of musical instruments.

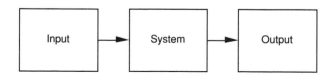

Figure 4.2 The input/system/output model applied to an instrument being played in a room.

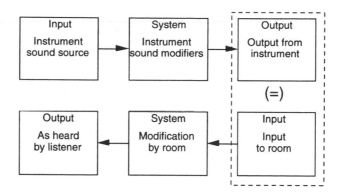

from the instrument which is not that which the listener hears since it is modified by the acoustics of the environment in which the instrument is being played. The input/system/output model can be extended to include the acoustic effects of the environment as follows.

If we are modelling the effect of an instrument being played in a room, then the output we require is the sound heard by the listener and not the output from the instrument itself. The environment itself acts as a sound modifier and therefore it too acts as a 'system' in terms of the input/system/output model. The input to the model of the environment is the output from the instrument being played. Thus the complete practical input/system/output model for an instrument being played in a room is shown in Figure 4.2. Here, the output from the instrument is equal to the input to the room.

In order to make use of the model in practice, acoustic details are required for the 'input' and 'system' boxes to enable the output(s) to be determined. The effects of the room are described in Chapter 6. In this chapter, the 'input' and 'system' characteristics for stringed, wind and percussion instruments as well as the singing voice are discussed. Such details can be calculated theoretically from first principles, or measured experimentally, in which case they must be carried out in an environment which either has no effect on the acoustic recording, or has a known effect which can be accounted for mathematically. An environment which has no acoustic effect is one where there are no reflections of sound—ideally this is known as 'free space'. In practice, free space is achieved in a laboratory in an anechoic ('no echo') room in which all sound reaching the walls, floor and ceiling is totally absorbed by large wedges of sound-absorbing material. However, anechoic rooms are rare, and a useful practical approximation to free space for experimental purposes is outside on grass during a windless day, with the experiment being conducted at a reasonable height above the ground.

This chapter considers the acoustics of stringed, wind, and percussion instruments. In each case, the sound source and the sound modifiers are discussed. These discussions are not intended to be exhaustive since the subject is a large one. Rather they focus on one or two instruments by way of examples as to how their acoustics can be described using the sound source and sound modifier model outlined above. References are included to other textbooks in which additional information can be found for those wishing to explore a particular area more fully.

Finally the singing voice is considered. It is often the case that budding music technologists are able to make good approximations with their voices to sounds they wish to synthesise electronically or acoustically, and a basic understanding of the acoustics of the human voice can facilitate this. As a starting point for the consideration of the acoustics of musical instruments the playing fundamental frequency ranges of a number of orchestral instruments as well as the organ, piano and singers are illustrated in Figure 4.3. A nine octave keyboard is provided for reference on which middle C is marked with a black square.

Figure 4.3 Playing fundamental frequency ranges of selected acoustic instruments and singers.

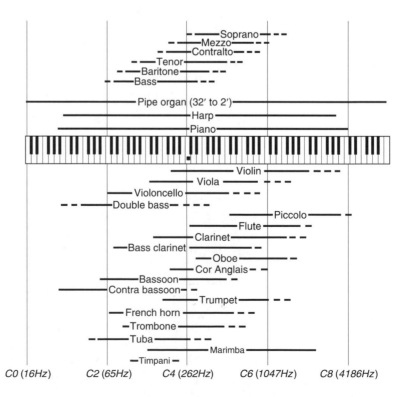

4.2 Stringed instruments

The string family of musical instruments includes the violin, viola, violoncello and double bass and all their predecessors, as well as keyboard instruments which make use of strings, such as the piano, harpsichord, clavichord and spinet. In each case, the acoustic output from the instrument can be considered in terms of an input sound source and sound modifiers as illustrated in Figure 4.1. A more detailed discussion on stringed instruments can be found in Hutchins (1975a,b), Benade (1976), Rossing (1989), Hall (1991) and Fletcher and Rossing (1999). The playing fundamental frequency (f_0) ranges of the orchestral stringed instruments are shown in Figure 4.3.

All stringed instruments consist of one or more strings stretched between two points, and the f_0 produced by the string is dependent on its mass per unit length, length and tension. For any practical musical instrument, the mass per unit length of an individual string is constant and changes are made to the tension and/or the length to enable different notes to be played. Figure 4.4 shows a string fixed at one end, supported on two single-point contact bridges and passed over a pulley with a variable mass hanging on the other end. The variable mass enables the string tension to be altered and the length of the string can be altered by moving the right-hand bridge. In a practical musical instrument, the tension of each string is usually altered by means of a peg with which the string can be wound or winched in to tune the string, and the position of one of the points of support is varied to enable different notes to be played except in instruments such as stringed keyboard instruments where individual strings are provided to play each note.

The string is set into vibration to provide the sound source to the instrument. A vibrating string on its own is extremely quiet because little energy is imparted to the surrounding air due to the small size of a string with respect to the air particle movement it can initiate. All practical stringed instruments have a body which is set in motion by the vibrations of the string(s) of the instrument, giving a large area from which vibration can be imparted to the surrounding air. The body of the instrument is the sound modifier. It imparts its own mechanical properties

Figure 4.4 Idealised string whose tension and length can be varied.

Figure 4.5
Input/system/output model
for a stringed instrument.

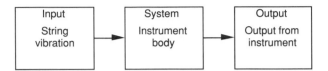

Input	System	Output
String vibration	Instrument body	Output from instrument

onto the acoustic input provided by the vibrating string (see Figure 4.5).

There are three main methods by which energy is provided to a stringed instrument. The strings are either 'plucked', 'bowed' or 'struck'. Instruments which are usually plucked or bowed include those in the violin family, instruments which are generally plucked only include the guitar, lute, and harpsichord, and the piano is an instrument whose strings are struck.

A vibrating string fixed at both ends, for example by being stretched across two bridge-like support as illustrated in Figure 4.4, has a unique set of standing waves (see Chapter 1). Any observed instantaneous shape adopted by the string can be analysed (and synthesised) as a combination of some or all of

Figure 4.6 The first ten possible modes of vibration of a string of length (*L*) fixed at both ends.

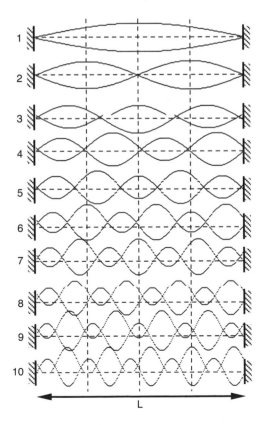

these standing wave modes. The first ten modes of a string fixed at both ends are shown in Figure 4.6. In each case the mode is illustrated in terms of the extreme positions of the string between which it oscillates. Every mode of a string fixed at both ends is constrained not to move, or it cannot be 'displaced', at the ends themselves, and these points are known as 'displacement nodes'. Points of maximum movement are known as 'displacement antinodes'. It can be seen in Figure 4.6 that the first mode has two displacement nodes (at the ends of the string) and one displacement antinode (in the centre). The sixth mode has seven displacement nodes and six displacement antinodes. In general, a particular mode (n) of a string fixed at both ends has ($n + 1$) displacement nodes and (n) displacement antinodes. The frequencies of the standing wave modes are related to the length of the string and the velocity of a transverse wave in a string (see Equation 1.7) by Equation 1.20.

4.2.1 Sound source from a plucked string

When a string is plucked it is pulled a small distance away from its rest position and released. The nature of the sound source it provides to the body of the instrument depends in part on the position on the string at which is plucked. This is directly related to the component modes a string can adopt. For example, if the string is plucked at the centre, as indicated by the central dashed vertical line in Figure 4.6, modes which have a node at the centre of the string (the 2nd, 4th, 6th, 8th, 10th, etc., or the even modes) are not excited, and those with an antinode at the centre (the 1st, 3rd, 5th, 7th, 9th, etc., or the odd modes) are maximally excited. If the string is plucked at a quarter of its length from either end (as indicated by the other dashed vertical lines in the Figure), modes with a node at the plucking point (the 4th, 8th etc.) are not excited and other modes are excited to a greater or lesser degree. In general, the modes that are not excited for a plucking point a distance (d) from the closest end of a string fixed at both ends, are those with a node at the plucking position. They are given by:

$$\text{Modes not excited} = m \left[\frac{L}{d} \right] \tag{4.1}$$

where m = 1, 2, 3, 4, ...
L = length of string
d = distance to plucking point from closest end of the string

Thus if the plucking point is a third of the way along the string, the modes not excited are the 3rd, 6th, 9th, 12th, 15th, etc. For a component mode not to be excited at all, it should be noted that the plucking distance has to be exactly an integer fraction of the length of the string in order that it exactly coincides with nodes of that component.

This gives the sound input to the body of a stringed instrument when it is plucked. The frequencies (f_n) of the component modes of a string supported at both ends can be related to the length, tension (T) and mass per unit length (μ) of the string by substituting Equation 1.7 for the transverse wave velocity in Equation 1.20 to give:

$$f_n = \left[\frac{n}{2L}\right]\sqrt{\frac{T}{\mu}} \tag{4.2}$$

where n = 1, 2, 3, 4, ...
L = length
T = tension
μ = mass per unit length

The frequency of the lowest mode is given by Equation 4.2 when (n=1):

$$f_1 = \left[\frac{1}{2L}\right]\sqrt{\frac{T}{\mu}}$$

This is the f_0 of the string which is also known as the first harmonic (see Table 3.1). Thus the first mode (f_1) in Equation 4.2 is the f_0 of string vibration. Equation 4.2 shows that the frequencies of the higher modes are harmonically related to f_0.

4.2.2 Sound source from a struck string

When a stringed instrument is struck such as in a piano, the same relationship exists between the point at which the strike occurs and the modes that will be missing in the sound source. There is, however, an additional effect that is particularly marked in the piano to consider. Piano strings are very hard and they are under enormous tension compared with the strings on plucked instruments. When a piano string is stuck, it behaves partly like a bar because it is not completely flexible since it has some stiffness. This results in a slight raising in frequency of all the component modes with the effect being greater for the higher modes, resulting in the modes no longer being exact integer multiples of the fundamental mode. This effect, known as 'inharmonicity', varies as the square of the component mode (n^2), or harmonic number, and as the fourth power of the string radius (R^4). Thus for a particular string, the third mode is shifted nine times (3^2) as much as the first, or fundamental, mode, and a doubling in string radius increases inharmonicity by a factor of sixteen (2^4). The effect would therefore be considerably greater for bass strings if they were simply made thicker to give them greater mass, and in many stringed instruments, including pianos, guitars and violins, the bass strings are wrapped with wire to increase their mass without increasing their stiffness. (A detailed discussion of the acoustics of pianos is given in Benade, 1976; Askenfelt, 1990; Fletcher and Rossing, 1999.)

The notes of a piano are usually tuned to equal temperament (see Chapter 3) and octaves are then tuned by minimising the beats between pairs of notes an octave apart. When tuning two notes an octave apart, the components which give rise to the strongest sensation of beats are the first harmonic of the upper note and the second harmonic of the lower note. These are tuned in unison to minimise the beats between the notes. This results in the f_0 of the lower note being slightly lower than half the f_0 of the higher note due to the inharmonicity between the first and second components of the lower note.

Example 4.1 If the f_0 of a piano note is 400 Hz and inharmonicity results in the second component being stretched to 801 Hz, how many cents sharp will the note an octave above be if it is tuned for no beats between it and the octave below?

Tuning for no beats will results in the f_0 of the upper note being 801 Hz, slightly greater than an exact octave above 400 Hz which would be 800 Hz. The frequency ratio (801/800) can be converted to cents using equation A2.4 in Appendix 2:

$$\text{Number of cents} = 3986.3137 \log_{10} \frac{801}{800} = 2.16 \text{ cents}$$

Inharmonicity on a piano increases as the strings become shorter and therefore the octave stretching effect increases with note pitch. The stretching effect is usually related to middle C and it becomes greater the further away the note of interest is in pitch. Figure 4.7 illustrates the effect in terms of the average deviation from equal-tempered tuning across the keyboard of a small piano. Thus high and low notes on the piano are tuned sharp and flat respectively to what they would have been if all octaves were tuned pure with a frequency ratio of 2:1. From the Figure it can be seen that this stretching effect amounts to approximately 35 cents sharp at C8 and 35 cents flat at C1 with respect to middle C.

Figure 4.7 Approximate form of the average deviations from equal temperament due to inharmonicity in a small piano. Middle C marked with a spot. (Data from Martin and Ward, 1961.)

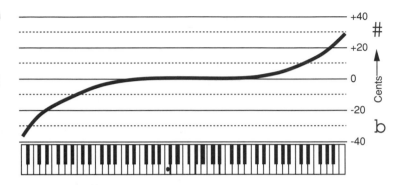

4.2.3 Sound source from a bowed string

The sound source that results from bowing a string is periodic and a continuous note can be produced while the bow travels in one direction. A bow supports many strands of hair, traditionally horsehair. Hair tends to grip in one direction but not in the other. This can be demonstrated with your own hair. Support the end of one hair firmly with one hand, and then grip the hair in the middle with the thumb and index finger of the other hand and slide that hand up and down the hair. You should feel the way the hair grips in one direction but slides easily in the other.

The hairs of a bow are laid out such that approximately half grip when the bow is moved in one direction and half when it is moved in the other. Rosin is applied to the hairs of a bow to increase its gripping ability. As the bow is moved across a string in either direction, the string is gripped and moved away from its rest position until the string releases itself, moving past its rest position until the bow hairs grip it again to repeat the cycle. One complete cycle of the motion of the string immediately under a bow moving in one direction is illustrated in the graph on the right-hand side of Figure 4.8. (When the bow moves in the other direction, the pattern is reversed.) The string moves at a constant velocity when it is gripped by the bow hairs and then returns rapidly through its rest position until it is gripped by the bow hairs again. If the minute detail of the motion of the bowed string is observed closely, for example by means of stroboscopic illumination, it is seen to consist of two straight-line segments joining at a point which moves at a constant velocity around the dotted track as shown in the snapshot sequence in Figure 4.8. The time taken to complete one cycle, or the fundamental period (T_0) is the time taken for the point joining the two line segments to travel twice the length of the string (2L):

$$T_0 = \frac{2L}{v}$$

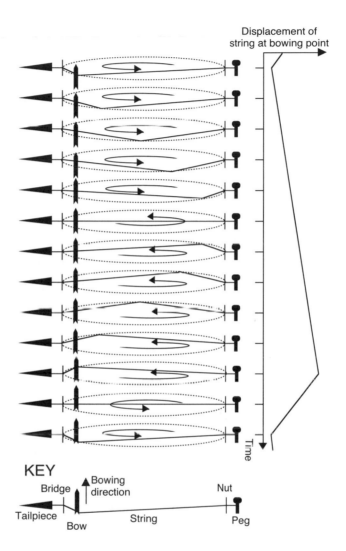

Figure 4.8 One complete cycle of vibration of a bowed string and graph of string displacement at the bowing point as a function of time. (Adapted from Rossing, 1989.)

Substituting equation 1.7 for the transverse wave velocity gives:

$$T_0 = 2L \sqrt{\frac{\mu}{T}}$$

The f_0 of vibration of the bowed string is therefore:

$$f_0 = \left[\frac{1}{T_0}\right] = \left[\frac{1}{2L}\right] \sqrt{\frac{T}{\mu}} \qquad (4.3)$$

Comparison with Equation 4.2 when ($n=1$) shows that this is the frequency of the first component mode of the string. Thus the f_0 for a bowed string is the frequency of the first natural mode of the string, and bowing is therefore an efficient way to excite the vibrational modes of a string

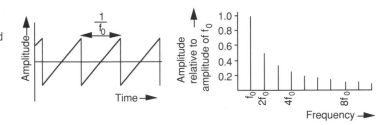

Figure 4.9 Idealised sound source sawtooth waveform and its spectrum for a bowed string.

The sound source from a bowed string is that of the waveform of string motion which excites the bridge of the instrument. Each of the snapshots in Figure 4.8 correspond to equal time instants on the graph of string displacement at the bowing point in the Figure, from which the resulting force acting on the bridge of the instrument can be inferred to be of a similar shape to that at the bowing point. In its ideal form, this is a sawtooth waveform (see Figure 4.9). The spectrum of an ideal sawtooth waveform contains all harmonics and their amplitudes decrease with ascending frequency as $(1/n)$, where n is the harmonic number. The spectrum of an ideal sawtooth waveform is plotted in Figure 4.9 and the amplitudes are shown relative to the amplitude of the f_0 component.

4.2.4 Sound modifiers in stringed instruments

The sound source provided by a plucked or bowed string is coupled to the sound modifiers of the instrument via a bridge. The vibrational properties of all elements of the body of the instrument play a part in determining the sound modification that takes place. In the case of the violin family, the components which contribute most significantly are the top plate (the plate under the strings which the bridge stands on and which has the f holes in it), the back plate (the back of the instrument), and the air contained within the main body of the instrument. The remainder of the instrument contributes to a very much lesser extent to the sound-modification process, and there is still lively debate in some quarters about the importance or otherwise of the glues, varnish, choice of wood and wood treatment used by highly regarded violin makers of the past.

Two acoustic resonances dominate the sound modification due to the body of instruments in the violin family at low frequencies: the resonance of the air contained within the body of the instrument or the 'air resonance', and the main resonance of the top plate or 'top resonance'. Hall (1991) summarises the important resonance features of a typical violin as follows:

- practically no response below the first resonance at approximately 273 Hz (air resonance)
- another prominent resonance at about 473 Hz (top resonance)
- rather uneven response up to about 900 Hz, with a significant dip around 600–700 Hz
- better mode overlapping and more even response (with some exceptions) above 900 Hz
- gradual decrease in response toward high frequencies.

Apart from the air resonance which is defined by the internal dimensions of the instrument and the shape and size of the *f* holes, the detailed nature of the response of these instruments is related to the main vibrational modes of the top and back plates. As these plates are being shaped by detailed carving, the maker will hold each plate at particular points and tap it to hear how the so-called 'tap tones' are developing to guide the shaping process. This ability is a vital part of the art of the experienced instrument maker in setting up what will become the resonant properties of the complete instrument when it is assembled.

The acoustic output from the instrument is the result of the sound input being modified by the acoustic properties of the instrument itself. Figure 4.10 (from Hall, 1991) shows the input spectrum for a bowed G3 (f_0 = 196 Hz) with a typical response curve for a violin, and the resulting output spectrum. Note that the frequency scales are logarithmic, therefore the harmonics in the input and output spectra bunch together at high frequencies. The output spectrum is derived by multiplying the amplitude of each component of the input spectrum by the response of the body of the instrument at that frequency. In the figure, this multiplication becomes an addition since the amplitudes are expressed logarithmically as dB values, and adding logarithms of numbers is mathematically equivalent to multiplying the numbers themselves.

There are basic differences between the members of the orchestral string family (violin, viola, cello and double bass) differ from each other acoustically in that the size of the body of each instrument becomes smaller relative to the f_0 values of the open strings (e.g. Hutchins, 1978). The air and tap resonances approximately coincide as follows: for the violin with f_0 of the D4 (2nd string) and A4 (3rd string) strings respectively, for the viola with f_0 values approximately midway between the G3 and D4 (2nd and 3rd strings) and D4 and A4 (3rd and 4th strings) strings respectively, for the cello with f_0 of the G2 string and approximately midway between the D3 and A3 (3rd and 4th

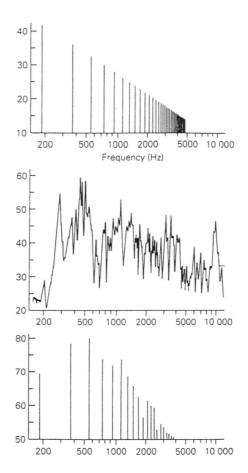

Figure 4.10 Sound source spectrum for: a bowed G3 ($f_0 = 196$ Hz); sound modifier response curve for a typical violin, and the resulting output spectrum. (From Figure 11.9 in *Musical Acoustics*, by Donald E. Hall, © 1991 Brooks/Cole Publishing Company, Pacific Grove, CA 93950, by permission of the publisher).

strings) respectively, and for the double bass with f_0 of the D2 (3rd string) and G2 (4th string) strings respectively. Thus there is more acoustic support for the lower notes of the violin than for those of the viola or the double bass, and the varying distribution of these two resonances between the instruments of the string family is part of the acoustic reason why each member of the family has its own characteristic sound.

Figure 4.11 shows waveforms and spectra for notes played on two plucked instruments, C3 on a lute and F3 on a guitar. The decay of the note can be seen on the waveforms, and in each case the note lasts just over a second. The pluck position can be estimated from the spectra by looking for those harmonics which are reduced in amplitude that are integer multiples of each other (see Equation 4.2). The lute spectrum suggests a pluck point at approximately one sixth of the string length due to the clear amplitude dips in the 6th and 12th harmonics, but there are also clear dips at the 15th and 20th harmonics. An important point

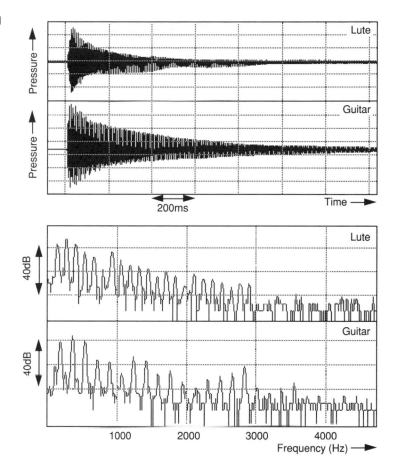

Figure 4.11 Waveforms and spectra for C3 played on a lute and F3 played on a six-string guitar.

to note is that this is the spectrum of the output from the instrument, and therefore it includes the effects of the sound modifiers (e.g. air and plate resonances), so harmonic amplitudes are affected by the sound modifiers as well as the sound source. Also, the 15th and 20th harmonics are nearly 40 dB lower than the low harmonics in amplitude and therefore background noise will have a greater effect on their amplitudes. The guitar spectrum also suggests particularly clearly a pluck point at approximately one sixth of the string length, given the dips in the amplitudes of the 6th, 12th and 18th harmonics.

Sound from stringed instruments does not radiate out in all directions to an equal extent and this can make a considerable difference if, for example, one is critically listening to or making recordings of members of the family. The acoustic output from any stringed instrument will contain frequency components across a wide range, whether it is plucked, struck or bowed. In general, low frequencies are radiated in essentially all directions, with the pattern of radiation becoming more directionally focused as frequency increases from the mid to high range. In the case of the violin, low frequencies in this context are those up to approximately 500 Hz, and high frequencies, which tend to radiate outwards from the top plate, are those above approximately 1000 Hz. The position of the listener's ear or a recording microphone is therefore an important factor in terms of the overall perceived sound of the instrument.

4.3 Wind instruments

The discussion of the acoustics of wind instruments involves similar principles to those used in the discussion of stringed instruments. However, the nature of the sound source in wind instruments is rather different but the description of the sound modifiers in wind instruments has much in common with that relating to possible modes on a string. This section concentrates on the acoustics of organ pipes to illustrate the acoustics of sound production in wind instruments. Some of the acoustic mechanisms basic to other wind instruments are given later in the section.

Wind instruments can be split into those without and those with reeds. This is the basis on which they are presented in this section. For convenience, organ pipes are used as the model. Organ pipes can also be split into two types based on the sound source mechanism involved: 'flues' and 'reeds'. Figure 4.12 shows the main parts of flue and reed pipes. Each is constructed of a particular material, usually wood or a tin–lead alloy, and has a resonator of a set shape, size depending on the sound that pipe is designed to produce (e.g. Audsley, 1965; Sumner, 1975; Norman and Norman, 1980). The sources of sound in the flue and the reed pipe will be considered first, followed by the sound modification that occurs due to the resonator.

4.3.1 Sound source in organ flue pipes

The source of sound in flue pipes is described in detail in Hall (1991) and his description is followed here. The important features of a flue sound source are a narrow slit (the flue) through which air flows, and a wedge-shaped obstacle placed in the airstream from the slit. Figure 4.13 shows the detail of this mechanism for a wooden organ flue pipe (the similarity with a metal organ flue pipe can be observed in Figure 4.12). A narrow slit exists between the lower lip and the languid, and this is known as the 'flue', and the wedge-shaped obstacle is the upper

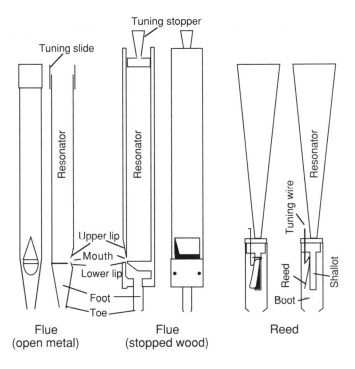

Figure 4.12 The main parts of flue (open metal and stopped wood) and reed organ pipes.

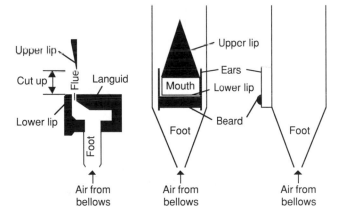

Figure 4.13 The main elements of the sound source in organ flue pipes based on a wooden flue pipe (left) and additional features found on some metal pipes (centre and left).

lip which is positioned in the airstream from the flue. This obstacle is usually placed off-centre to the airflow.

Air enters the pipe from the organ bellows via the foot and a thin sheet of air emerges from the flue. If the upper lip were not present, the air emerging from the flue would be heard as noise. This indicates that the airstream is turbulent. A similar effect can be observed if you form the mouth shape for the 'ff' in *far*, in which the bottom lip is placed in contact with the upper teeth and produce the 'ff' sound. The airflow is turbulent, producing the acoustic noise which can be clearly heard. If the airstream flow rate is reduced, there is an air velocity below which turbulent flow ceases and acoustic noise is no longer heard. At this point the airflow has become smooth or 'laminar'. Turbulent airflow is the mechanism responsible for the non-pitched sounds in speech such as the 'sh' in *shoe* and the 'ss' in *sea* for which waveforms and spectra are shown in Figure 3.9.

When a wedge-like obstruction is placed in the airstream emerging from the flue a definite sound is heard known as an 'edgetone'. Hall suggests a method for demonstrating this by placing a thin card in front of the mouth and blowing on its edge. Researchers are not fully agreed on the mechanism which underlies the sound source in flues. The preferred explanation is illustrated in Figure 4.14, and it is described in relation to the sequence of snapshots in the figure as follows. Air flows to one side of the obstruction, causing a local increase in pressure on that side of it. This local pressure increase causes air in its local vicinity to be moved out of the way, and some finds its way in a circular motion into the pipe via the mouth. This has the effect of 'bending' the main stream of air increasingly, until it flips into the pipe. The process repeats itself, only this time the local pressure increase causes air to move in a circular motion out of the pipe via the mouth, gradually bending the main airstream until it flips outside the pipe again. The cycle then repeats providing a periodic sound source to the pipe itself. This process is sometimes referred to as a vibrating 'air reed' due to the regular flipping to and fro of the airstream.

Figure 4.14 Sequence of events to illustrate the sound source mechanism in a flue organ pipe.

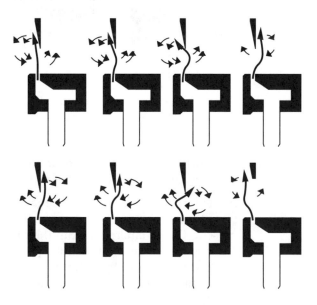

The f_0 of the pulses generated by this air reed mechanism in the *absence* of a pipe resonator is directly proportional to the airflow velocity from the flue, and inversely proportional to the length of the cut-up:

$$f_0 \propto \frac{v_j}{L_{cut\text{-}up}} \qquad\qquad (4.4)$$

where $\propto$ means 'is proportional to'
$\quad\quad f_0$ = fundamental frequency of air reed oscillation in absence of pipe resonator
$\quad\quad v_j$ = airflow velocity
$\quad\quad L_{cut\text{-}up}$ = length of cut-up

In other words, f_0 can be raised by either increasing the airflow velocity or reducing the cut-up. As the airflow velocity is increased or the cut-up size is decreased, there comes a point where the f_0 jumps up in value. This effect can be observed in the presence of a resonator with respect to increasing the airflow velocity by blowing with an increasing flow rate into a recorder (or if available, a flue organ pipe). It is often referred to as an 'overblown' mode.

The acoustic nature of the sound source in flues is set by the pipe voicer, whose job it is to determine the overall sound from individual pipes and to establish an even tone across complete ranks of pipes. The following comments on the voicer's art in relation to the sound source in flue pipes are summarised from Norman and Norman (1980), who give the main modifications made by the voicer in order of application as:

* adjusting the cut-up
* 'nicking' the languid and lower lip
* adjusting languid height with respect to that of the lower lip

Adjusting the cut-up needs to be done accurately to achieve an even tone across a rank of pipes. This is achieved on metal pipes by using a sharp, short thick bladed knife. A high cut-up produces a louder and more 'hollow' sound, and a lower cut-up gives a softer and 'edgier' sound. The higher the cut-up, the greater the airflow required from the foot. However, the higher the air flow, the less prompt the speech of the pipe.

Nicking relates to a series of small nicks that are made in the approximating edges of the languid and the upper lip. This has the effect of reducing the high-frequency components in the sound source spectrum and giving the pipe a smoother, but slower, onset to its speech. More nicking is customarily applied to pipes which are heavily blown. A pipe which is not nicked has a characteristic consonantal attack to its sound, sometimes referred to as a 'chiff'. A current trend in organ voicing is the use of less or no nicking in order to take advantage of the onset chiff musically to give increased clarity to notes particularly in contrapuntal music (e.g. Hurford, 1994).

The height of the languid is fixed at manufacture for wooden pipes, but it can be altered for metal pipes. The languid controls, in part, the direction of the air flowing from the flue. If it is too high, the pipe will be slow to speak and may not speak at all if the air misses the upper lip completely. If it is too low the pipe will speak too quickly, or speak in an uncontrolled manner. A pipe is adjusted to speak more rapidly if it is set to speak with a consonantal chiff by means of little or no nicking. Narrow scaled pipes (small diameter compared with the length), usually having a 'stringy' tone colour often have ears added (see Figure 4.12) which stabilise air reed oscillation, and some bass pipes also have a wooden roller or 'beard' placed between the ears to aid prompt pipe speech.

4.3.2 Sound modifiers in organ flue pipes

The sound modifier in a flue organ pipe is the main body of the pipe itself, or its 'resonator' (see Figure 4.12). Organ pipe resonators are made in a variety of shapes developed over a number of years to achieve subtleties of tone colour, but the most straightforward to consider are resonators whose dimensions do not vary along their length, or resonators of 'uniform cross-section'. Pipes made of metal are usually round in cross-section and those made of wood are generally square (some builders make triangular wooden pipes, partly to save on raw material). These shapes arise mainly from ease of construction with the material involved.

There are two basic types of organ flue pipes, those that are open and those that are stopped at the end farthest from the flue itself (see Figure 4.12). The flue end of the pipe is acoustically equivalent to an open end. Thus the open flue pipe is acoustically open at both ends, and the stopped flue pipe is acoustically open at one end and closed at the other. The air reed sound source mechanism in flue pipes as illustrated in Figure 4.14 launches a pulse of acoustic energy into the pipe When a compression (positive amplitude) pulse of sound pressure energy is launched into a pipe, for example at the instant in the air reed cycle illustrated in the lower-right snapshot in Figure 4.14, it travels down the pipe at the velocity of sound as a compression pulse.

Figure 4.15 The reflected pulses resulting from a compression (upper) and rarefaction (lower) pulse arriving at an open (left) and a stopped (right) end of a pipe of uniform cross-section. Note: Time axes marked in equal arbitrary units.

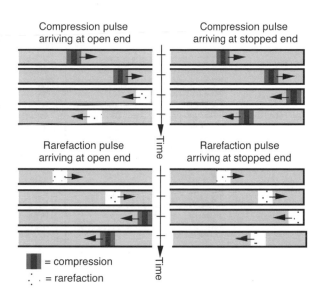

When the compression pulse reaches the far end of the pipe, it is reflected in one of the two ways described in the 'standing waves' section of Chapter 1, depending (Section 1.5.7) on whether the end is open or closed. At a closed end there is a pressure antinode and a compression pulse is reflected back down the pipe. At an open end there is a pressure node and a compression pulse is reflected back as a rarefaction pulse to maintain atmospheric pressure at the open end of the pipe. Similarly, a rarefaction pulse arriving at a closed end is reflected back as a rarefaction pulse, but as a compression pulse when reflected from an open end. All four conditions are illustrated in Figure 4.15.

When the action of the resonator on the air reed sound source in a flue organ pipe is considered (see Figure 4.14), it is found that the f_0 of air reed vibration is entirely controlled by: (a) the length of the resonator, and (b) whether the pipe is open or stopped. This dependence of the f_0 of the air reed vibration can be appreciated by considering the arrival and departure of pulses at each end of the open and the stopped pipe.

Figure 4.16 shows a sequence of snapshots of pressure pulses generated by the air reed travelling down an open pipe of length L_o (left) and a stopped pipe of length L_c (right) and how they drive the vibration of the air reed. (Air reed vibration is illustrated in a manner similar to that used in Figure 4.14.) The figure shows pulses moving from left to right in the upper third of each pipe, those moving from right to left in the centre third, and the summed pressure in the lower third. A time axis with

Figure 4.16 Pulses travelling in open (left) and stopped (right) pipes when they drive an air reed sound source. Note: Time axis marked in equal arbitrary time units; pulses travelling left to right are shown in the upper part of each pipe, those going right to left are shown in the centre and the sum is shown in the lower part.

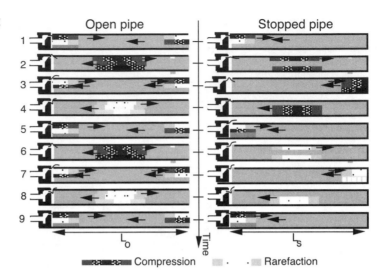

arbitrary but equal units is marked in the figure to show equal time intervals. The pulses travel an equal distance in each frame of the figure since an acoustic pulse moves at a constant velocity. The flue end of the pipe acts as an open end in terms the manner in which pulses are reflected (see Figure 4.15). At every instant when a pulse arrives and is reflected from the flue end, the air reed is flipped from inside to outside when a compression pulse arrives and is reflected as a rarefaction pulse, and vice versa when a rarefaction pulse arrives. This can be observed in Figure 4.16.

For the open pipe, the sequence in the figure begins with a compression pulse being launched into the pipe, and another compression pulse just leaving the open end (the presence of this second pulse will be explained shortly). The next snapshot shows the instant when these two pulses reach the centre of the pipe, their summed pressure being a maximum at this point. The pulses effectively travel through each other and emerge with their original identities due to 'superposition' (see Chapter 1). In the third snapshot the compression pulse is being reflected from the open end of the pipe as a rarefaction pulse, and the air reed flips outside the pipe, generating a rarefaction pulse. (This may seem strange at first, but it is a necessary consequence of the event happening in the fifth snapshot.) The fourth snapshot shows two rarefactions at the centre giving a summed pressure which is a minimum at this instant of twice the rarefaction pulse amplitude. In the fifth snapshot, when the rarefaction pulse is reflected from the flue end as a compression pulse, the air reed is flipped from outside to inside the pipe. One cycle is complete at this point since events in the fifth and first snapshots are similar. (A second cycle is illustrated to enable comparison with events in the stopped pipe.)

The fundamental period for the open pipe is the time taken to complete a complete cycle (i.e. the time between a compression pulse leaving the flue end of the pipe and the next compression pulse leaving the flue end of the pipe). In terms of Figure 4.16, it is four time frames (snapshot one to snapshot five), being the time taken for the pulse to travel down to the other end and back again (see Figure 4.15), or twice the open pipe length:

$$T_{0(open)} = \left[\frac{2L_o}{c} \right]$$

where $T_{0(open)}$ = fundamental period of open pipe
        L_o = length of the open pipe
        c = velocity of sound

The f_0 value for the open pipe is therefore:

$$f_{0(open)} = \left[\frac{1}{T_{0(open)}} \right] = \left[\frac{c}{2L_o} \right] \tag{4.5}$$

In the stopped pipe, the sequence in Figure 4.15 again begins with a compression pulse being launched into the pipe, but there is no second pulse. Snapshot two shows the instant when the pulses reach the centre of the pipe, and the third snapshot the instant when the compression pulse is reflected from the stopped end as a compression pulse (see Figure 4.15) and the summed pressure is a maximum for the cycle of twice the amplitude of the compression pulse. The fourth snapshot shows the compression pulse at the centre and in the fifth, the compression pulse is reflected from the flue end as a rarefaction pulse, flipping the air reed from inside to outside the pipe. The sixth snapshot shows the rarefaction pulse half way down the pipe and the seventh shows its reflection as a rarefaction pulse from the stopped end when the summed pressure there is the minimum for the cycle of twice the amplitude of the rarefaction pulse. The eighth snapshot shows the rarefaction pulse half-way back to the flue end, and by the ninth, one cycle is complete, since events in the ninth and first snapshots are the same.

It is immediately clear that one cycle for the stopped pipe takes twice as long as one cycle for the open pipe if the pipe lengths are equal (ignoring a small end correction which has to be applied in practice), its fundamental period is therefore double that for the open pipe, and its f_0 is therefore half that for the open pipe, or an octave lower. This can be quantified by considering the time taken to complete a complete cycle is the time required for the pulse to travel to the other end of the pipe and back twice, or four times the stopped pipe length (see Figure 4.15):

$$T_{0(stopped)} = \left[\frac{4L_s}{c}\right]$$

where $T_{0(stopped)}$ = fundamental period of stopped pipe
L_s = length of the stopped pipe
c = velocity of sound

Therefore:

$$f_{0(stopped)} = \left[\frac{1}{T_{0(stopped)}}\right] = \left[\frac{c}{4L_s}\right] \tag{4.6}$$

Example 4.2 If an open pipe and a stopped pipe are the same length, what is the relationship between their f_0 values?
Let $(L_s = L_o = L)$ and substitute into Equations 4.5 and 4.6:

$$f_{0(open)} = \left[\frac{c}{2L}\right]$$

$$f_{0(stopped)} = \left[\frac{c}{4L}\right]$$

Therefore:

$$f_{0(stopped)} = \frac{1}{2}\left[\frac{c}{2L}\right] = \left[\frac{f_{0(open)}}{2}\right]$$

Therefore $f_{0(stopped)}$ is an octave lower than $f_{0(open)}$ (frequency ratio 1:2).

The natural modes of a pipe are constrained as described in the 'standing waves' section of Chapter 1. Equation 1.20 gives the frequencies of the modes of an open pipe and Equation 1.21 gives the frequencies of the modes of a stopped pipe. In both equations, the velocity is the velocity of sound (c).

The frequency of the first mode of the open pipe is given by Equation 1.20 when ($n = 1$):

$$f_{open(1)} = \left[\frac{c}{2L_o}\right] \tag{4.7}$$

which is the same value obtained in Equation 4.5 by considering pulses in the open pipe. Using Equation 1.20, the frequencies of the other modes can be expressed in terms of its f_0 value as follows:

$$f_{open(2)} = \left[\frac{2c}{2L_o}\right] = 2f_{open(1)}$$

$$f_{\text{open}(3)} = \left[\frac{3c}{2L_o}\right] = 3f_{\text{open}(1)}$$

$$f_{\text{open}(4)} = \left[\frac{4c}{2L_o}\right] = 4f_{\text{open}(1)}$$

In general:

$$f_{\text{open}(n)} = nf_{\text{open}(1)} \tag{4.8}$$

The modes of the open pipe are thus all harmonically related and all harmonics are present. The musical intervals between the modes can be read from Figure 3.3.

The frequency of the fundamental mode of the stopped pipe is given by equation 1.21 when ($n=1$):

$$f_{\text{stopped}(1)} = \left[\frac{c}{4L_s}\right] \tag{4.9}$$

This is the same value obtained in Equation 4.6 by considering pulses in the stopped pipe. The frequencies of the other stopped pipe modes can be expressed in terms of its f_{stopped} using Equation 1.21 as follows:

$$f_{\text{stopped}(2)} = \left[\frac{3c}{4L_s}\right] = 3f_{\text{stopped}(1)}$$

$$f_{\text{stopped}(3)} = \left[\frac{5c}{4L_s}\right] = 5f_{\text{stopped}(1)}$$

$$f_{\text{stopped}(4)} = \left[\frac{7c}{4L_s}\right] = 7f_{\text{stopped}(1)}$$

In general:

$$f_{\text{stopped}(n)} = (2n-1)\, f_{\text{stopped}(1)} \tag{4.10}$$

where $n = 1, 2, 3, 4, \ldots$

Thus the modes of the stopped pipe are harmonically related, but only the odd-numbered harmonics are present. The musical intervals between the modes can be read from Figure 3.3.

In open and stopped pipes the pipe's resonator acts as the sound modifier and the sound source is the air reed. The nature of the spectrum of the air reed source depends on the detailed shape of the pulses launched into the pipe, which in turn depends on the pipe's voicing summarised above. If a pipe is overblown, its f_0 jumps to the next higher mode that the resonator can support: up one octave to the second harmonic for an open pipe and up an octave and a fifth to the third harmonic for the stopped pipe.

Figure 4.17 Waveforms and spectra for middle C (C4) played on a gedackt 8' (stopped flue) and a principal 8' (open flue).

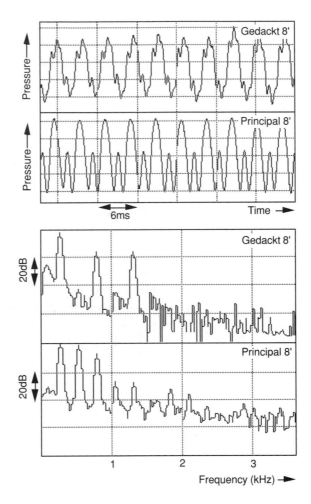

Figure 4.17 Waveforms and spectra for middle C (C4) played on a gedackt 8' (stopped flue) and a principal 8' (open flue).

The length of the resonator controls the f_0 of the air reed (see Figure 4.15) and the natural modes of the pipe are the frequencies that the pipe can support in its output. The amplitude relationship between the pipe modes is governed by the material from which the pipe is constructed and the diameter of the pipe with respect to its length. In particular, wide pipes tend to be weak in upper harmonics. Organ pipes are tuned by adjusting the length of their resonators. In open pipes this is usually done nowadays by means of a tuning slide fitted round the outside of the pipe at the open end, and for a stopped pipes by moving the stopper (see Figure 4.12).

A stopped organ pipe has a f_0 value which is an octave below that of an open organ pipe (Example 4.2), and where space is limited in an organ, stopped pipes are often used in the bass register and played by the pedals. However, the trade-off is between the physical space saved and the acoustic result in that only the odd-numbered harmonics are supported. Figure 4.17 illustrates this with waveforms and spectra for middle C played on a gedackt 8' and a principal 8' (Section 5.4 describes organ stop footages: 8', 4' etc.). The gedackt stop has stopped wooden pipes, and the spectrum clearly shows the presence of odd harmonics only, in particular, the first, third and fifth. The principal stop consists of open metal pipes, and odd and even harmonics exist in its output spectrum. Although the pitch of these stops is equivalent, and they are therefore both labelled 8', the stopped gedackt pipe is half the length of the open principal pipe.

4.3.3 Woodwind flue instruments

Other musical instruments which have an air reed sound source include the recorder and the flute. Useful additional material on woodwind flue instruments can be found in Benade (1976) and Fletcher and Rossing (1999). The air reed action is controlled by oscillatory changes in flow of air in and out of the flue (see Figure 4.16), often referred to as a *flow-controlled valve*, and therefore there must be a displacement antinode and a pressure node. Hence the flue end of the pipe is acting as an open end and woodwind flue instruments act acoustically as pipes open at both ends (see Figure 4.18).

Figure 4.18 The first four pressure and displacement modes of an open and a stopped pipe of uniform cross-section. Note: The plots show maximum and minimum amplitudes of pressure and displacement.

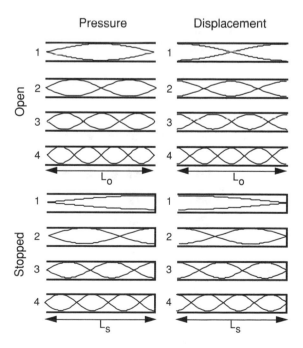

Players are able to play a number of different notes on the same instrument by changing the effective acoustic length of the resonator. This can be achieved, for example, by means of the sliding piston associated with a swanee whistle or more commonly when particular notes are required by covering and uncovering holes in the pipe walls known as 'finger holes'. A hole in a pipe will act in an acoustically similar manner to an open pipe end (pressure node, displacement antinode). The extent to which it does this is determined by the diameter of the hole with respect to the pipe diameter. When these are large with respect to the pipe diameter, as in the flute, they equal the uncovered hole acts acoustically as if the pipe had an open end at that position. Smaller finger holes result acoustically in the effective open end being further (away from the flue end) down the pipe. This is an important factor in the practical design of instruments with long resonators since it can enable the finger holes to be placed within the physical reach of a player's hands. It does, however, have an importance consequence on the frequency relationship between the modes, and this is explored in detail below in connection with woodwind reed instruments. The other way to give a player control over finger holes which are out of reach, for example on a flute, is by providing each hole with a pad controlled by a key mechanism of rods and levers operated by the player's fingers to close or open the hole (depending whether the hole is normally open or closed by default).

In general, a row of finger holes is gradually uncovered to effectively shorten the acoustic length of the resonator as an ascending scale is played. Occasionally some cross-fingering is used in instruments with small holes or small pairs of holes such as the recorder as illustrated in Figure 4.19. Here, the pressure node is

Figure 4.19 Fingering chart for recorders in C (descants and tenors).

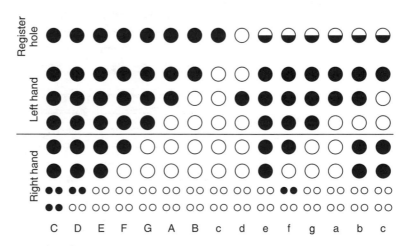

further away from the flue than the first uncovered hole itself such that the state of other holes beyond it will affect its position. The figure shows typical fingerings used to play a two octave C major scale on a descant or tenor recorder. Hole fingerings are available to enable notes to be played which cover a full chromatic scale across one octave. To play a second octave on woodwind flue instruments such as the recorder or flute, the flue is overblown. Since these instruments are acoustically open at both ends, the overblown flue jumps to the second mode which is one octave higher than the first (see Equation 4.8 and Figure 3.3). The finger holes can be reused to play the notes of the second octave.

Once an octave and a fifth above the bottom note has been reached, the flue can be overblown to the third mode (an octave and a fifth above the first mode) and the fingering can be started again to ascend higher. The fourth mode is available at the start of the third octave and so on. Overblowing is supported in instruments such as the recorder by opening a small 'register' or 'vent' hole which is positioned such that it is at the pressure antinode for unwanted modes and these modes will be suppressed. The register hole marked in Figure 4.19 is a small hole on the back of the instrument which is controlled by the thumb of the left hand which either covers it completely, half covers it by pressing the thumb nail end-on against it, or uncovers it completely. To suppress the first mode in this way without affecting the second, this hole should be drilled in a position where the undesired mode has a pressure maximum. When all the tone holes are covered, this would be exactly half-way down the resonator, a point where the first mode has a pressure maximum and is therefore reduced, but the second mode has a pressure node and is therefore unaffected (see Figure 4.18). Register holes can be placed at other positions to enable overblowing to different modes. In practice, register holes may be set in compromise positions because they have to support all the notes available in that register, for which the effective pipe length is altered by uncovering tone holes.

A flute has a playing range between B3 and D7 and the piccolo sounds one octave higher between B4 and D8 (see Figure 4.3). Flute and piccolo players can control the stability of the overblown modes by adjusting their lip position with respect to the embouchure hole as illustrated in Figure 4.20. The air reed mechanism can be compared with that of flue organ pipes illustrated in Figures 4.13 and 4.14 as well as the associated discussion relating to organ pipe voicing. The flautist is able to adjust the distance between the flue outlet (the player's lips) and the

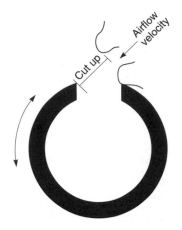

Figure 4.20 Illustration of lip to embouchure adjustments available to a flautist.

edge of the mouthpiece, marked as the 'cut-up' in the Figure, a term borrowed from organ nomenclature (see Figure 4.13), by rolling the flute as indicated by the double-ended arrow. In addition, the airflow velocity can be varied as well as the fine detailed nature of the airstream dimensions by adjusting the shape, width and height of the opening between the lips. The flautist therefore has direct control over the stability of the overblown modes (Equation 4.4).

4.3.4 Sound source in organ reed pipes

The basic components of an organ reed pipe is shown in Figure 4.12. The sound source results from the vibrations of the reed, which is slightly larger than the shallot opening, against the edges of the shallot. Very occasionally, organ reeds make use of 'free reeds' which are cut smaller than the shallot opening and they move in and out of the shallot without coming into contact with its edges. In its rest position as illustrated in Figure 4.12, there is a gap between the reed and shallot, enabled by the slight curve in the reed itself. The vibrating length of the reed is governed by the position of the 'tuning wire', or 'tuning spring' which can be nudged up or down to make the vibrating length longer or shorter, accordingly lowering or raising the f_0 of the reed vibration.

The reed vibrates when the stop is selected and a key on the appropriate keyboard is pressed. This causes air to enter the boot and flow past the open reed via the shallot to the resonator. The gap between the reed and shallot is narrow and for air to flow, there must be a higher pressure in the boot than the shallot. The higher pressure in the boot than the shallot tends to close the reed fractionally, resulting in the gap between the reed and shallot being narrowed. When the gap is narrowed, the airflow rate is increased and the pressure difference which supports this increased airflow increases. The increase in pressure difference exerts a slightly greater closing force on the reed, and this series of events continues, accelerating the reed towards the shallot until it hits the edge of the shallot, closing the gap completely and rapidly.

The reed is springy and once the gap is closed and the flow has dropped to zero, the reed's restoring force causes the reed to spring back towards its equilibrium position, opening the gap. The reed overshoots its equilibrium position, stops and returns towards the shallot, in a manner similar to its vibration if it had been displaced from its equilibrium position and released by hand. Air flow is restored via the shallot and the cycle repeats.

In the absences of a resonator, the reed wound vibrate at its natural frequency. This is the frequency at which it would vibrate if it were plucked, or displaced from its equilibrium position and released by hand. If a plucked reed continues to vibrate for a long time, then it has a strong tendency to vibrate at a frequency within a narrow range, but if it vibrates for a short time, there is a wide range of frequencies over which it is able to vibrate. This effect is illustrated in Figure 4.21. This difference

Figure 4.21 Time (left) and frequency (right) responses of hard (upper) and soft (lower) reeds when plucked. Natural frequency (F_N) and natural period (T_N) are shown.

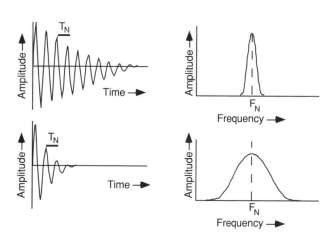

is exhibited depending on the material from which the reed is made and how it is supported. A reed which vibrates over a narrow frequency range is usually made from brass and supported rigidly, and is known as a 'hard' reed. A reed which vibrates over a wide range might be made from cane or plastic and held in a pliable support is known as a 'soft' reed. As shown in the figure, the natural period (T_N) is related to the natural frequency (F_N) as:

$$F_N = \frac{1}{T_N} \qquad\qquad (4.11)$$

A reed vibrating against a shallot shuts of the flow of air rapidly and totally, and the consequent acoustic pressure variations are the sound source provided to the resonator. The rapid shutting off of the airflow produces a rapid, instantaneous drop in acoustic pressure within the shallot (as air flowing fast into the shallot is suddenly cut off). A rapid amplitude change in a waveform indicates a relatively high proportion of high harmonics are present. The exact nature of the sound source spectrum depends on the particular reed, shallot and bellows pressure being considered. Free reeds which do not make contact with a shallot, as found, for example, in a harmonica or harmonium, do not produce as high a proportion of high harmonics since the airflow is never completely shut off.

4.3.5 Sound modifiers in organ reed pipes

All reed pipes have resonators. The effect of a resonator has already been described and illustrated in Figure 4.16 in connection with air reeds. The same principles apply to reed pipes, but there is a major difference in that the shallot end of the resonator acts as a stopped end (as opposed to an open end as in the case of a flue). This is because during reed vibration, the pipe is either closed completely at the shallot end (when the reed is in contact with the shallot) or open with a very small aperture compared with the pipe diameter.

Organ reed pipes have hard reeds, which have a narrow natural frequency range (see Figure 4.21). Unlike the air reed, the presence of a resonator does not control the frequency of vibration of the hard reed. The sound-modifying effect of the resonator is based on the modes it supports (e.g. see Figure 4.18), bearing in mind the closed end at the shallot. Because the reed itself fixes the f_0 of the pipe, the resonator does not need to reinforce the fundamental and fractional length resonators are sometimes used to support only the higher harmonics. Figure 4.22 shows waveforms and spectra for middle C (C4) played on a hautbois 8′, or oboe 8′, and a trompette 8′, or trumpet 8′. Both spectra exhibit an overall peak around the sixth/seventh harmonic. For the trompette this peak is quite broad with the odd harmonics dominating the even ones up to the tenth harmonic, probably a feature of its resonator shape. The hautbois

Figure 4.22 Waveform and spectra for middle C (C4) played on a hautbois 8' and a trompette 8'.

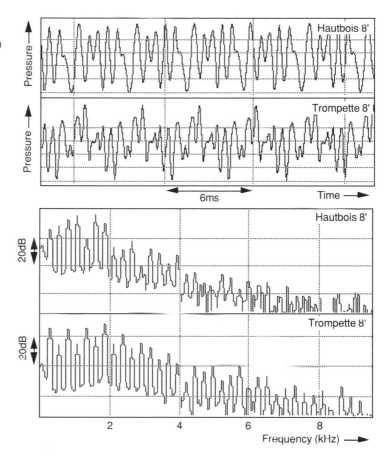

spectrum exhibits more dips in the spectrum than the trompette – these are all features which characterise the sounds of different instruments as being different.

4.3.6 Woodwind reed instruments

Woodwind reed instruments make use of either a single or a double vibrating reed sound source which controls the flow of air from the player's lungs to the instrument. The action of a vibrating reed at the end of a pipe is controlled as a function of the relative air pressure on either side of it in terms of when it opens and closes. It is therefore usually described as a *pressure-controlled valve*, and the reed end of the pipe acts as a *stopped end* (pressure antinode and displacement node—see Figure 4.18). Note that although the reed opens and closes such that airflow is not always zero, the reed opening is very much smaller than the pipe diameter elsewhere making a stopped end reasonable.

This is in direct contrast to the air reed in woodwind flue instruments such as the flute and recorder (see above), which is a flow-controlled valve which provides a displacement antinode and a pressure node and hence the flue end of the pipe is acting as an open end (see Figure 4.18).

Soft reeds are employed in woodwind reed instruments which can vibrate over a wide frequency range (see Figure 4.21). The reeds in clarinets and saxophones are single reeds which can close against the edge of the mouthpiece as in organ reed pipes where they vibrate against their shallots. The oboe and bassoon on the other hand use double reeds, but the basic opening and closing action of the sound source mechanism is the same.

Woodwind reed instruments have resonators whose modal behaviour is crucial to the operation of the instrument and provide the sound modifier function. Woodwind instruments incorporate finger holes to enable chromatic scales to be played from the first mode to the second mode when the fingering can be used again as the reed excites the second mode. These mode changes continue up the chromatic scale to cover the full playing range of the instrument (see Figure 4.3). Clearly it is essential that the modes of the resonator retain their frequency ratios relative to each other as the tone holes are opened or else the instrument's tuning will be adversely affected as higher modes are reached. Benade (1976) summarises this effect and indicates the resulting constraint as follows:

> Preserving a constant frequency ratio between the vibrational modes as the holes are opened is essential in all woodwinds and provides a limitation on the types of air column (often referred to as the bore) that are musically useful.

The musically useful bores in this context are based on tubing that is either cylindrical as in the clarinet, or conical as in the oboe, cor Anglais, and members of the saxophone and bassoon families. The cylindrical resonator of a clarinet acts as a pipe that is stopped at the reed end (see above) that is open at the other. Odd-numbered modes only are supported by such a resonator (see Figure 4.18), and its f_0 is an octave lower (see Example 4.2) than that of an instrument with a similar length pipe which is open at both ends, such as a flute (see Figure 4.3). The first overblown mode of a clarinet is therefore the third mode, an interval of an octave and a fifth (see Figure 3.3), and therefore unlike a flute or recorder, it has to have sufficient holes to enable at least 19 chromatic notes to be fingered within the first mode prior to transition to the second.

Conical resonators that are stopped at the reed end and open at the other support all modes in a harmonically related manner. Taylor (1976) gives a description of this effect as follows:

> Suppose by some means we can start a compression from the narrow end; the pipe will behave just as our pipe open at both ends until the rarefaction has returned to the start. Now, because the pipe has shrunk to a very small bore, the speed of the wave slows down and no real reflection occurs. ... The result is that we need only consider one journey out and one back regardless of whether the pipe is open or closed at the narrow end. ... The conical pipe will behave something like an pipe open at both ends as far as its modes are concerned.

The conical resonator therefore supports all modes, and the overblown modes of instruments with conical resonators, such as the oboe, cor Anglais, bassoon and saxophone family, is therefore to the second mode, or up an octave. Sufficient holes are therefore required for at least 12 chromatic notes to be fingered to enable the player to arrive at the second mode from the first.

The presence of a sequence of open tone holes in a pipe resonator of any shape is described by Benade (1976) as a *tone-hole lattice*. The effective acoustical end-point of the pipe varies slightly as a function of frequency when there is a tone-hole lattice and therefore the effective pipe length is somewhat different for each mode. A pipe with an tone-hole lattice is acoustically shorter for low frequency standing wave modes compared with higher-frequency modes, and therefore the higher-frequency modes are increasingly lowered slightly in frequency (lengthening the wavelength lowers the frequency). Above a particular frequency, described by Benade (1976) the *open-holes lattice cut-off frequency* (given as around 350–500 Hz for quality bassoons, 1500 Hz for quality clarinets and between 1100 and 1500 Hz for quality oboes), sound waves are not reflected due to the presence of the lattice. Benade notes that this has a direct effect on the perceived timbre of woodwind instruments, correlating well with descriptions such as *bright* or *dark* given to instruments by players. It should also be noted that holes that are closed modify the acoustic properties of the pipe also, and this can be effectively modelled as a slight increase in pipe diameter at the position of the tone hole. The resulting acoustic change is considered below.

In order to compensate for these slight variations in the frequencies of the modes produced by the presence of open and closed tone holes, alterations can be made to the shape of the pipe. These might include flaring the open end, adding a tapered section, or small local voicing adjustments by enlarging or constricting the pipe which on a wooden instrument can be achieved by reaming out or adding wax respectively (e.g. Nederveen, 1969). The acoustic effect on individual pipe mode frequencies of either enlarging or constricting the size of the pipe depends directly on the mode's distribution of standing wave pressure nodes and antinodes (or displacement antinodes and nodes respectively). The main effect of a constriction in relation to pressure antinodes (displacement nodes) is as follows (Kent and Read, 1992):

- a constriction near a pressure node (displacement antinode) lowers that mode's frequency
- a constriction near a pressure antinode (displacement node) raises that mode's frequency.

A constriction at a pressure node (displacement antinode) has the effect of reducing the flow at the constriction since the local pressure difference across the constriction has not changed. Benade (1976) notes that this is equivalent to raising the local air density, and the discussion in Chapter 1 indicates that this will result in a lowering of the velocity of sound (see Equation 1.1) and therefore a lowering in the mode frequency (see Equations 4.7 and 4.9). A constriction at a pressure antinode (displacement node), on the other hand, provides a local rise in acoustic pressure which produces a greater opposition to local air flow of the sound waves that combine to produce the standing wave modes. This is equivalent to raising the local springiness in the medium, which is shown in Chapter 1 to be equivalent for air of Young's modulus, which raises the velocity of sound (see Equation 1.5) and therefore raises the mode frequency (see Equations 4.7 and 4.9). By the same token, the effect of locally enlarging a pipe will be exactly opposite to that of constricting it.

Knowledge of the position of the pressure and displacement nodes and antinodes for the standing wave modes in a pipe therefore allows the effect on the mode frequencies of a local constriction or enlargement of a pipe to be predicted. Figure 4.23 shows the potential mode frequency variation for the first three modes of a cylindrical stopped pipe that could be caused by a constriction or enlargement at any point along its length. (The equivalent diagram for a cylindrical pipe open at both ends could be readily produced with reference to Figures 4.18 and 4.23; this is left as an exercise for the interested reader.)

The upper part of Figure 4.23 (taken from Figure 4.18) indicates the pressure and displacement node and antinode positions for the first three standing wave modes. The lower part of the Figure

Figure 4.23 The effect of locally constricting or enlarging a stopped pipe on the frequencies of its first three modes: '+' indicates raised modal frequency, '–' indicates lowered modal frequency, and the magnitude of the change is indicated by the size of the '+' or '–' signs. The first three pressure and displacement modes of a stopped pipe are shown for reference: 'N' and 'A' indicate node and antinode positions respectively .

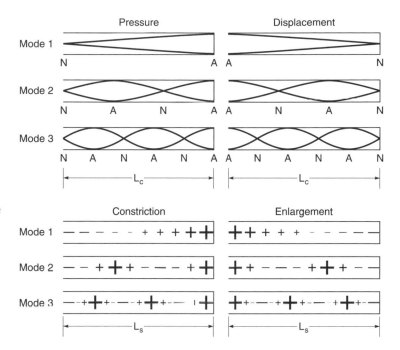

exhibits plus and minus signs to indicate where that particular mode's frequency would be raised or lowered respectively by a local constriction or enlargement at that position in the pipe. The size of the signs indicate the sensitivity of the frequency variation based on how close the constriction is to the mode's pressure/displacement nodes and antinodes shown in the upper part of the Figure. For example, a constriction close to the closed end of a cylindrical pipe will raise the frequencies of all modes since there is a pressure antinode at a closed end, whereas an enlargement at that position would lower the frequencies of all modes. However, if a constriction or enlargement were made one third the way along a stopped cylindrical pipe from the closed end, the frequencies of the first and third modes would be raised somewhat, but that of the second would be lowered maximally. By creating local constrictions or enlargements, the skilled maker is able to set up a woodwind instrument to compensate for the presence of tone holes such that the modes remain close to being in integer frequency ratios over the playing range of the instrument.

Figure 4.24 shows waveforms and spectra for the note middle C played on a clarinet and a tenor saxophone. The saxophone spectrum contains all harmonics since its resonator is conical. The clarinet spectrum exhibits the odd harmonics clearly as its resonator is a cylindrical pipe closed at one end (see Figure 4.18),

Figure 4.24 Waveforms and spectra for middle C (C4) played on a clarinet and a tenor saxophone.

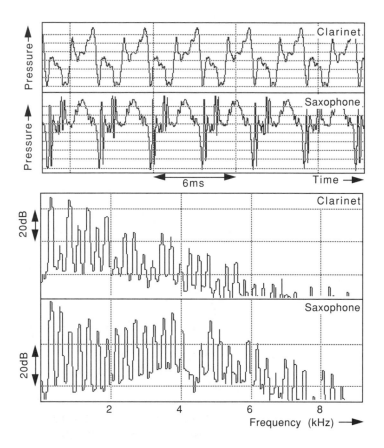

but there is also energy clearly visible in some of the even harmonics. Although the resonator itself does not support the even modes, the spectrum of the sound source does contain all harmonics (the saxophone and the clarinet are both single reed instruments). Therefore some energy will be radiated by the clarinet at even harmonics.

Sundberg (1989) summarises this effect for the clarinet as follows:

> This means that the even numbered modes are not welcome in the resonator ... A common misunderstanding is that these partials are all but missing in the spectrum. The truth is that the second partial may be about 40 dB below the fundamental, so it hardly contributes to the timbre. Higher up in the spectrum there differences between odd- and even-numbered neighbours are smaller. Further ... the differences can be found only for the instruments' lower tones.

This description is in accord with the spectrum in Figure 4.24, where the amplitude of the second harmonic is approximately

40 dB below that of the fundamental, and the odd/even differences become less with increased frequency.

4.3.7 Brass instruments

The brass instrument family has an interesting history from early instruments derived from natural tube structures such as the horns of animals, seashells and plant stems, through a variety of wooden and metal instruments to today's metal brass orchestral family (e.g. Fletcher and Rossing, 1999; Campbell and Greated, 1998). The sound source in all brass instruments is the vibrating lips of the player in the mouthpiece. They form a double soft reed, but the player has the possibility of adjusting the physical properties of the double reed by lip tension and shape. The lips act as a pressure-controlled valve in the manner described in relation to the woodwind reed sound source, and therefore the mouthpiece end of the instrument acts acoustically as a stopped end (pressure antinode and displacement node—see Figure 4.18).

The double reed action of the lips can be illustrated if the lips are held slightly apart, and air is blown between them. For slow airflow rates, nothing is heard, but as the airflow is increased, acoustic noise is heard as the airflow becomes turbulent. If the flow is increased further, the lips will vibrate together as a double reed. This vibration is sustained by the physical vibrational properties of the lips themselves, and an effect known as the 'Bernoulli effect'.

As air flows past a constriction, in this case the lips, its velocity increases. The Bernoulli effect is based on the fact that at all points, the sum of the energy of motion, or 'kinetic' energy, plus the pressure energy, or 'potential' energy, must be constant at all points along the tube. Figure 4.25 illustrates this effect in a tube with a flexible constriction. Airflow direction is represented

Figure 4.25 An illustration of the Bernoulli effect (potential energy + kinetic energy = a constant) in a tube with a constriction. Note: Lines with arrows represent airflow direction, and the distance between them is proportional to the airflow velocity. PE = potential energy; KE = kinetic energy.

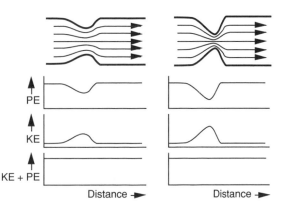

by the lines with arrows, and the velocity of airflow is represented by the distance between these lines. Since airflow increases as it flows through the constriction, the kinetic energy increases. In order to satisfy the Bernoulli principle that the total energy remains constant, the potential energy or the pressure at the point of constriction must therefore reduce. This means that the force on the tube walls is lower at the point of constriction.

If the wall material at the point of constriction is elastic and the force exerted by the Bernoulli effect is sufficient to move their mass (such as the brass players lips) from its rest (equilibrium) position, then the walls are sucked together a little (compare the right- and left-hand illustrations in the figure). Now the kinetic energy (airflow velocity) becomes greater because the constriction is narrower, thus the potential energy (pressure) must reduce some more to compensate (compare the graphs in the Figure), and the walls of the tube are sucked together with greater force. Therefore the walls are accelerated together as the constriction narrows until they smack together, cutting off the airflow. The air pressure in the tube tends to push the constriction apart, as does the natural tendency of the walls to return to their equilibrium position. Like two displaced pendulums, the walls move past their equilibrium position, stop and return towards each other and the Bernoulli effect accelerates them together again. The oscillation of the walls will be sustained by the airflow, and the vibration will be regular if the two walls at the point of constriction have similar masses and tensions, such as the lips.

The lip reed vibration is supported by the resonator of the brass instrument formed by a length of tubing attached to a mouthpiece. Some mechanism is provided to enable the player to vary the length of the tube, originally, for example, in the horn family by adding different lengths of tubing or 'crooks' by hand. Nowadays this is accomplished by means of a sliding section as in the trombone or by adding extra lengths of tubing by means of valves. The tube profile in the region of the trombone slide or tuneable valve mechanism has to be cylindrical in order for slides to function.

All brass instruments consist of four sections (see Figure 4.26): mouthpiece, a tapered mouthpipe, a main pipe fitted with slide or valves which is cylindrical (e.g. trumpet, French horn, trombone) or conical (e.g. cornet, fluegelhorn, baritone horn, tuba), and a flared bell (Benade, 1976; Hall, 1991). If a brass instrument consisted only of a conical main pipe, all modes would be supported (see discussion on woodwind reed instruments above),

Figure 4.26 Basic sections of a brass instrument.

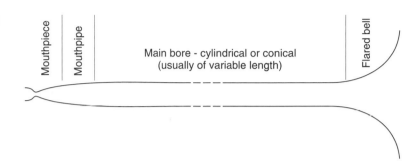

but if it were cylindrical, it acts as a stopped pipe due to the pressure-controlled action of the lip reed and therefore only odd-numbered modes would be supported (see Figure 4.18). However, instruments in the brass family support almost all modes which are essentially harmonically related due to the acoustic action of the addition of the mouthpiece and bell.

The bell modifies as a function of frequency the manner in which the open end of the pipe acts as a reflector of sound waves arriving there from within the pipe. A detailed discussion is provided by Benade (1976) from which a summary is given here. Lower-frequency components are reflected back into the instrument from the narrower part of the bell whilst higher frequency components are reflected from the wider regions of the bell. Frequencies higher than a cut-off frequency determined by the diameter of the outer edge of the bell (approximately 1500 Hz for a trumpet) are not reflected appreciably by the bell. Adding a bell to the main bore of the instrument has the effect of making the effective pipe length longer with increasing frequency. The frequency relationship between the modes of the stopped cylindrical pipe (odd-numbered modes only: $1f$, $3f$, $5f$, $7f$, etc.) will therefore be altered such that they are brought closer together in frequency. This effect is greater for the first few modes of the series.

The addition of a mouthpiece at the other end of the main bore also affects the frequency of some of the modes. The mouthpiece consists of a cup-shaped cavity which communicates via a small aperture with a short conical pipe. The mouthpiece has a resonant frequency associated with it, which is generally in the region of 850 Hz for a trumpet, which is otherwise known as the *popping frequency* since it can be heard by slapping its lip contact end on the flattened palm of one hand (Benade, 1976). The addition of a mouthpiece effectively extends the overall pipe length by an increasing amount. Benade notes that this effect *'is a steady increase nearly to the top of the instrument's playing range'*, and that a mouthpiece with a *'lower popping frequency will show*

191

a greater total change in effective length as one goes up in frequency' (Benade, 1976, p.416). This pipe length extension caused by adding a mouthpiece therefore has a greater downwards frequency shifting effect on the higher compared with the lower modes.

In a complete brass instrument, it is possible through the use of an appropriately shaped bell, mouthpiece and mouthpipe to construct an instrument whose modes are frequency shifted from the odd only modes of a stopped cylindrical pipe to being very close to a complete harmonic series. In practice, the result is a harmonic series where all modes are within a few per cent of being integer multiples of a common lower-frequency value except for the first mode itself, which is well below that lower frequency value common to the higher modes, and therefore it is not harmonically related to them. The effects of the addition of the bell and mouthpiece/mouthpipe on the individual lowest six modes are broadly as summarised in Figure 4.27. Here the odd-numbered modal frequencies of the stopped cylindrical pipe are denoted as integer multiples of frequency 'f', and the resulting brass instrument modal frequencies are shown as multiples of another frequency 'F'.

Figure 4.27 Brass instrument mode frequency modification to stopped cylindrical pipe by the addition of mouthpiece/mouthpipe and bell.

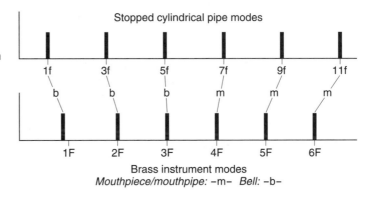

The second mode is therefore the lowest musically usable mode available in a brass instrument (note that the lowest mode does not correspond with 1F). Overblowing from the second mode to the third mode results in a pitch jump of a perfect fifth, or seven semitones. The addition of three valves to brass instruments (except the trombone), each of which adds a different length of tubing when it is depressed, enables six semitones to be played, sufficient to progress from the first to the second mode. Assuming this is from the written notes C4 to G4, the six required semitones are: C#4, D4, D#4, E4, F4, and F#4. Figure 4.28 shows how this is achieved. The centre (or second) valve lowers the pitch by one

Figure 4.28 The basic valve combinations used on brass instruments to enable seven semitones to be fingered. Note: Black circle = valve depressed; white circle = valve not depressed; on a trumpet, first valve is nearest mouthpiece, second in the middle and third nearest the bell.

Semitones	−1	−2	−3	−3	−4	−5	−6
1st valve (−2 semitones)	○	●	●	○	○	●	●
2nd valve (−1 semitone)	●	○	●	○	●	○	●
3rd valve (−3 semitones)	○	○	○	●	●	●	●

semitone, the first valve (nearest the mouthpiece) by two semitones, and the third valve by three semitones. Combinations of these valves therefore in principle enable the required six semitones to be played. It may at first sight seem odd that there are two valve fingerings for a lowering of three semitones (third valve alone or first and second valves together) as shown in the Figure. This relates to a significant problem in relation to the use of valves for this purpose which is described below.

Assuming equal-tempered tuning for the purposes of this section, it was shown in Chapter 3 that the frequency ratio for one semitone (1/12 of one octave) is:

$$r = \sqrt[12]{2} = 1.0595$$

The decrease in frequency required to lower a note by one semitone is therefore 5.95%, and this is also the factor by which a pipe should be lengthened by the second valve on a brass instrument. Depressing the first valve only should lower the f_0 and hence lengthen the pipe by 12.25% since the frequency ratio for two semitones is the square of that for one semitone ($1.0595^2 = 1.1225$). Depressing the first and second valve together will lengthen the pipe by 18.2% (12.25% + 5.95%), which is not sufficient for three semitones since this requires the pipe to be lengthened by 18.9% ($1.0595^3 = 1.1893$). The player must lip notes using this valve combination down in pitch. The third valve is also set nominally to lower the f_0 by three semitones, but because of the requirement to add a larger length the further down that is progressed, it is set to operate with the first valve to produce an accurate lowering of five semitones. Five semitones is equivalent to 33.51% ($1.0595^5 = 1.3351$), and subtracting the lowering produced by the first valve gives the extra pipe length required from the third valve as 21.26% (33.51%−12.25%), which is rather more than both the 18.2% available from the combination of the first and second valves as well as the 18.9% required for an accurate three-semitone lowering. In practice on a trumpet, for example, the third valve is often fitted with a

Figure 4.29 Waveforms and spectra for C3 played on a trombone and a tuba.

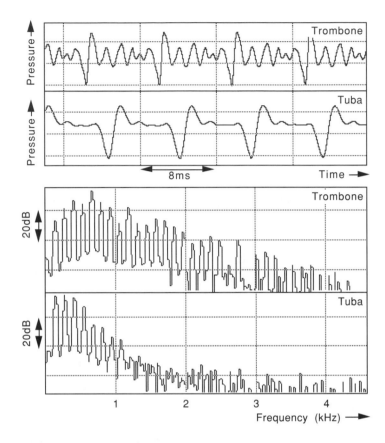

tuning slide so that the player can alter the added pipe length while playing. No such issues arise for the trombonist who can alter the slide position accurately to ensure the appropriate additional pipe lengths are added for accurate tuning of the intervals.

Figure 4.29 shows waveforms and spectra for the note C3 played on a trombone and a tuba. The harmonics in the spectrum of the trombone extend far higher in frequency than those of the tuba. This effect can be seen by comparing the shape of their waveforms where the trombone has many more oscillations during each cycle than the tuba. In these examples, the first three harmonics dominate the spectrum of the tuba in terms of amplitude and eight harmonics can be readily seen, whereas the fifth harmonic dominates the spectrum of the trombone, and harmonics up to about the 29th can be identified.

4.4 Percussion instruments

The percussion family is an important class of instruments which can also be described acoustically in terms of the 'black box' model. Humans have always struck objects, whether to draw attention or to imbue others and themselves with rhythm. Rhythm is basic to all forms of music in all cultures and members of the percussion family are often used to support it. Further reading in this area can be found in Benade, 1976; Rossing, 1990; Hall, 1991; and Fletcher and Rossing, 1998.

4.4.1 Sound source in percussion instruments

The sound source in percussion instruments usually involves some kind of striking. This is most often by means of a stick, but not, for example, in a cymbal crash. Such a sound source is known as an 'impulse'. The spectrum of a single impulse is continuous since it is non-periodic (i.e. it never repeats), and all frequency components are present. Therefore any instrument which is struck is excited by an acoustic sound source of short duration in which all frequencies are present. All modes that the instrument can support will be excited, and each will respond in the same way that the plucked reed vibrates as illustrated in Figure 4.21. The narrower the frequency band of the mode, the longer it will 'ring' for. (One useful analogy is the impulse provided if a parent pushes a child on a swing just once. The child will swing back and forth at the natural frequency of the swing and child, and the amplitude of the swinging will gradually diminish. A graph of swing position against time would be similar to the time response for the hard reed plotted in Figure 4.21.)

4.4.2 Sound modifiers in percussion instruments

Percussion instruments are characterised acoustically by the modes of vibration that they are able to support, and the position of the strike point with respect to the node and antinode points of each mode (e.g. see the discussion on plucked and struck strings earlier in this chapter). Percussion instruments can be considered in three classes: those that make use of bars (e.g. xylophone, glockenspiel, celeste, triangle); membranes (e.g. drums) or plates (e.g. cymbals). In each case, the natural mode frequencies are not harmonically related, with the exception of longitudinal modes excited in a bar which is stimulated by stroking with a cloth or glove coated with rosin whose mode frequencies are given by Equation 1.20 if the bar is free to move (unfixed) at both ends, and 1.21 is it is supported at one end and free at the other.

Transverse modes are excited in bars that are struck, as for example when playing a xylophone or triangle, and these are not harmonically related. The following equations (adapted from Fletcher and Rossing, 1999) relate the frequencies of higher modes to that of the first mode:

For transverse modes in a bar resting on supports (e.g. glockenspiel, xylophone):

$$f_n = 0.11030 \, ((2n)+1)^2 \, f1 \qquad (4.12)$$

where n = mode numbers from 2 (i.e. 2, 3, 4, ...)
      f_1 = frequency of first mode

For transverse modes in a bar clamped at one end (e.g. celeste):

$$f_2 = 0.70144 \, (2.988)^2 \, f_1$$
$$f_n = 0.70144 \, (2n+1)^2 \, f_1 \qquad (4.13)$$

where n = mode numbers from 3 (i.e. 3, 4, 5, ...)
      f_1 = frequency of first mode

The frequencies of the transverse modes in a bar are inversely proportional to the square of the length of the bar:

$$f_{transverse} \propto \left[\frac{1}{L^2} \right] \qquad (4.14)$$

whereas those of the longitudinal modes are inversely proportional to the length (from Equations 1.20 and 1.21):

$$f_{longitudinal} \propto \left[\frac{1}{L} \right] \qquad (4.15)$$

Therefore halving the length of a bar will raise its transverse mode frequencies by a factor of four, or two octaves, whereas the longitudinal modes will be raised by a factor of two, or one octave. The transverse mode frequencies vary as the square of the mode number, apart from the second mode of the clamped bar (see Equation 4.13) whose factor (2.988) is very close to (3). Table 4.1 shows the frequencies of the first five modes relative to the frequency of the first mode as a ratio and in equal-tempered

Table 4.1 Frequency ratios (Equations 4.20 and 4.21) and semitone spacings (see Appendix 2) of the first five theoretical modes relative to the first mode for a bar clamped at one end and a bar resting on supports

Transverse mode of bar	Bar resting on supports		Bar clamped at one end	
	Ratio	Semitones	Ratio	Semitones
1 (rel. 1st mode)	1.000	0.00	1.000	0.00
2 (rel. 1st mode)	2.758	17.56	6.267	31.77
3 (rel. 1st mode)	5.405	29.21	17.536	49.58
4 (rel. 1st mode)	8.934	37.91	34.371	61.23
5 (rel. 1st mode)	13.346	44.86	56.817	69.93

semitones (Appendix 2 gives a frequency ratio to semitone conversion equation) for a bar resting on supports (Equation 4.12) and one clamped at one end (Equation 4.13). None of the higher modes are a whole number of equal-tempered semitones above the fundamental and none form an interval available within a musical scale. The intervals between the modes are very wide compared to harmonic spacing as they are essentially related by the square of odd integers (i.e. $3^2, 5^2, 7^2, 9^2, ...$). The relative excitation strength of each mode is in part governed by the point at which the bar is hit.

Benade (1976) notes that often the measured frequencies of the vibrating modes of instruments which use bars differs somewhat from the theoretical modes (in Table 4.1) due to the effect of 'mounting hole(s) drilled in the actual bar and the grinding away of the underside of the center of the bar which is done for tuning purposes'.

In order that notes can be played which have a clearly perceived pitch on percussion instruments such as the xylophone, marimba, and vibraphone, the bars are shaped with an arch on their undersides to tune the modes to be close to harmonics of the first mode. In the marimba and vibraphone the second mode is tuned to two octaves above the first mode, and in the xylophone it is tuned to a twelfth above the first mode. These instruments have resonators, which consist of a tube closed at one end, mounted under each bar. The first mode of these resonators is tuned to the f_0 of the bar to enhance its loudness, and therefore the length of the resonator is a quarter of the wavelength of f_0 (see Equation 1.21).

In percussion instruments which make use of membranes and plates, the modal patterns which can be adopted by the membrane or plate themselves govern the frequencies of the modes that are supported. The membrane in a drum and the plate of a cymbal are circular, and the first five mode patterns which they can adopt in terms of where displacement nodes and antinodes can occur are shown in Figure 4.30. Displacement nodes occur in circles and/or diametrically across and these are shown in the figure. They are identified by the numbers given in brackets as follows: (number of diametric modes, number of circular modes). The drum membrane always has at least one circular mode where there is a displacement node, which is the clamped edge.

The frequencies of the modes can be calculated mathematically, but the result is rather more complicated than for the bars. Table 4.2 gives the frequencies of each mode relative to the first mode (Fletcher and Rossing, 1999) and the equivalent number of

Figure 4.30 The first five modes of an ideal drum membrane (upper) and a cymbal plate (lower). Notes: (1) the edge of the cymbal plate is shown dotted as it is unclamped (displacement antinode) and the edge of the drum is shown undotted as it is clamped (displacement node), (2) the mode numbers are given in brackets as (number of diametric modes, number of circular modes). (Adapted from Fletcher and Rossing, 1999.)

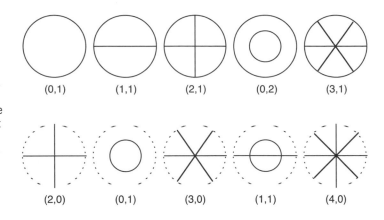

Table 4.2 Modes and frequency ratios (from Fletcher and Rossing, 1999) as well as semitone spacings (see Appendix 2) of the first five theoretical modes relative to the first mode for an ideal circular membrane and plate

Mode	Ideal circular membrane			Ideal circular plate		
	Mode	Ratio	Semitones	Mode	Ratio	Semitones
1 (rel. 1st mode)	(0,1)	1.000	0.00	(2,0)	1.000	0.00
2 (rel. 1st mode)	(1,1)	1.593	8.06	(0,1)	2.092	12.77
3 (rel. 1st mode)	(2,1)	2.136	13.13	(3,0)	3.427	21.32
4 (rel. 1st mode)	(0,2)	2.296	14.38	(1,1)	3.910	23.60
5 (rel. 1st mode)	(3,1)	2.653	16.89	(4,0)	6.067	31.21

semitones (calculated using the equation given in Appendix 2). As with the bars, none of the modes are an exact number of equal tempered semitones apart or in an integer ratio and therefore they are not harmonically related. These frequencies are for an 'ideal' membrane since they will change when the membrane is mounted on a drum body. In the case of the tympani or kettledrum, Rossing (1989) notes that the air loading of the air enclosed in the drum body is 'mainly responsible for establishing the harmonic relationship of kettledrum modes'.

4.5 The speaking and singing voice

The singing voice is probably the most versatile of all musical instruments. Anyone who can speak is capable of singing, but we are not all destined to be opera or pop stars. Whilst considerable mystique surrounds the work of some singing teachers and how they achieve their results, the acoustics of the singing voice is now established as a research topic in its own right. Issues such as the following are being considered:

- the differences between adult male and adult female voices
- the effects of singing training
- the development of pitching skills by children
- 'tone deafness'
- the acoustic nature of different singing styles
- the effect of different acoustics on choral singing
- electronic synthesis of the singing voice.

Knowledge of the acoustics of the singing and speaking voice can be helpful to music technologists when they are developing synthetic sounds since humans are remarkably good at vocalising the sound they desire. In such cases, knowledge of the acoustics of the singing and speaking voice can help in the development of synthesis strategies. This section discusses the human singing voice in terms of the input/system/output model and points to some of the key differences between the speaking and singing voice. The discussion presented in this section is necessarily brief. A number of texts are available which consider the acoustics of the speaking voice (e.g. Fant, 1960; Fry, 1979; Borden and Harris, 1980; Baken, 1987; Baken and Danilof, 1991; Kent and Read, 1992; Howard, 1998; Howard and Angus, 1998), and the acoustics of singing voice (e.g. Benade, 1876, Sundberg, 1987; Bunch, 1993; Dejonckere *et al.*, 1995; Howard, 1999).

4.5.1 Sound source in singing

The sound source in singing is the acoustic result of the vocal folds vibrating in the larynx which is sustained by air flowing from the lungs. The sound modifiers in singing are the spaces between the larynx and the lips and nostrils, known as the 'vocal tract', which can be changed in shape and size by moving the 'articulators', for example the jaw, tongue and lips (see Figure 4.31). As we sing or

Figure 4.31 A cross-section of the vocal tract.

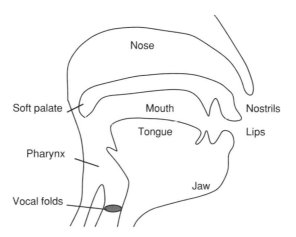

speak, the shape of the vocal tract is continually changing to produce different sounds. The soft palate acts as a valve to shut off and open the nasal cavity (nose) from the airstream.

Vocal fold vibration in a healthy larynx is a cyclic sequence in which the vocal folds close and open regularly when a note is being sung. Thus the vocal folds of a soprano singing A4 (f_0 = 440.0 Hz) will complete this vocal fold closing and opening sequence 440 times a second. Singers have two methods by which they can change the f_0 of vocal fold vibration: they alter the stiffness of the folds themselves by changing the tension of the fold muscle tissue or by altering the vibrating mass by supporting an equal portion of each fold in an immobile position. Adjustments of the physical properties of the folds themselves allows many trained singers to sing over a pitch range of well over two octaves.

The vocal folds vibrate as a result of the Bernoulli effect in much the same way as the lips of a brass player. A consequence of this is that the folds close more rapidly than they open. An acoustic pressure pulse is generated at each instant when the vocal folds snap together, rather like a hand clap. As these closures occur regularly during singing, the acoustic input to the vocal tract consists of a regular series of pressure pulses (see Figure 4.32), the number per second depending on the note being sung. The pressure pulses are shown as negative going in the Figure since the rapid closure of the vocal folds suddenly causes the air flow from the lungs to stop, resulting in a pressure *drop* immediately above the vocal folds. The time between each pulse is the fundamental period. Benade (1976) notes though that the analogy between the lip vibration of brass players and vocal fold vibration speakers and singers should not be taken too far because the vocal folds can vibrate with little influence being exerted by the presence of the vocal tract, whereas the brass player's lip vibration is very strongly influenced by the presence of the instrument's pipe.

Figure 4.32 Idealised waveform (left) and spectrum (right) of acoustic excitation due to normal vocal fold vibration.

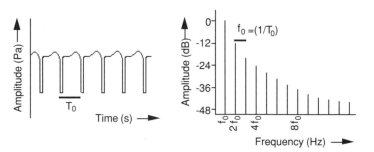

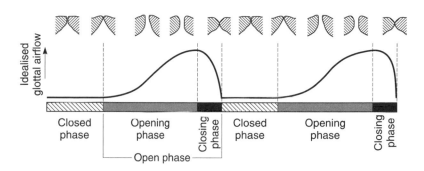

Figure 4.33 Schematic sequence for two vocal fold vibration cycles to illustrate vocal fold vibration sequence as if viewed from the front and idealised glottal airflow waveform. Vocal fold opening, closing, open and closed phases are indicated.

Figure 4.33 shows a schematic vocal fold vibration sequence as if viewed from the front associated with an idealised airflow waveform between the vibrating vocal folds. This is referred to as 'glottal' airflow since the space between the vocal folds is known as the 'glottis'. Three key phases of the vibration cycle are usefully identified: closed phase (vocal folds together), opening phase (vocal folds parting), closing phase (vocal folds coming together). The opening and closing phases are often referred to as the 'open phase' as shown in the Figure, because this is the time during which air flows. It should also be noted that airflow is not necessarily zero during the closed phase since there are vocal fold vibration configurations for which the vocal folds do not come together over their whole length (e.g. Sundberg, 1987; Howard, 1998, 1999).

The nature of vocal fold vibration changes with voice training, whether for oratory, acting or singing. The time for which the vocal folds are in contact in each cycle, known as 'larynx closed quotient' or 'CQ', has been investigated as a possible means by which trained adult male (Howard *et al.*, 1990) and female (Howard, 1995) singers are helped in producing a more efficient acoustic output. Experimental measurements on trained and untrained singers suggest that CQ is higher at all pitches for trained adult males, and that it tends to increase with pitch for trained adult females in a patterned manner. Howard *et al.* suggest that the higher CQ provides the potential for a more efficient voice output by three means: (i) the time in each cycle during which there is an acoustic path via the open vocal folds to the lungs where sound is essentially completely absorbed is reduced, (ii) longer notes can be sustained since less air is lost via the open vocal folds in each cycle, and (iii) the voice quality is less breathy since less air flows via the open vocal folds in each cycle.

The frequency spectrum of the regular pressure pulses generated by the vibrating vocal folds during speech and singing consists of all harmonics with an amplitude change on average of –12 dB per octave rise in frequency (see the illustration on the right in Figure 4.32). Thus for every doubling in frequency, equivalent to an increase of one octave the amplitude reduces by 12 dB. The amplitudes of the first, second, fourth and eighth harmonics (which are separated by octaves) in the figure illustrate this effect.

The shape of the acoustic excitation spectrum remains essentially constant while singing, although the amplitude change of –12 dB per octave is varied for artistic effect, singing style and to aid voice projection by professional singers (e.g. Sundberg, 1987). The spacing between the harmonics will change as different notes are sung, and Figure 4.34 shows three input spectra for sung notes an octave apart. Trained singers, particularly those with Western operatic voices, exhibit an effect known as 'vibrato' in which their f_0 is varied at a rate of approximately 5.5–7.5 Hz with a range of between ± 0.5 and ± 2 semitones (Dejonckere *et al.*, 1995).

Figure 4.34 Idealised vocal tract response plots for the vowels in the words *fast* (left), *feed* (centre), and *food* (right).

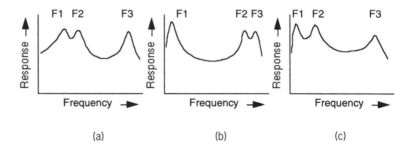

(a)   (b)   (c)

4.5.2 Sound modifiers in singing

The regular series of pulses from the vibrating vocal folds are modified by the acoustic properties of the vocal tract (see Figure 4.21). In acoustic terms, the vocal tract can be considered as a stopped tube (closed at the larynx which operates as a *flow-controlled reed*, open at the lips) which is approximately 17.5 cm in length for an adult male. When the vowel at the end of *announcer* is produced, the vocal tract is set to what is referred to as a neutral position, in which the articulators are relaxed, and the soft palate (see Figure 4.31) is raised to cut off the nose; the vowel is termed 'non-nasalised'. The neutral vocal tract approximates quite closely to a tube of constant diameter throughout its length and therefore the equation governing modal frequencies in a cylindrical stopped pipe can be used to find the vocal tract standing wave mode frequencies for this vowel.

Example 4.3 Calculate the first three mode frequencies of the neutral adult male vocal tract. (Take the velocity of sound in air as 344 ms⁻¹.)

The vocal tract length is 17.5 cm, or 0.175 m.
From Equation 4.9, the fundamental or first mode:

$$F_{\text{stopped}(1)} = \left[\frac{c}{4L_s} \right] = \left[\frac{344}{4 * 0.175} \right] 491.4 \text{ Hz}$$

From Equation 4.10, the higher mode frequencies are:

$$f_{\text{stopped}(n)} = (2n - 1) f_{\text{stopped}(1)}$$

where $n = 1, 2, 3, 4, \ldots$

Thus the second mode frequency ($n = 2$) is: $3 * 491.4 = 1474$ Hz and the third mode frequency ($n = 3$) is: $5 * 491.4 = 2457$ Hz

Example 4.3 gives the frequencies for the neutral vowel, and these are often rounded to 500 Hz, 1500 Hz and 2500 Hz for convenience. When considering the acoustics of speech and singing, the standing wave modes are generally referred to as 'formants'. Idealised frequency response curves for a vocal tract set to produce the vowels in the words *fast*, *feed* and *food* are shown in Figure 4.33. and the centre frequency of each formant is labelled starting with 'F1' or 'first formant' for the peak that is lowest in frequency, continuing with 'F2' (second formant) and 'F3' (third formant) as shown in the figure. The formants are acoustic resonances of the vocal tract itself resulting from the various dimensions of the vocal tract spaces. These are modified during speech and singing by movements of the articulators. When considering the different sounds produced during speech, usually just the first, second and third formants are considered since these are the only formants whose frequencies tend to vary. Six or seven formants can often be identified in the laboratory and the higher formants are thought to contribute to the individual identity of a speaking or singing voice. However, in singing important contributions to the overall projection of sound are believed to be made by formants higher than the third.

In order to produce different sounds, the shape of the vocal tract shape is altered by means of the articulators to change its acoustic properties. The *perturbation theory* principles explored in the context of woodwind reed instruments (see Figure 4.23) can be employed here also (Kent and Read, 1992). Figure 4.35 shows the displacement nodes and antinode positions for the first three formants of the vocal tract during a neutral non-nasalised vowel,

Figure 4.35 Displacement nodes and antinode positions for the first three modes (or formants: F1, F2, F3) of the vocal tract during a neutral non-nasalised vowel.

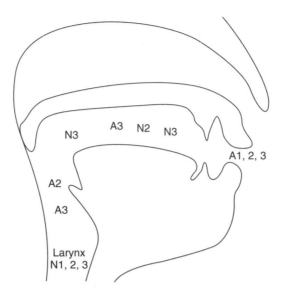

which can be confirmed with reference to the upper-right-hand part of Figure 4.23. Following the same line of reasoning as that presented in the context of Figure 4.23, the effect of constrictions (and therefore enlargements) on the first three formants of the vocal tract can be predicted as shown in Figure 4.36. For example, all formants have a volume velocity antinode at the lips, and a lip constriction therefore lowers the frequencies of all formants. (It should be noted that there are two other means of lowering all formant frequencies by means of vocal tract lengthening either by protruding lip or by lowering the larynx.)

A commonly referenced set of average formant frequency values for men, women and children for a number of vowels , taken from Peterson and Barney (1952), is shown in Table 4.3. Formant frequency values for these vowels can be predicted with reference to their articulation. For example, the vowel in *beat* has a constriction towards the front of the tongue in the region of both N2 and N3 (see Figure 4.35), and reference to Figure 4.36 suggests that F1 is lowered in frequency and F2 and F3 are raised from the values one would expect for the neutral vowel. The vowel in *part*, on the other hand, has a significant constriction in the region of both A2 and A3 (see Figure 4.35) resulting in a raising of F1, and a lowering of both F2 and F3 from their neutral vowel values. The vowel in *boot* has a constriction at the lips which are also rounded so as to extend the length of the vocal tract and thus all formant frequencies are lowered from their neutral vowel values. These changes can be confirmed from Table 4.3.

Figure 4.36 Formant frequency modification with position of vocal tract of constriction.

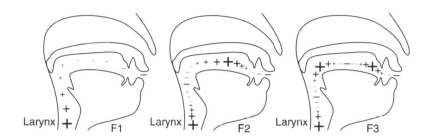

Table 4.3 Average formant frequencies in Hz for men, women and children for a selection of vowels. (From Peterson and Barney, 1952)

Vowel in	Men			Women			Children		
	F1	F2	F3	F1	F2	F3	F1	F2	F3
beat	270	2300	3000	300	2800	3300	370	3200	3700
bit	400	2000	2550	430	2500	3100	530	2750	3600
bet	530	1850	2500	600	2350	3000	700	2600	3550
bat	660	1700	2400	860	2050	2850	1000	2300	3300
part	730	1100	2450	850	1200	2800	1030	1350	3200
pot	570	850	2400	590	900	2700	680	1050	3200
boot	440	1000	2250	470	1150	2700	560	1400	3300
book	300	850	2250	370	950	2650	430	1150	3250
but	640	1200	2400	760	1400	2800	850	1600	3350
pert	490	1350	1700	500	1650	1950	560	1650	2150

The input/system/output model for singing consists of the acoustic excitation due to vocal fold vibration (input), the vocal tract response (system) to give the output. These are usually considered in terms of their spectra, and both the input and system change with time during singing. Figure 4.37 shows the model for the vowel in *fast* sung on three different notes. This is to allow one of the main effects of singing at different pitches to be illustrated.

The input in each case is the acoustic spectrum resulting from vocal fold vibration (see Figure 4.32). The output is the result of the response of the vocal tract for the vowel in *fast* acting on the input vocal fold vibration. The effect of this is to multiply the amplitude of each harmonic of the input by the response of the vocal tract at that frequency. This effectively imparts the formant peaks of the vocal tract response curve onto the harmonics of the input spectrum. In this example, there are three formant peaks shown, and it can be seen that in the cases of the lower two notes, the formant structure can be readily seen in the output, but that in the case of the highest note the formant peaks cannot be identified in the output spectrum because the harmonics of the input are too far apart to represent clearly the formant structure.

Figure 4.37 Singing voice input/system/output model idealised for the vowel in *fast* sung on three notes an octave apart.

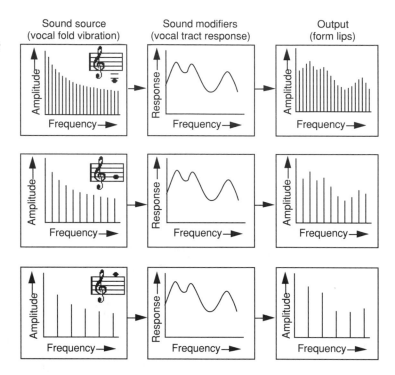

Sound source (vocal fold vibration) | Sound modifiers (vocal tract response) | Output (form lips)

The representation of the formant structure in the output spectrum is important if the listener is to identify different vowels. Figure 4.37 suggests that somewhere between the G above middle C and the G an octave above, vowel identification will become increasingly difficult. This is readily tested by asking a soprano to sing different vowels on mid and top G as shown in the figure and listening to the result. In fact, when singing these higher notes, professional sopranos adopt vocal tract shapes which place the lower formants over individual harmonics of the excitation so that they are transmitted via the vocal tract with the greatest amplitude. In this way, sopranos can produce sounds of high intensity which will project well. This effect is used from approximately the C above middle C where the vocal tract is, in effect, being 'tuned-in' to each individual note sung, but at the expense of vowel clarity.

This tuning-in effect is not something that tenors need to do since the ratio between the formant frequencies and the f_0 of the tenor's range is higher than that for sopranos. However, all singers who do not use amplification need to project above accompaniment, particularly when this is a full orchestra and the performance is in a large auditorium. The way in which profes-

Figure 4.38 Idealised spectra for a singer speaking the text of an opera aria (a), the orchestra playing the accompaniment to the aria (b), and aria being sung with orchestral accompaniment (c) (adapted from Sundberg 1987).

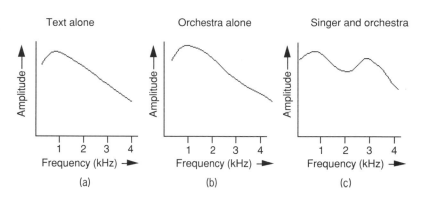

Text alone

Orchestra alone

Singer and orchestra

(a)

(b)

(c)

sional opera singers achieve this can be seen with reference to Figure 4.38 which shows idealised spectra for the following:

- a professional opera singer speaking the text of an operatic aria
- the orchestra playing the accompaniment to the aria and
- the aria sung by the singer with the orchestral accompaniment.

It should be noted that the amplitude levels cannot be directly compared between (A) and (B) in the Figure (i.e. the singer does not speak as loudly as the orchestral accompaniment!) since they have been normalised for comparison.

The idealised spectrum for the text read alone has the same general shape as that for the orchestra playing alone. When the professional singer sings the aria with orchestra accompaniment, it can be seen that this combined response curve has a shape similar to both the speech and orchestral accompaniment at low frequencies, but with an additional broad peak between approximately 2.5 kHz and 4 kHz and centred at about 3 kHz. This peak relates to the acoustic output from the singer when singing but not when speaking, since it is absent for the read text and also in the orchestral accompaniment alone. This peak has similar characteristics to the formants in the vocal tract response, and for this reason it is known as the 'singer's formant'. The presence of energy in this peak enables the singer to be heard above an accompanying orchestra because it is a section of the frequency spectrum in which the singer's output prevails. This is what gives the professional singing voice its characteristic 'ring', and it is believed to be the result of lowering the larynx and widening the pharynx (see Figure 4.31) which is adopted by trained Western operatic singers. (The lower plot in Figure 5.5 is an analysis of a CD recording of a professional tenor whose singer's formant is very much in evidence.)

Singing teachers set out to achieve these effects from pupils by suggesting that pupils: 'sing on the point of the yawn', or 'sing as if they have swallowed an apple which has stuck in their throat'. Sundberg (1987) discusses the articulatory origin of the singer's formant as follows: '... it shows a strong dependence on the larynx tube ...', concluding that: '... it is necessary, however, that the pharynx tube be lengthened and that the cross-sectional area in the pharynx at the level of the larynx tube opening be more than six times the area of that opening'.

Professional singing is a complex task which extends the action of the instrument used for speech. It is salutary to note that the prime function of the vocal folds is to act as a valve to protect the lungs, and not to provide the sound source basic to human communication by means of speech and song.

References

Askenfelt, A. (ed.) (1990). *Five Lectures on the Acoustics of the Piano*, with compact disc, Publication No. 64. Stockholm: Royal Swedish Academy of Music.

Audsley, G.A. (1965). *The Art of Organ-Building*, 2 Vols. New York: Dover (reprint of 1905 edition, New York: Dodd, Mead and Company).

Backus, J. (1977). *The Acoustical Foundations of Music*. New York: Norton.

Baken, R.J. (1987). *Clinical Measurement of Speech and Voice*. London: Taylor and Francis.

Baken, R.J. and Danilof, R.G. (1991). *Readings in Clinical Spectrography of Speech*. San Diego: Singular Publishing Group.

Benade, A.H. (1976). *Fundamentals of Musical Acoustics*. New York: Oxford University Press.

Borden, G.J. and Harris, K.S. (1980). *Speech Science Primer*. Baltimore: Williams and Wilkins.

Bunch, M. (1993). *Dynamics of the Singing Voice*. New York: Springer-Verlag.

Campbell, M. and Greated, C. (1998). *The Musician's Guide to Acoustics*. Oxford: Oxford University Press.

Dejonckere, P.H., Hirano, M. and Sundberg, J. (eds) (1995). *Vibrato*. San Diego: Singular Publishing Group.

Fant, C.G.M. (1960). *Acoustic Theory of Speech Production*. The Hague: Mouton.

Fletcher, N.H. and Rossing, T.D. (1999). *The Physics of Musical Instruments* (2nd end). New York: Springer-Verlag.

Fry, D.B. (1979). *The Physical of Speech*. Cambridge: Cambridge University Press.

Hall, D.E. (1991). *Musical Acoustics: An introduction* (2nd edn). Belmont, CA: Wadsworth Publishing Company.

Howard, D.M. (1998). Practical voice measurement, In: *The Voice Clinic Handbook*. (T. Harris, S. Harris, J.S. Rubin and D.M. Howard, eds). London: Whurr Publishing Company.

Howard, D.M. (1999). The human singing voice, In: *Killers in the Brain* (P. Day, ed.). Oxford: Oxford University Press, 113–134.

Howard, D.M. and Angus, J.A.S. (1998). Introduction to human speech production, human hearing and speech analysis, In: *Speech Technology for Telecommunications* (F.A. Westall, R.D. Johnson and A.V. Lewis, eds). London: Chapman and Hall, 30–72.

Howard, D.M. (1995). Variation of electrolaryngographically derived closed quotient for trained and untrained adult female singers. *Journal of Voice*, **9**, 163–172.

Howard, D.M., Lindsey, G.A. and Allen, B. (1990). Towards the quantification of vocal efficiency. *Journal of Voice*, **4**, 205–21. (See also errata: (1991). *Journal of Voice*, **5**, 93–95.)

Hurford, P. (1994). *Making Music on the Organ* (6th edn). Oxford: Oxford University Press.

Hutchins, C.M. (ed.) (1975a). *Musical Acoustics, part 1: Violin Family Components*. Pennsylvania: Dowden, Hutchinson and Ross Inc.

Hutchins, C.M. (ed.) (1975b). *Musical Acoustics, part II: Violin Family Functions*. Pennsylvania: Dowden, Hutchinson and Ross Inc.

Hutchins, C.M. (ed.) (1978). *The Physics of Music*. Reprints from Scientific American, San Francisco: W.H. Freeman and Company.

Kent, R.D. and Read, C. (1992). *The Acoustic Analysis of Speech*. San Diego: Singular Publishing Group.

Martin, D. and Ward, D. (1961). Subjective evaluation of musical scale temperament in pianos. *Journal of the Acoustical Society of America*, **33**, 582-585.

Nederveen, C.J. (1969). *Acoustical Aspects of Woodwind Instruments*. Amsterdam: Frits Knuf.

Norman, H. and Norman, H.J. (1980). *The Organ Today*. London: David and Charles.

Peterson, G.E. and Barney, H.E. (1952). Control methods used in the study of vowels. *Journal of the Acoustical Society of America*, **24**, 175–184.

Proctor, D.F. (1980). *Breathing, Speech and Song*. New York: Springer-Verlag.

Rossing, T.D. (1989). *The Science of Sound*. New York: Addison-Wesley.

Sumner, W.L. (1975). *The Organ*, (5th edn). London: Macdonald and Company.

Sundberg, J. (1987). *The Science of the Singing Voice*. DeKalb: Illinois University Press.

Sundberg, J. (1989). *The Science of Musical Sounds*. San Diego: Academic Press.

Taylor, C.A. (1976). *Sounds of Music*. London: Butler and Tanner Ltd.

5 Hearing timbre and deceiving the ear

5.1 What is timbre?

Pitch and loudness are two of three important descriptors of musical sounds commonly used by musicians; the other being 'timbre'. Pitch relates to issues such as notes on a score, key, melody, harmony, tuning systems, and intonation in performance. Loudness relates to matters such as musical dynamics (e.g. pp, p, mp, mf, f, ff, etc.), the balance between members of a musical ensemble (e.g. between individual parts, choir and orchestra, or soloist and accompaniment). Timbre to sound quality descriptions include: mellow, rich, covered, open, dull, bright, dark, strident, grating, harsh, shrill, sonorous, sombre, colourless and lacklustre. Timbral descriptors are therefore used to indicate the perceived quality or tonal nature of a sound which can have a particular pitch and loudness also.

There is no subjective rating scale against which timbre judgements can be made, unlike pitch and loudness which can, on average, be reliably rated by listeners on scales from 'high' to 'low'. The commonly quoted American National Standards Institute formal definition of timbre reflects this: 'Timbre is that attribute of auditory sensation in terms of which a listener can judge two sounds similarly presented and having the same loudness and pitch as being dissimilar' (ANSI, 1960). In other words, two sounds that are perceived as being different but which have the same perceived loudness and pitch differ by virtue of their timbre. The timbre of a note is the aspect by which a listener recognises the instrument which is playing a note when, for example, instruments play notes with the same pitch, loudness and duration. The definition given by Scholes (1970) encompasses some timbral descriptors: 'Timbre means tone quality–coarse or smooth, ringing or more subtly penetrating, "scarlet" like that of a trumpet, "rich brown" like that of a cello, or "silver" like that of the flute. These colour analogies come naturally to every mind ... The one and only factor in sound production which conditions timbre is the presence or absence, or relative strength or weakness, of overtones'. (Table 3.1 gives the relationship between overtones and harmonics.) Whilst his colour analogies might not come naturally to every mind, Scholes' later comments about the acoustic nature of sounds which have different timbres are a useful contribution to the acoustic discussion of the timbre of musical sounds.

When considering the notes played on pitched musical instruments, timbre relates to those aspects of the note which can be varied without affecting the pitch, duration or loudness of the note as a whole, such as the spectral components present and the way in which their frequencies and amplitudes vary during the sound. In Chapter 4 the acoustics of musical instruments is considered in terms of the output from the instrument as a consequence of the effect of the sound modifiers on the sound input (e.g. Figure 4.2). What is not considered, due to the complexity of modelling, is the acoustic development from silence at the start of the note and back to silence at the end. It is then, convenient to consider a note in terms of three phases: the 'onset' or 'attack' (the build-up from silence at the start of the note), the 'steady state' (the main portion of the note), and the 'offset' or 'release' (the return to silence at the end of the note after the energy source is stopped). The onset and offset portions of a note tend to last for a short time of the order of a few tens of milliseconds (or a few hundredths of a second). Changes that occur during the onset and offset phases, and in particular during the onset, turn out to have a very important role in defining the timbre of a note.

In this chapter, timbre is considered in terms of the acoustics of sounds which have different timbres, and the psychoacoustics of how sounds are perceived. Finally, the pipe organ is reviewed in terms of its capacity to synthesize different timbres.

5.2 Acoustics of timbre

The description of the acoustics of notes played on musical instruments presented in Chapter 4 was in many cases, supported by plots of waveforms and spectra of the outputs from some instruments (Figures 4.17, 4.22, 4.24 and 4.29). Except in the plots for the plucked notes on the lute and guitar (Figure 4.11) where the waveforms are for the whole note and the spectra are for a single spectral analysis, the waveform plots show a few cycles from the steady-state phase of the note concerned and the spectral plots are based on averaging together individual spectral measurements taken during the steady-state phase. The number of spectra averaged together depend on how long the steady-state portion of the note lasts. For the single notes illustrated in Chapter 4, spectral averaging was carried out over approximately a quarter to three quarters of a second, depending on the length of the note available. An alternative way of thinking about this is in terms of the number of cycles of the waveform over which the averaging takes place, which would be 110 cycles for a quarter of a second to 330 cycles for three quarters of a second for A4 (f_0 = 440 Hz), or 66 cycles to 198 cycles for C4 (f_0 = 261.6 Hz). Such average spectra are commonly used in for analysing the frequency components of musical notes, and they are known as 'long-term average spectra' or 'LTAS'. One main advantage of using LTAS is that the spectral features of interest during the steady state portion of the note are enhanced in the resulting plot by the averaging process with respect to competing acoustic sounds such as background noise which change over the period of the LTAS and thus average towards zero.

LTAS cannot, however, be used to investigate acoustic features that change rapidly such as the onset and offset of musical notes, because these will also tend to average towards zero. In terms of the timbre of the note, it is not only the variations that occur during the onset and offset that are of interest, but also how they change with time. Therefore an analysis method is required in which the timing of acoustic changes during a note is preserved in the result. One analysis technique commonly used for the acoustic analysis of speech is a plot of amplitude, frequency and time known as a 'spectrogram'. Frequency is plotted on the vertical scale, time on the horizontal axis and amplitude is plotted as the darkness on a grey scale, or in some cases the colour, of the spectrogram.

The upper plot in Figure 5.1 shows a spectrogram and acoustic pressure waveform of C4 played on a principal 8' (open flue), the same note for which an LTAS is presented in Figure 4.17. The LTAS plot in Figure 4.17 showed that the first and second harmonics dominate the spectrum, the amplitude of the third harmonic being approximately 8 dB lower than the first harmonic, and with energy also clearly visible in the fourth, fifth,

Figure 5.1 Waveform and spectrogram of whole note (upper) and onset phase (lower) for C4 played on the principal 8' organ stop (open flue) for which an LTAS is shown in Figure 4.17. Note onset, steady-state and offset phases are marked.

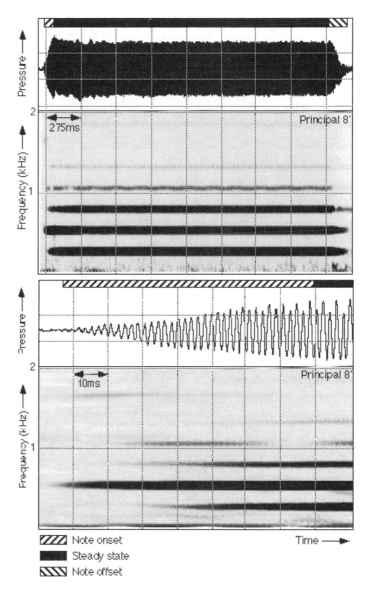

seventh and eighth harmonics whose amplitudes are at least 25 dB lower than that of the first harmonic.

A spectrogram shows which frequency components are present (measured against the vertical axis), at what amplitude (blackness of marking) and when (measured against the horizontal axis). Thus harmonics are represented on spectrograms as horizontal lines, where the vertical position of the line marks the frequency and the horizontal position shows the time for which that harmonic lasts. The amplitudes of the harmonics are plotted as the blackness of marking of the lines. The frequency and time axes on the spectrogram are marked and the amplitude is shown as the blackness of the marking. The spectrogram shown in Figure 5.1 shows three black horizontal lines which are the first three harmonics of the principal note (since the frequency axis is linear, they are equally spaced). The first and second harmonics are slightly blacker (and thicker) than the third, reflecting the amplitude difference as shown in Figure 4.17. The fourth, fifth and seventh harmonics are visible and their amplitude relative to the first harmonic is reflected in the blackness with which they are plotted.

5.2.1 Note envelope

The onset, steady-state, and offset phases of the note are indicated above the waveform in the Figure, and these are determined mainly with reference to the spectrogram because they relate to the changes in spectral content at the start and end of the note, leaving the steady portion in between. However, 'steady state' does not mean that no aspect of the note varies. The timbre of a principal organ stop sounds 'steady' during a prolonged note such as that plotted, which lasts for approximately 2 seconds, but it is clear from the acoustic pressure waveform plot in Figure 5.1 that the amplitude, or 'envelope', varies even during the so-called 'steady-state' portion of this note. This is an important aspect of musical notes to be aware of when, for example, synthesising notes of musical instruments; particularly if using looping techniques on a sampling synthesiser.

For the principal pipe, the end of the note begins when the key is released and the air flowing from the organ bellows to drive the air reed is stopped. In the note offset for this example which lasts approximately 200 ms, the high harmonics die away well before the first and second. However, interpretation of note offsets is rather difficult if the recording has been made in an enclosed environment as opposed to free space (see Chapter 4), since any reverberation due to the acoustics of the space are also being analysed (see Chapter 6). It is difficult to see the details of the note onset in this example, due to the time scale required to view the complete note.

The note onset phase is particularly important to perceived timbre. Since listeners can reliably perceive the timbre of notes *during* the steady-state phase, it is clear that the offset phase is rather less important to the perception of timbre than the onset and steady state phases. The onset phase is also more acoustically robust from the effects of the local environment in which the notes are played, since colouration of the direct sound by the first reflection (see Chapter 6) may occur *after* the onset phase is complete (and therefore transmitted uncoloured to the listener). By definition the first reflection certainly occurs after part of the note onset has been heard uncoloured. The onset phase is therefore a vital element and the offset phase an important factor in terms of timbre perception. Spectrograms whose time scales are expanded to cover the time of the note onset phase are particularly useful when analysing notes acoustically.

The lower plot in Figure 5.1 shows an expanded timescale version of the upper plot in the figure, showing the note onset phase which lasts approximately 70 ms, and start of the steady-state phase. It can be seen that the detail of the onset instant of each of the harmonics is clearly visible, with the second harmonic starting approximately 30 ms before the first and third harmonics. This is a common feature of organ pipes voiced with a chiff or consonantal onset which manifests itself acoustically in the onset phase as a initial jump to the first, or sometimes higher, overblown mode. The first overblown mode for an open flue pipe is to the second harmonic (see Chapter 4). Careful listening to pipes voiced with a chiff will reveal that open pipes start briefly an octave high since their first overblown mode is the second harmonic, and stopped pipes start an octave and a fifth high since their first overblown mode is the third harmonic. The fourth harmonic in the figure starts with the third and its amplitude briefly drops 60 ms into the sound when the fifth starts, and the seventh starts almost with the second and its amplitude drops 30 ms later. The effect of the harmonic build-up on the acoustic pressure waveform can be observed in the figure in terms of the changes in its shape, particularly the gradual increase in amplitude, during the onset phase. The onset phase for this principal organ pipe is a complex series of acoustic events, or acoustic 'cues', which are available as potentially contributors to the listener's perception of the timbre of the note.

5.2.2 Note onset

In order to provide some data to enable appreciation of the importance of the note onset phase for timbre perception, Figures 5.2 to 5.4 are presented in which the note onset and start of the steady-state phases for four organ stops, four woodwind instruments and four brass instruments respectively are presented for the note C4 (except for the trombone and tuba for which the note is C3). By way of a caveat it should be noted that

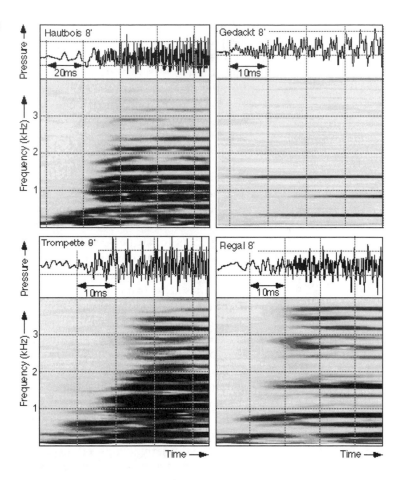

Figure 5.2 Waveform (upper) and spectrogram (lower) of the note onset phase for C4 played on the following pipe organ stops: hautbois 8' (reed), trompette 8' (reed), gedackt 8' (stopped flue) and regal 8' (reed). LTAS for the hautbois and trompette notes are shown in Figure 4.22, and for the gedackt in Figure 4.17.

these figures are presented to provide only examples of the general nature of the acoustics of the note onset phases for these instruments. Had these notes been played at a different loudness, by a different player, on a different instrument, in a different environment, or even just a second time by the same played on the same instrument in the same place while attempting to keep all details constant, the waveforms and spectra would probably be noticeably different.

The organ stops for which waveforms and spectra are illustrated in Figure 5.2 are three reed stops: hautbois and trompette (LTAS in Figure 4.22) and a regal, and gedackt (LTAS in Figure 4.17) which is an example of a stopped flue pipe. The stopped flue supports only the odd modes (see Chapter 4) , and during the onset phase of this particular example, the fifth harmonic sounds first, which is the second overblown mode sounding two octaves and a major third above the fundamental (see Figure 3.3),

Figure 5.3 Waveform (upper) and spectrogram (lower) of the note onset phase for C4 played on a clarinet, flute, oboe and tenor saxophone. LTAS for the clarinet and tenor saxophone are shown in Figure 4.24.

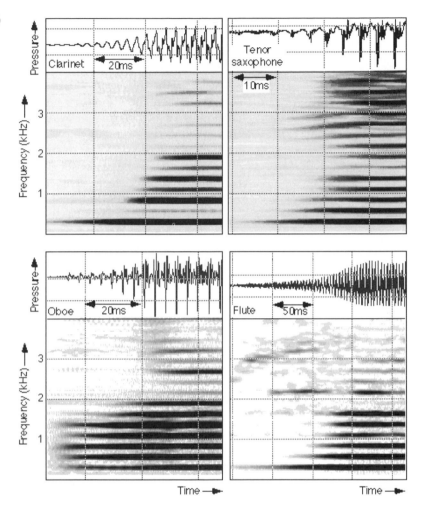

followed by the fundamental and then the third harmonic, giving a characteristic chiff to the stop. The onset phase for the reed stops is considerably more complicated since many more harmonics are present in each case. The fundamental for the hautbois and regal is evident first, and the second harmonic for the trompette. In all cases, the fundamental exhibits a frequency rise at onset during the first few cycles of reed vibration. The staggered times of entry of the higher harmonics forms part of the acoustic timbral characteristic of that particular stop, the trompette having all harmonics present up to the 4 kHz upper frequency bound of the plot, the hautbois having all harmonics up to about 2.5 kHz, and the regal exhibiting little or no evidence (at this amplitude setting of the plot) of the fourth or eighth harmonics.

Figure 5.3 shows plots for four woodwind instruments: clarinet, oboe, tenor saxophone and flute. For these particular examples of the clarinet and tenor saxophone, the fundamental is apparent first and the oboe note begins with the second harmonic, followed by the third and fourth harmonics after approximately 5 ms, and then the fundamental some 8 ms later. The higher harmonics of the clarinet are apparent nearly 30 ms after the fundamental; the dominance of the odd harmonics is discussed in Chapter 4. This particular example of C4 from the flute begins with a notably 'breathy' onset just prior to and as the fundamental component starts. This can be seen in the frequency region of the spectrogram that is above 2 kHz lasting some 70 ms. The higher harmonics enter one by one approximately 80 ms after the fundamental starts. The rather long note onset phase is characteristic of a flute note played with some deliberation. The periodicity in the waveforms develops gradually, and in all cases, there is an appreciable time over which the amplitude of the waveform reaches its steady state.

Figure 5.4 shows plots for four brass instruments. The notes played on the trumpet and French horn are C4 and those for the trombone and tuba are C3. The trumpet is the only example with energy in high harmonic components in this particular example, with the fourth, fifth and sixth harmonics having the highest amplitudes. The other instruments in this figure do not have energy apparent above approximately the fourth harmonic (French horn and tuba) or the sixth harmonic for the trombone. The note onset phase for all four instruments starts with the fundamental (noting that this is rather weak for the trombone) and continues with increasing frequency. The waveforms in all cases become periodic almost immediately.

Waveforms and spectrograms are presented in the upper plot of Figure 5.5 for C4 played with a bow on a violin. Approximately 250 ms into the violin note, vibrato is apparent as a frequency variation particularly in the high harmonics. This is a feature of using a linear frequency scale, since a change of x Hz in f_0 will manifest itself as a change of $2x$ Hz in the second harmonic, $3x$ Hz in the third harmonic and so on. In general the frequency change in the nth harmonic will be nx Hz, therefore the frequency variation in the upper harmonics during vibrato will be greater than that for the lower harmonics when frequency is plotted on a linear scale as in the figure. Vibrato often has a delayed start, as in this example, as the player makes subtle intonation adjustments to the note. This particular bowed violin note has an onset phase of approximately 160 ms and an offset phase of some 250 ms.

Figure 5.4 Waveform (upper) and spectrogram (lower) of the note onset phase for C4 played on a: trumpet and French horn, and C3 played on a trombone and tuba. LTAS for the trombone and tuba are shown in Figure 4.29.

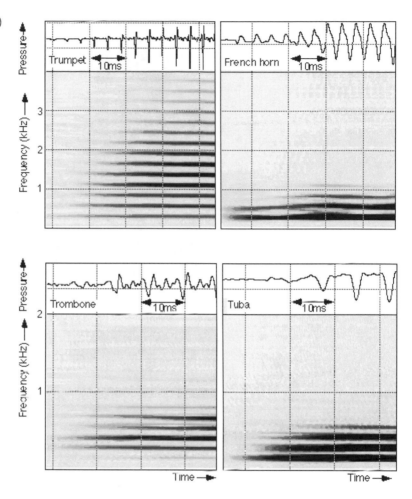

Finally in this section, a note analyses from a CD recording of a professional tenor singing the last three syllables of the word 'vittoria' (i.e. 'toria') on Bb4 from the second act of *Tosca* by Puccini (lower plot in Figure 5.5). This is a moment in the score when the orchestra stops playing and leaves the tenor singing alone. The orchestra stops approximately 500 ms into this example, its spectrographic record can be seen particularly in the lower left-hand corner of the spectrogram, where it is almost impossible to analyse any detailed acoustic patterning. This provides just a hint at the real acoustic analysis task facing the hearing system when listening to music. The spectrogram of the tenor shows the harmonics and the extent of the vibrato very clearly, and his singer's formant (compare with Figure 4.38) can be clearly seen in the frequency region between 2.4 kHz and 3.5 kHz. The first and third of the three syllables ('toria') are

Figure 5.5 Waveform (upper) and spectrogram (lower) of the C4 played on a violin and analysed from a CD recording of the last three syllables of the word 'Vittoria' from act 2 of *Tosca* by Puccini sung by a professional tenor (f_0=Bb4).

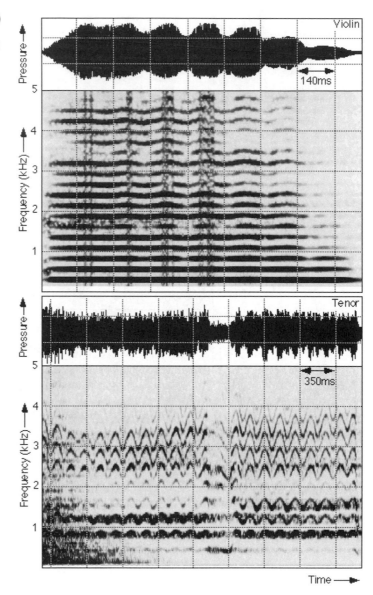

long, and the second ('ri') is considerably shorter in this particular tenor's interpretation. The second syllable manifests itself as the dip in amplitude of all harmonics just over half-way through the spectrogram.

5.3 Psychoacoustics of timbre

A number of psychoacoustic experiments have been carried out to explore listeners' perceptions of the timbre of musical instruments and the acoustic factors on which it depends. Such experiments have demonstrated, for example, that listeners cannot reliably identify musical instruments if the onset and offset phases of notes are removed. For example, if recordings of a note played on a violin open string and the same note played on a trumpet are modified to remove their onset and offset phases in each case, it becomes very difficult to tell them apart. The detailed acoustic nature of a number of example onset phases is provided in Figures 5.1 to 5.5, from which differences can be noted. Thus, for example the initial scraping of the bow on a stringed instrument, the consonant-like onset of a note played on a brass instrument, the breath noise of the flautist, the initial flapping of a reed, the percussive thud of a piano hammer and the final fall of the jacks of a harpsichord back onto the strings are all vital acoustic cues to the timbral identity of an instrument. Careful attention must be paid to such acoustic features, for example when synthesising acoustic musical instruments if the resulting timbre is to sound convincingly natural to listeners.

5.3.1 Critical bands and timbre

A psychoacoustic description of timbre perception must be based on the nature of the critical bandwidth variation with frequency since this describes the nature of the spectral analysis carried out by the hearing system. The variation in critical bandwidth is such that it becomes wider with increasing frequency, and the general conclusion was drawn in the section on pitch perception in Chapter 3 (Section 3.2) that no harmonic above about the fifth to seventh is resolved no matter what the value of f_0. Harmonics below the fifth to seventh are therefore resolved separately by the hearing system (e.g. see Figure 3.11), which suggests that these harmonics might play a distinct and individual rôle in timbre perception. Harmonics above the fifth or seventh, on the other hand, which are not separately isolated by the hearing system are not likely to have such a strong individual effect on timbre perception, but could affect it as groups that lie within a particular critical band. Based on this notion, the perceived timbre is reviewed of instruments for which the results of acoustic analysis are presented in this book, bearing in mind that these analyses are for single examples of notes played on these instruments by a particular player on a particular instrument at a particular loudness and pitch in a particular acoustic environment.

Instruments amongst those for which spectra have been presented that have significant amplitudes in harmonics above the fifth or seventh during their steady-state phases include organ reed stops (see Figures 4.22 and 5.2), the tenor saxophone (see Figures 4.24 and 5.3), the trumpet (see Figure 5.4), the violin and professional singing voice (see Figure 5.5). The timbres of such instruments might be compared with those of other instruments using descriptive terms such as 'bright', 'brilliant', or 'shrill'. Instruments which do not exhibit energy in harmonics above the fifth or seventh during their steady-state phases include the principal 8' (see Figures 4.17 and 5.1), the gedackt 8' (see Figures 4.17 and 5.2), the clarinet, oboe and flute (see Figures 4.24 and 5.3), and the trombone, French horn and tuba (see Figures 4.29 and 5.4). In comparison with their counterpart organ stops or other instruments of their category (woodwind or brass), their timbres might be described as being: 'less bright' or 'dark', 'less brilliant' or 'dull', or 'less shrill' or 'bland'.

Within this latter group of instruments there is an additional potential timbral grouping between those instruments which exhibit all harmonics up to the fifth or seventh, such as the clarinet, oboe, flute, compared with those which just have a few low harmonics such as the principal 8', gedackt 8', trombone, French horn and tuba. It may come as a surprise to find the flute in the same group as the oboe and clarinet, but the lack of the seventh harmonic in the flute spectrum compared to the clarinet and oboe (see Figure 5.3) is crucial. Notes excluding the seventh harmonic sound considerably less 'reedy' than those with it, the seventh harmonic is one of the lowest which is not resolved by the hearing system (provided the sixth and/or eighth are/is also present). This last point is relevant to the clarinet where the seventh harmonic is present but both the sixth and eighth are weak. The clarinet has a particular timbre of its own due to the dominance of the odd harmonics in its output, and it is often described as being 'nasal'. Organists who are familiar with the effect of the tierce (1⅗) and the rarely found septième (1⅐) stops (see Section 5.4) will appreciate the particular timbral significance of the fifth and seventh harmonics respectively and the 'reediness' they tend to impart to the overall timbre when used in combination with other stops.

Percussion instruments which make use of bars, membranes or plates as their vibrating system (described in Section 4.4) which are struck have a distinct timbral quality of their own. This is due to the non-harmonic relationship between the frequencies of their natural modes which provides a clear acoustic cue to their family identity. It gives the characteristic 'clanginess' to this class of instruments which endows them with a timbral quality of their own.

5.3.2 Acoustic cues and timbre perception

Timbre judgements are highly subjective and therefore individualistic. Unlike pitch or loudness judgements, where listeners might be asked to rate sounds on scales of low to high or soft to loud respectively, there is no 'right' answer for timbre judgements. Listeners will usually be asked to compare the timbre of different sounds and rate each numerically between two opposite extremes of descriptive adjectives, for example on a one to ten scale between 'bright' (1)—'dark' (10) or 'brilliant' (1)—'dull' (10), and a number of such descriptive adjective pairs could be rated for a particular sound. The average of ratings obtained from a number of listeners is often used to give a sound an overall timbral description. Hall (1991) suggests that it is theoretically possible that one day up to five specific rating scales could be 'sufficient to accurately identify almost any timbre'.

Researchers have attempted to identify relationships between particular features in the outputs from acoustic musical instruments and their perceived timbre. A significant experiment in this field was conducted by Grey (1977). Listeners were asked to rate the similarity between recordings of pairs of synthesised musical instruments on a numerical scale from one to thirty. All sounds were equalised in pitch, loudness and duration. The results were analysed by 'multidimensional scaling' which is a computational technique that places the instruments in relation to each other in a multidimensional space based on the similarity ratings given by listeners. In Grey's experiment, a three-dimensional space was chosen and each dimension in the resulting three-dimensional representation was then investigated in terms of the acoustic differences between the instruments lying along it 'to explore the various factors which contributed to the subjective distance relationships'. Grey identified the following acoustic factors with respect to each of the three axes: (1) 'spectral energy distribution' observed as increasing high-frequency components in the spectrum; (2) 'synchronicity in the collective attacks and decays of upper harmonics' from sounds with note onsets in which all harmonics enter in close time alignment to those in which the entry of the harmonics is tapered; and (3) from sounds with 'precedent high-frequency, low-amplitude energy, most often inharmonic energy, during the attack phase' to those without high-frequency attack energy. These results serve to demonstrate that (a) useful experimental work can and has been carried out on timbre, and (b) that acoustic conclusions can be reached which fit in with other observations, for example the emphasis of Grey's axes (2) and (3) on the note onset phase.

The sound of an acoustic musical instrument is always changing, even during the rather misleadingly so-called 'steady-state' portion of a note. This is clearly shown, for example in the waveforms and spectrograms for the violin and sung notes in Figure 5.5. Pipe organ notes are often presented as being 'steady' due to the inherent air flow regulation within the instrument, but Figure 5.1 shows that even the acoustic output from a single organ pipe has an amplitude envelope that is not particularly steady. This effect manifests itself perceptually extremely clearly when attempts are made to synthesise the sounds of musical instruments electronically and no attempt is made to vary the sound in any way during its steady state. Variation of some kind is needed during any sound in order to hold the listener's attention. The acoustic communication of new information to a listener, whether speech, music, environmental sounds or warning signals from a natural or person-made source, requires that the input signal varies in some way, with time. Such variation may be of the pitch, loudness or timbre of the sound. The popularity of post-processing effects, particularly chorus (see Chapter 7), either as a feature on synthesisers themselves or as a studio effects unit reflects this. However, whilst these can make sounds more interesting to listen to by time variation imposed by adding post-processing, such an addition rarely does anything to improve the overall *naturalness* of a synthesised sound.

A note from any acoustic musical instrument typically changes dynamically throughout in its pitch, loudness and timbre. Pitch and loudness have one dimensional subjective scales from 'low' to 'high' which can be related fairly directly to physical changes which can be measured, but timbre has no such one-dimensional subjective scale. Methods have been proposed to track the dynamic nature of timbre based on the manner in which the harmonic content of a sound changes throughout. The 'tristimulus diagram' described by Pollard and Jansson (1982) is one such method in which the time course of individual notes is plotted on a triangular graph such as the example plotted in Figure 5.6. The graph is plotted based on the proportion of energy in (1) the second, third and fourth harmonics or 'mid' frequency components (Y axis); and (2) the high-frequency partials, which here are the fifth and above, or 'high' frequency components (X axis); and (3) the fundamental or f_0 (where X and Y tend towards zero). The corners of the plot in Figure 5.6 are marked: 'mid', 'high' and 'f_0' to indicate this. A point on a tristimulus diagram therefore indicates the relationship between f_0, harmonics which are resolved and harmonics which are not resolved.

The tristimulus diagram enables the dynamic relationship between high, mid and f_0 to be plotted as a line, and a number

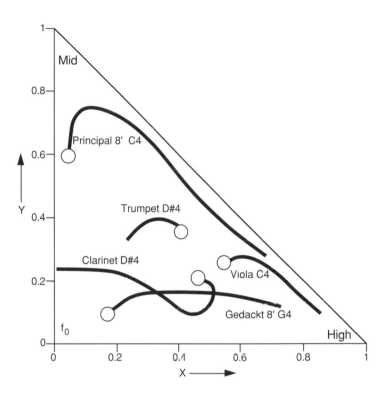

Figure 5.6 Approximate timbre representation by means of a tristimulus diagram for note onsets of notes played on a selection of instruments. In each case, the note onset tracks along the line towards the open circle which represents the approximate steady-state position. 'Mid' represents 'strong mid frequency partials'; 'High' represents 'strong high frequency partials'; 'f_0' represents 'strong fundamental'. (Data from Pollard and Jansson, 1982.)

are shown in the figure for the note onset phases of notes from a selection of instruments (data from Pollard and Jansson, 1982). The time course is not even and is not calibrated here for clarity. The approximate steady-state position of each note is represented by the open circle, and the start of the note is at the other end of the line. The note onsets in these examples lasted as follows: gedackt (10–60 ms); trumpet (10–100 ms); clarinet (30–160 ms); principal (10–150 ms); and viola (10–65 ms). The tracks taken by each note is very different and the steady-state positions lie in different locations. Pollard and Jansson present data for additional notes on some of these instruments which suggest that each instrument maintains its approximate position on the tristimulus representation as shown in the figure. This provides a method for visualising timbral differences between instruments which is based on critical band analysis. It also provides a particular representation which gives an insight as to the nature of the patterns which could be used to represent timbral differences perceptually.

There is still much work to be done on timbre to understand it more fully. Whilst there are results and ideas which indicate what acoustic aspects of different instruments contribute to the perception of their timbre differences, such difference are far too coarse to explain how the experienced listeners are able to tell apart the timbre differences between, for example violins made by different makers. The importance of timbre in music performance has been realised for many hundreds of years as manifested in the so-called 'king' of instruments—the pipe organ —well before there was any detailed knowledge of the function of the human hearing system.

5.4 The pipe organ as a timbral synthesiser

There are references to the existence of some form of pipe organ since at least 250 BC (e.g. Sumner, 1975), and it is one of the earliest forms of an acoustic timbral synthesiser based on the 'harmonic additive synthesis' principle for the production of sound. In harmonic additive synthesis, the timbre of the output sound is manipulated by means of adding harmonics together, and the stops of a pipe organ provide the means for this process.

An organ stop which has the same f_0 values as on a piano (i.e. f_0 for its A4 is 440 Hz—see Figure 3.21) is known as an 'eight foot' (8') rank on the manuals and 'sixteen foot' (16') rank on the pedals, because eight and sixteen feet are the approximate lengths of open pipes of the bottom note of a manual (C2) and the pedals (C1) respectively. A 4' rank and a 2' rank would sound one and two octaves higher than an 8' rank respectively, and a 32' rank would sound one octave lower than a 16' rank. It should be noted that the footage terminology is used to denote the sounding pitch of the rank and give no indication as to whether open or stopped pipes are employed. Thus the bottom pipes of a stopped rank on a manual sounding a pitch equivalent to a rank of 8' open pipes would be four foot long physically but its stop knob would be labelled 8'.

Organs have a number of stops on each manual of various footages, most of which are flues. Some are voiced to be used alone as solo stops usually as 8' stops, but the majority are voiced to blend together, allowing variations in loudness and timbre to be achieved by acoustic synthesis involving drawing different combinations of stops. The timbral changes are controlled by reinforcing the natural harmonics of the 8' harmonic series on the manuals (16 foot harmonic series for the pedals). The following equation relates the footage of a stop to the member of the 8' natural harmonic series which its f_0 reinforces:

$$\text{Stop footage} = \frac{8}{N} \qquad (5.1)$$

where N = harmonic number (1, 2, 3, ...)

Example 5.1 Find the footage of pipe organ stops which reinforce the third and sixth natural harmonics of the 8' harmonic series.

The third harmonic is reinforced by a stop of $\dfrac{8}{3} = 2\tfrac{2}{3}'$

The sixth harmonic is reinforced by a stop of $\dfrac{8}{6} = 1\tfrac{1}{3}'$

However, it is important to note that a single 8 foot principal stop, the foundation tone of the organ, produces a sound which is itself rich in harmonics (see Figure 5.1). Therefore the addition of a 4' principal will enhance not only the second harmonic of the 8' stop, but it will also enhance all other even harmonics. The odd harmonics of the 8' pipe are not members of the harmonic series of the 4' pipe. In general, when a stop is added whose f_0 is set to reinforce a member ($n = 1, 2, 3, 4, ...$) of the natural harmonic series at 8' pitch on the manuals (16' pitch on the pedals), it enhances the ($2n, 3n, 4n, ...$) members also. Those stops which reinforce harmonics which are not in unison (1:1) with, or a whole number of octaves (i.e. 2:1, 4:1, 8:1, ... $2n$:1) away from the first harmonic are known as 'mutation' stops.

There is a basic pipe organ timbral problem when tuning the instrument to equal temperament (see Chapter 3). Stops have to be tuned in their appropriate integer frequency ratio (see Figure 3.3) to reinforce harmonics appropriately, but as a result of this those which therefore introduce beats when chords are played. For example, supposing two stops are drawn, an 8' and a 2⅔'. The 2⅔' stop sounds an octave and a third above the 8' stop, and reinforces the third harmonic of the 8' harmonic series and therefore it must be exactly in tune with the third harmonic of the 8' stop. Thus if middle C is played with these two stops drawn, the f_0 of the C on the 2⅔' rank will be exactly in tune with the third harmonic of the C on the 8' rank. If the organ is tuned in equal temperament and the G above middle C is also played to form a two-note chord, the second harmonic of the G on the 8' rank will beat with the f_0 of the C on the 2⅔' rank as well as with the third harmonic of the C on the 8' rank. Equal-tempered tuning thus colours with beats the desired effect of adding mutation stops to build up the timbre of the organ. Mutation stops therefore tended to go out of fashion with the introduction of equal-tempered tuning on pipe organs (Padgham, 1986). Recent revivals in authentic performance of early music has extended to the pipe organ with the use of non-equal-tempered tuning systems and increased use of mutation stops. This gives new life particularly to contrapuntal music.

5.5 Deceiving the ear

This section concerns sounds which in some sense could be said to 'deceive' the ear. Such sounds have a psychoacoustic realisation which is not what might be expected from knowledge of their acoustic components. In other words, the subjective and objective realisations of sounds cannot be always directly matched up. Whilst some of the examples given may be of no obvious musical use to the performer or composer, they may in the future find musical application in electronically produced music for particular musical effects where control over the acoustic components of the output sound is exact.

5.5.1 Perception of pure tones

When two pure tones are played simultaneously, they are not always perceived as two separate pure tones. The discussion relating to Figure 2.7 introducing critical bandwidth in Chapter 2 provides a first example of sounds which in some sense, deceive the ear. These two pure tones are only perceived as separate pure tones when their frequency difference is greater than the critical bandwidth. Otherwise they are perceived as a single fused tone which is 'rough' or as beats depending on the frequency difference between the two pure tones.

When two pure tones are heard together, other tones with frequencies lower than the frequencies of either of the two pure tones themselves may be heard also. These lower tones are not acoustically present in the stimulating signal and they occur as a result of the stimulus consisting of a 'combination' of at least two pure tones and they are known as 'combination tones'. The frequency of one such combination tone which is usually quite easily perceived is the difference (higher minus the lower) between the frequencies of the two tones, and this is known as the 'difference tone':

$$f_d = f_h - f_l \tag{5.2}$$

where f_d = frequency of the difference tone

$ f_h$ = frequency of the higher frequency pure tone

$ f_l$ = frequency of the lower frequency pure tone

Notice that this is the beat frequency when the frequency difference is less than approximately 12.5 Hz (see Chapter 2). The frequencies of other possible combination tones that can result from two pure tones sounding simultaneously can be calculated as follows:

$$f_{(n)} = f_l - [n \, (f_h - f_l)] = f_l - [n \, f_d] \qquad (5.3)$$

where $f_{(n)}$ = frequency of the nth combination tone

$\quad n$ = (1, 2, 3, 4, ...)

$\quad f_l$ = frequency of the pure tone with the lower frequency

$\quad f_h$ = frequency of the pure tone with the higher frequency

These tones are always below the frequency of the lower pure tone, and occur at integer multiples of the difference tone frequency below the lower tone. No listeners hear all and some hear none of these combination tones. The difference tone and the combination tones for n = 1 and n = 2, known as the 'second-order difference tone' and the 'third-order difference tone', are those that are perceived most readily (e.g. Rasch and Plomp, 1982).

Example 5.2 Calculate the difference tone and first four combination tones which occur when pure tones of 1200 Hz and 1100 Hz sound simultaneously.

Equation 5.2 gives the difference tone frequency = $f_h - f_l$ = 1200 − 1100 = 100 Hz

Equation 5.3 gives other combination tone frequencies, and the first three are for n = 1, 2, 3 and 4.

for n = 1: $f_{(1)}$ = 1100 − (1 * 100) = 1000 Hz
for n = 2: $f_{(2)}$ = 1100 − (2 * 100) = 900 Hz
for n = 3: $f_{(3)}$ = 1100 − (3 * 100) = 800 Hz
for n = 4: $f_{(4)}$ = 1100 − (4 * 100) = 700 Hz

When the two pure tone frequencies are both themselves adjacent harmonics of some f_0 (in Example 5.2 the tones are the 11th and 12th harmonics of 100 Hz), then the difference tone is equal to f_0 and the other combination tones form 'missing' members of the harmonic series. When the two tones are not members of a harmonic series, the combination tones have no equivalent f_0, but they will be equally spaced in frequency.

Combination tones are perceived quite easily when two musical instruments which produce fairly pure tone outputs, such as the descant recorder, baroque flute or piccolo, whose f_0 values are high and close in frequency.

When the two notes played are themselves both exact and adjacent members of the harmonic series formed on their difference tone, the combination tones will be consecutive members of the harmonic series adjacent and below the lower played note (i.e. the f_0 values of both notes and their combination tones would be exact integer multiples of the difference frequency between the notes themselves). The musical relationship of combination tones to notes played therefore depends on the tuning system in use. Two notes played using a tuning system which results in the interval between the notes never being pure, such as the equal-tempered system, will produce combinations tones which are close but not exact harmonics of the series formed on the difference tone.

Example 5.3 If two descant recorders are playing the notes A5 and B5 simultaneously in equal tempered tuning, which notes on the equal tempered scale are closest to the most readily perceived combination tones?

The most readily perceived combination tones are the difference tone and the combination tones for $n = 1$ and $n = 2$ in Equation 5.3. Equal-tempered f_0 values for notes are given in Figure 3.21. Thus A5 has an f_0 of 880.0 Hz and for B5, f_0=987.8 Hz.

The difference tone frequency = 987.8 − 880.0 = 107.8 Hz; closest note is A2 (f_0=110.0 Hz).

The combination tones are:

for $n = 1$: 880.0 − 107.8 = 772.2 Hz; closest note is G5 (f_0=784.0 Hz)
for $n = 2$: 880.0 − 215.6 = 664.4 Hz; closest note is E5 (f_0=659.3 Hz)

These combination tones would beat with the f_0 component of any other instruments in an ensemble playing a note close to a combination tone. This will not be as marked as it might appear at first, due to an effect known as 'masking', which is described in the next section.

5.5.2 Masking of one sound by another

When we listen to music, it is very rare that it consists of just a single pure tone. Whilst it is possible and relatively simple to arrange to listen to a pure tone of a particular frequency in a laboratory or by means of an electronic synthesiser (a useful, important and valuable experience) such a sound would not sustain any prolonged musical interest. Almost every sound we hear in music consists of at least two frequency components.

When two or more pure tones are heard together an effect known as 'masking' can occur, where each individual tone can become more difficult or impossible to perceive, or it is partially or completely 'masked', due to the presence of another tone. In such a case the tone which causes the masking is known as the 'masker' and the tone which is masked is known as the 'maskee'. These tones could be individual pure tones, but given the rarity of such sounds in music, they are more likely to be individual frequency components of a note played on one instrument which either mask other components in that note, or frequency components of another note. The extent to which masking occurs depends on the frequencies of the masker and maskee and their amplitudes.

As is the case with most psychoacoustic investigations, masking is usually discussed in terms of the masking effect one pure tone can have on another, and the result is extended to complex sounds by considering the masking effect in relation to individual components. (This is similar, for example, to the approach adopted in the section on consonance and dissonance in Chapter 3, Section 3.3.2.) In psychoacoustic terms, the threshold of hearing of the maskee is shifted when in the presence of the masker, which gives the basis on which masking can be measured as the shift of a listener's threshold hearing curve caused by the presence of the masker.

The dependence of masking on the frequencies of masker and maskee can be illustrated by reference to Figure 2.9 in which an idealised frequency response curve for an auditory filter is plotted. The filter will respond to components in the input acoustic signal which fall within its response curve whose bandwidth is given by the critical bandwidth for the filter's centre frequency. The filter will respond to components in the input whose frequencies are lower than its centre frequency to a greater degree than components which are higher in frequency than the centre frequency due to the asymmetry of the response curve. Masking can be thought of as the filter's effectiveness in analysing a component at its centre frequency (maskee) being reduced to some degree by the presence of another component (masker) whose frequency falls within the filter's response curve. The degree to which the filter's effectiveness is reduced is usually measured as a shift in hearing threshold, or 'masking level', as illustrated in Figure 5.7. The figure shows that the asymmetry of the response curve results in the masking effect being considerably greater for maskees which are above rather than those below the frequency of the masker. This effect is often referred as:

Figure 5.7 Idealised masking level to illustrate the 'low masks high', or 'upward spread of masking', effect for a masker of frequency f_{masker} Hz.

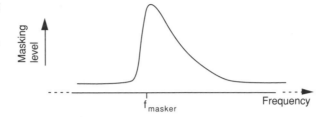

- the upward spread of masking, *or*
- low masks high.

The dependence of masking on the amplitudes of masker and maskee is illustrated in Figure 5.8 in which idealised masking level curves are plotted for different amplitude levels of a masker of frequency f_{masker}. At low amplitude levels, the masking effect tends to be similar for frequency above and below f_{masker}. As the amplitude of the masker is raised the low masks high effect increases and the resulting masking level curve becomes increasingly asymmetric. Thus the masking effect is highly dependent on the amplitude of the masker. This effect is illustrated in Figure 5.9 which is taken from Sundberg (1991). The frequency scale in this Figure is plotted such that each critical bandwidth occupies the same distance. Sundberg summarises this Figure in terms of a three straight-line approximation to the threshold of hearing in the presence of the masker, or 'masked threshold', as follows:

- the masked threshold above the critical band in which the masker falls off at about 5–13 dB per critical band
- the masked threshold in the critical band in which the masker falls is approximately 20 dB below the level of the masker itself
- the masked threshold below the critical band in which the masker falls off considerably more steeply than it does above the critical band in which the masker falls.

Figure 5.8 Idealised change in masking level with different levels of masker of frequency f_{masker} Hz.

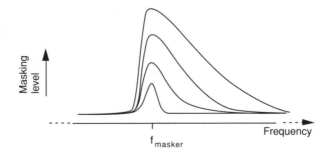

Figure 5.9 Idealised masked thresholds for masker pure tones at 300 Hz, 350 Hz and 400 Hz at 50 dBSPL, 70 dBSPL and 90 dBSPL respectively plotted on a critical band spaced frequency scale. (From Sundberg, 1991.)

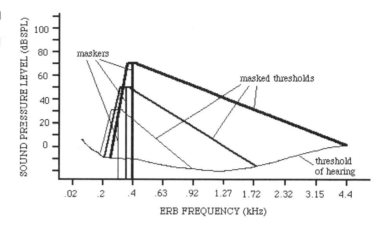

The masking effect of individual components in musical sounds which are complex with many spectral components can be determined in terms of the masking effect of individual components on other components in the sound. If a component is completely masked by another component in the sound, the masked component makes no contribution to the perceived nature of the sound itself and is therefore effectively ignored. If the masker is broadband noise, or 'white noise', then components at all frequencies are masked in an essentially linear fashion (i.e. a 10 dB increase in the level of the noise increases the masking effect by 10 dB at all frequencies). This can be the case, for example, with background noise or a brushed snare drum (see Figure 3.6) which have spectral energy generally spread over a wide frequency range and this can mask components of other sounds that fall within that frequency range.

The masking effects considered so far are known as 'simultaneous masking' because the masking effect on the maskee by the masker occurs when both sound together (or simultaneously). Two further masking effects are important for the perception of music where the masker and maskee are not sounding together, and these are referred to as 'non-simultaneous masking'. These are 'forward masking' or 'post-masking' and 'backward masking' or pre-masking. In forward masking, a pure tone masker can mask another tone (maskee) which starts after the masker itself has ceased to sound. In other words the masking effect is 'forward' in time from the masker to the maskee. Forward masking can occur for time intervals between the end of the masker and the start of the maskee of up to approximately 30 ms. In backward masking a maskee can be masked by a masker which follows it in time, starting up to approximately 10 ms after the maskee itself has ended. It should be noted,

Figure 5.10 Idealised illustration of simultaneous and non-simultaneous masking.

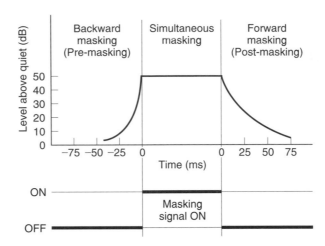

however, that considerable variation exists between listeners in terms of the time intervals over which forward and backward masking takes place.

Simultaneous and non-simultaneous masking are summarised in an idealised graphical format in Figure 5.10, which gives an indication of the masking effect in the time domain. The instant at which the masker starts and stops is indicated at the bottom of the figure, and it is assumed that the simultaneous masking effect is such that the threshold is raised by 50 dB. The potential spreading in time of masking as non-simultaneous pre- and post-masking effects is also shown. Moore (1996) makes the following observations about non-simultaneous masking:

- backward masking is considerably lessened (to zero in some cases) with practice
- recovery rate from forward masking is greater at higher masking levels
- the forward masking effect disappears 100–200 ms after the masker ceases
- the forward masking effect increases for masker durations up to about 50 ms.

Masking is exploited practically in digital systems that store and transmit digital audio in order to reduce the amount of information that has to be handled, and therefore reduce the transmission resource, or *bandwidth*, and memory, disk or other storage medium required. Such systems are generally referred to as *perceptual coders* because they exploit knowledge of human perception. For example, perceptual coding is the operating basis of the MP3 system that is used to transmit music over the Internet, MP3 players that store many hours of such music in a pocket-sized device, multi-channel sound in digital audio broadcasting and satellite television systems, MiniDisk recorders (Maes, 1996), and the now obsolete digital compact cassette (DCC).

There are international standards that define perceptual coding schemes for the encoding (recording) and decoding (playback) parts of these systems which enable different manufacturers to produce equipment, and the Moving Pictures Expert Group (MPEG) was set up in 1988. Their task was then and still is now to develop international standards for the coding of moving pictures and associated audio, and their work has resulted in standards such as MPEG-1, MPEG-2 and MPEG-4, each of which includes three layers: 1, 2, and 3. MP3 itself is based on MPEG-1, layer III (not MPEG-3 as this does not exist!). The basic principles of perceptual coding schemes for audio are outlined below and more details can be found in Watkinson (1994, 1999), Gilchrist and Grewin (1996) and Rumsey (1996).

Figure 5.11 shows the basic block structure of all audio perceptual coders. The input signal is first split into a number of frequency bands (box 1), generally by means of a bank of band-pass filters, and these are sometimes referred to as *sub-bands* giving some coders the often used name *sub-band coders*. The extent to which this process matches the human peripheral hearing system critical band analysis (see Chapter 2) depends on the complexity of the particular coding scheme itself. The energy in each of these sub-bands is used with reference to the original signal to calculate the simultaneous (and in some cases also the non-simultaneous) masking effects for that instant of input signal (box 2). Those elements of the sub-bands that the system decides would not be masked are then digitally coded (box 3) for transmission and/or storage. At the receiving end there is an encoder which reverses this process to reproduce the original audio material, which is not of course an exact copy of the original input since masking predictions have been employed to

Figure 5.11 Basic schematic of a perceptual coding system.

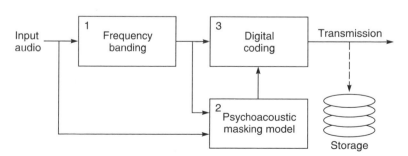

remove material that the listener would not have heard in that context.

Demonstrations of masking effects are available on the CD recording of Houtsma *et al.* (1987).

5.5.3 Note grouping illusions

There are some situations when the perceived sound is unexpected, as a result of either what amounts to an acoustic illusion or the way in which the human hearing system analyses sounds. Whilst some of these sounds will not be found in traditional musical performances using acoustic instruments since they can only be generated electronically, some of the effects have a bearing on how music is performed. The nature of the illusion and its relationship with the acoustic input which produced it can give rise to new theories of how sound is perceived, and in some cases, the effect might have already or could in the future be used in the performance of music.

Diana Deutsch describes a number of note grouping acoustic illusions, some of which are summarised below with an indication of their manifestation in music perception and/or performance. Deutsch (1974) describes an 'octave illusion' in which a sequence of two tones an octave apart with high (800 Hz) and low (400 Hz) f_0 values are alternated between the ears as illustrated in the upper part of Figure 5.12, Most listeners report hearing a high tone in the right ear alternating with a low tone in the left ear as illustrated in the figure, no matter which way round the headphones are placed. She further notes that right-handed listeners tend to report hearing the high tone in the right ear alternating with a low tone in the left ear whilst left-handed listeners tend to hear a high tone alternating with a low tone but it is equally likely that the high tone is heard in the left or right ear. This illusion persists when the stimuli are played over loudspeakers. This stimulus is available on the CD recording of Houtsma *et al.* (1987).

In a further experiment (Deutsch, 1975) played an ascending and descending C major scale simultaneously with alternate notes being switched between the two ears as shown in the lower part of Figure 5.12. The most commonly perceived response is also shown in the figure. Once again the high notes tend to be heard in the right ear and the low notes in the left ear, resulting in a snippet of a C major scale being heard in each ear. Such effects are known as 'grouping' or 'streaming', and by way of explanation, Deutsch invokes some of the grouping principles of the

Figure 5.12 A schematic representation of the stimulus for and most common response to the 'octave illusion' (upper) described by Deutsch (1974), and scale illusion (lower) described by Deutsch (1975).

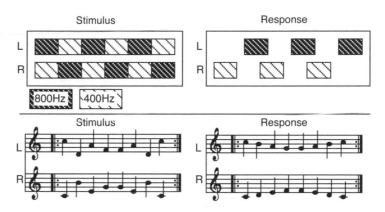

'Gestalt school' of psychology known as 'good continuation', 'proximity' and 'similarity'. She describes these (Deutsch, 1982) as follows:

• grouping by good continuation—'elements that follow each other in a given direction are perceived as blending together'.

• grouping by proximity—'nearer elements are grouped together in preference to elements that are spaced farther apart'.

• grouping by similarity—'like elements are grouped together'.

In each case the 'elements' referred to are the individual notes in these stimuli. Applying these principles to the stimuli shown in the figure, Deutsch suggests that the principle of proximity is important, grouping the higher tones (and lower tones) together, rather than good continuation which would suggest that complete ascending and/or descending scales of C major would be perceived. Deutsch (1982) describes other experiments which support this view. Music in which grouping of notes together by frequency proximity produces the sensation of a number of parts being played even though only a single line of music is being performed, includes works for solo instruments such as the Partitas and Sonatas for solo violin by J.S. Bach. An example of this effect is shown in Figure 5.13 from the Preludio from Partita number III in E major for solo violin by J.S. Bach. The score (upper stave) and three parts usually perceived (lower stave) are shown, where the perceived parts are grouped by frequency proximity.

The rather extraordinary string part writing in the final movement of Tchaikovsky's 6th symphony in the passage shown in Figure

Figure 5.13 Bars 45 to 50 of the *Preludio* from Partita III in E major for solo violin by J.S. Bach showing the notes scored for the violin (upper stave) and the three parts normally perceived by streaming (lower three staves).

5.14 is also often noted in this context because it is generally perceived as the four-part passage shown. This can again be explained by the principle of grouping by frequency proximity. The effect can be considered in terms of stereo listening if the strings are heard in a contemporaneous orchestral positioning in the following order (as viewed from left to right): first violins, double basses, cellos, violas, second violins. This is as opposed to the more common arrangement today (as viewed from left to right): first violins, second violins, violas, double basses and cellos.

Other illusions can be produced which are based on timbral proximity streaming. Pierce (1992) describes an experiment

Figure 5.14 Snippet of the final movement of Tchaikovsky's sixth symphony showing the notes scored for the strings and the four parts normally perceived.

'described in 1978 by David L. Wessel' and illustrated in Figure 5.15. In this experiment the rising arpeggio shown as response (A) is perceived as expected for stimulus (A) when all the note timbres are the same. However, the response changes to two separate falling arpeggii shown as response (B) if note timbres are alternated between two timbres represented by the different notehead shapes and 'the difference in timbres is increased' as shown for stimulus (B). This is described as timbral streaming (e.g. Bregman, 1990). A variation on this effect is shown in Figure 5.16 in which the pattern of notes shown is produced with four different timbres represented by the different notehead shapes. (This forms the basis of one of our laboratory exercises

Figure 5.15 Stimulus and usually perceived responses for timbral streaming experiment of Wessel described by Pierce (1992). Different timbres in (B) are represented by open square and filled diamond. (Adapted from Pierce, 1992.)

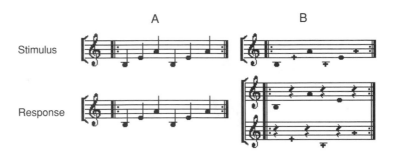

for music technology students.) The score is repeated indefinitely and the speed can be varied. Ascending or descending scales are perceived depending on the speed at which this sequence is played. For slow speeds (less than one note per second) an ascending sequence of scales is perceived (stave B in the figure). The streaming is based on 'note order'. When the speed is increased, for example to greater than ten notes per second, a descending sequence of scales of different timbres is perceived (staves C–F in the figure). The streaming is based on

Figure 5.16 Stimuli (stave A) used in timbre and note-order streaming experiment in which notehead shapes represent different timbres. At slow speeds, note order streaming is perceived (stave B) and at higher speeds timbre streaming is perceived (staves C–F).

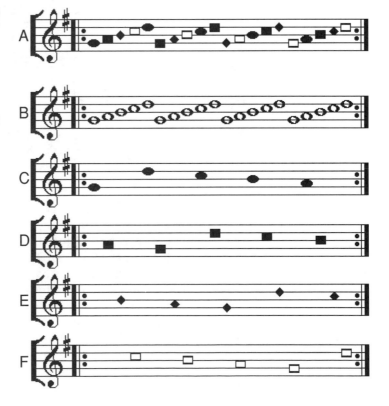

Figure 5.17 Traditional performer and audience layout in a concert situation showing treble/bass bias in the ears of performers and listeners.

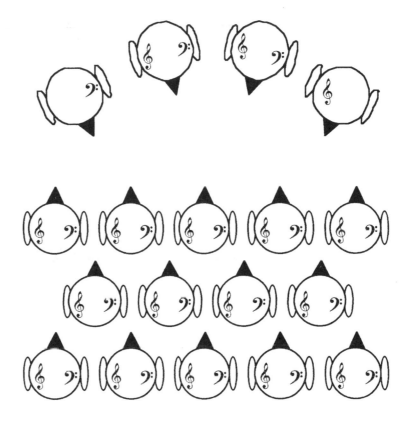

timbre. The ear can switch from one descending stream to another between those shown in staves (C–F) in the figure by concentrating on another timbre in the stimulus.

The finding that the majority of listeners to the stimuli shown in Figure 5.12 hear the high notes in the right ear and the low notes in the left ear may have some bearing on the natural layout of groups of performing musicians. For example, a string quartet will usually play with the cellist sitting on the left of the viola player who is sitting on the left of the second violinist who in turn is sitting on the left of the first violinist as illustrated in Figure 5.17. This means that each player has the instruments playing parts lower than their own on their left-hand side, and those instruments playing higher parts on their right-hand side. Vocal groups tend to organise themselves such that the sopranos are on the right of the altos, and the tenors are on the right of the basses if they are in two or more rows. Small vocal groups such as a quartet consisting of a soprano, alto, tenor and bass will tend to be in a line with the bass on the left and the soprano on the right. In orchestras, the treble instruments tend to be placed with the highest pitched instruments within their section (e.g. first violin, piccolo, trumpet etc.) on the left and bass instruments on the right. Such layouts have become traditional and moving players or singers around such that they are not in this physical position with respect to other instruments or singers is not welcomed. This tradition of musical performance layout may well be in part due to a right-ear preference for the higher notes.

However, whilst this may work well for the performers, it is back-to-front for the audience. When an audience faces a stage to watch a live performance (see Figure 5.17), the instruments or singers producing the treble notes are on the left and the bass instruments or singers are on the right. This is the wrong way round in terms of the right-ear treble preference, but the correct way round for observing the performers themselves. It is interesting to compare the normal concert hall layout as a listener with the experience of sitting in the audience area behind the orchestra which is possible in halls such as the Royal Festival Hall in London. Unfortunately this is not a test that can be carried out very satisfactorily since it is not usually possible to sit far enough behind the players to gain as good an overall balance as can be obtained from the auditorium in front of the orchestra. It is, however, possible to experience this effect by turning round when listening to a good stereo recording over loudspeakers.

5.5.4 Pitch illusions

A pitch illusion, which has been compared with the continuous staircase pictures of Maurits Escher, has been demonstrated by Shepherd (1964) and is often referred to as a 'Shepherd tone'. This illusion produces the sensation of an endless scale which ascends in semitone steps. After 12 semitone steps when the pitch has risen by one octave, the spectrum is identical to the starting spectrum so the scale ascends but never climbs acoustically more than one octave. This stimulus is available on the CD recording of Houtsma *et al.* (1987). Figure 5.18 illustrates the spectral nature of the Shepherd tone stimuli. Only the fundamental and harmonics which are multiple octaves above the fundamental are employed in the stimuli. The component frequencies of the Shepherd tone can be represented as:

$$f_{(Shepherd)} = (2_n \, f_0) \qquad (5.4)$$

where $f_{(Shepherd)}$ = frequencies of Shepherd tone components
$$n = 0, 1, 2, 3, ...)$$

The amplitudes of the components are constrained within the curved envelope shown. Each time the tone moves up one

Figure 5.18 Illustration of the spectra of stimuli which would produce the Shepherd continuous ascending scale illusion.

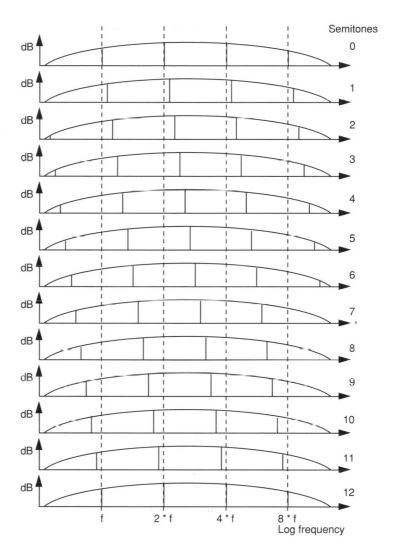

semitone, the partials all move up by a twelfth of an octave, or one semitone. The upper harmonics become weaker and eventually disappear and new lower harmonics appear and become stronger.

A musical example relating to the Shepherd tone effect in which some pitch ambiguity is perceived by some listeners can be found in the pedal line starting at bar 31 of the *Fantasia in G minor* (BWV 542) for organ by J.S. Bach. These bars are reproduced in Figure 5.19 as an organ score in which the lower of the three staves is played on the pedals while the upper two staves are played with the left and right hands. The pedal line consists of a sequence of five descending scales with eight notes in each except the last.

Figure 5.19 An extract from bar 31 of the *Fantasia in G minor* (BWV 542) for organ by J.S. Bach.

Each scale ends with an upward leap of a minor seventh and the exact moment where the upward leap occurs is often perceived with some ambiguity, even when listeners have the score in front of them. The strength of this effect depends on the particular stops used. This ambiguity is particularly common amongst listeners in the third bar of the extract where the upward leap is often very strongly perceived as occurring one or even two notes later. This could be due to the entry of a new part in the left hand playing f3 which starts as the pedal part jumps up to written Bb3. (Reference to 'written' Bb3 is made since the 16' rank provides the fundamental on the pedals which sounds an octave lower than written pitch as discussed in Section 5.4.) The f_0 components of these two notes, i.e. F3 and Bb2, form the second and third harmonics of Bb1, which would have been the next sounding note of the pedal descending scale had it not jumped up an octave. At all the upward leaps in the descending pedal scales the chord in the manual part changes from minor to major. In bar 32 the left hand change from Eb to E natural adds a member of the harmonic series (see Figure 3.3) of what would have been the next note (written C3) in the pedal scale had it not risen up the octave. Eb is not a member of that harmonic series. Similarly for the D natural in the right-hand in the third bar of the extract with the entry of the left-hand F3, and the left-hand C natural in the fourth bar. These entries of notes which are members of the harmonic series of what would have been the next note in the descending pedal scale had it not jumped up the octave serve to provide the perceived ambiguity in definition as to the exact instant at which the upward leap occurs.

The illusion produced by the combination of organ pipes to produce a sensation of pitch lower than any note actually sounding is also used in organ pedal resultant bass stops. These sound at 32′ (and very occasionally 64′), and their f_0 values for bottom C is 16.25 Hz and 8.175 Hz respectively. A resultant bass at 32′ pitch is formed by sounding together stops of 16′ and 10⅔′ which form the second and third harmonics of the 32′ harmonic series (see section 5.4). A 32′ stop perhaps labelled 'acoustic bass' is a mutation stop of 10⅔′ which when sounded with a 16′ rank produces a perceived pitch at 32′ (place theory of pitch perception from the second and third harmonics—see Chapter 3). Similarly, a 64′ stop perhaps labelled 'resultant bass' works similarly, sounding a 22⅓′ rank with a 32′ rank. The f_0 value of the middle C of a 32′ stop (C2) is 65.4 Hz and thus its bottom note is two octaves below this (C0) with an f_0 of (65¼) or 16.35 Hz. The f_0 for the bottom note of a 64′ stop (C-1) is 8.175 Hz which is below the human hearing range but within the frequency range of difference frequencies that are perceived as beats (see Figure 2.6). Harmonics that are within the human hearing range will contribute to a perception of pitch at these f_0 values which are themselves below the frequency range of the hearing system. Organists will sometimes play fifths in the pedals to imitate this effect, particularly on the last note of a piece. However, the effect is not as satisfactory as that obtained with a properly voiced resultant bass stop because the third harmonic (e.g. 10⅔′) should be softer than the second harmonic (16′) for best effect.

Roederer (1975) describes an organ-based example to illustrate residue pitch which constitutes a pitch illusion. The solo line of a chorale prelude, he suggests chorale number 40 from the *Orgelbuchlein* by J.S. Bach, is played using a number of mutation stops (see Section 5.5) if available (e.g. 8′, 4′, 2⅔′, 2′, 1⅗′, 1⅓′, 1′) accompanied by 8′, 4′ in the left hand and 16′, 8′ in the pedal. A musically trained audience should be asked to track the pitch of the melody and warned that timbre changes will occur. After playing a short snippet, play some more without the 8′, then without the 4′, then without the 2′ and finally without the 1′. What remains in the solo part is only mutation stops (i.e. those with a non-unison or non-octave pitch relationship to the fundamental). Roederer suggests making: 'the audience aware of what was left in the upper voice and point out that the pitch of the written note was absent altogether (in any of its octaves)—they will find it hard to believe! A repetition of the experiment is likely to fail—because the audience will redirect their pitch processing strategies!' Experience shows that such an experiment relies on pitch context being established when using such stimuli, usually through the use of a known or continuing musical melody.

A musical illusion only works by virtue of establishing a strong expectation in the mind's ear of the listener.

References

ANSI (1960). *American Standard Acoustical Terminology*. New York: American National Standards Institute.

Bregman, A.S. (1990). *Auditory Scene Analysis*. Cambridge: MIT Press.

Deutsch, D. (1974). An auditory illusion. *Nature*, **251**, 307–309.

Deutsch, D. (1975). Musical illusions. *Scientific American*, **233**, 92–104.

Deutsch, D. (1982). Grouping mechanisms in music. In *The Psychology of Music*. (J. Deutsch, ed.). London: Academic press.

Deutsch, D. (1983). *Journal of the Audio Engineering Society*, **31**, (9), 607–622.

Gilchrist, N. and Grewin, C. (eds) (1996). *Collected Papers on Digital Audio Bit-rate Reduction*. The Audio Engineering Society.

Grey, J. (1977). Timbre discrimination in musical patterns, *Journal of the Acoustical Society of America*, **64**, 467–472.

Hall, D.E. (1991). *Musical Acoustics: An introduction* (2nd edn). Belmont, CA: Wadsworth Publishing Company.

Houtsma, A.J.M., Rossing, T.D. and Wagenaars, W.M. (1987). *Auditory Demonstrations*. (Philips compact disc No.1126–061 and text.) New York: Acoustical Society of America.

Maes, J. (1996). *The Minidisk*. Oxford: Focal Press.

Moore, B.C.J. (1996). Masking and the human auditory system, In *Collected Papers on Digital Audio Bit-rate Reduction*. (N. Gilchrist and C. Grewin, (eds).) The Audio Engineering Society, 9–22.

Padgham, C.A. (1986). *The Well-tempered Organ*. Oxford: Positif Press.

Pierce, J.R. (1992). *The Science of Musical Sound*. Scientific American Books (2nd edn). New York: W.H. Freeman and Company.

Pollard, H.F. and Jansson, E.V. (1982). A tristimulus method for the specification of musical timbre. *Acustica*, **51**, 162–171.

Rasch, R.A. and Plomp, R. (1982). The perception of musical tones. In *The Psychology of Music*. (D. Deutsch, ed.) London: Academic Press.

Roederer, J.G. (1975). *Introduction to the Physics and Psychophysics of Music*. New York: Springer-Verlag.

Rumsey, F. (1996). *The Audio Workstation Handbook*. Oxford: Focal Press.

Scholes, P.A. (1970). *The Oxford Companion to Music*. London: Oxford University Press.

Shepherd, R.N. (1964). Circularity in pitch judgement. *Journal of the Acoustical Society of America*, **36**, 2346–2353.

Sumner, W.L. (1975). *The Organ*. London: MacDonald and Company.

Sundberg, J. (1991). *The Science of Musical Sounds*. San Diego: Academic Press.

Watkinson, J. (1994). *The Art of Digital Audio*. (2nd edn). Oxford: Focal Press.

Watkinson, J. (1999). *MPEG 2*. Oxford: Focal Press.

6 Hearing music in different environments

In this chapter we will examine the behaviour of the sound in a room with particular reference to how the room's characteristics affect the quality of the perceived sound. We will also examine strategies for analysing and improving the acoustic quality of a room. Finally we will look at how we hear sound direction and consider how this affects the design of listening rooms, studios and control rooms to give good stereo listening environment.

6.1 Acoustics of enclosed spaces

In Chapter 1 the concept of a wave propagating without considering any boundaries was discussed. However most music is listened to within a room, and is therefore influenced by the presence of boundaries, and so it is important to understand how sound propagates in such an enclosed space. Figure 6.1 shows an idealised room with a starting pistol and a listener; assume that at some time ($t = 0$) that the gun is fired. There are three main aspects to how the sound of a gun behaves in the room which are as follows.

6.1.1 The direct sound

After a short delay the listener in the space will hear the sound of the starting pistol, which will have travelled the shortest distance between it and the listener. The delay will be a function

Figure 6.1 An idealised room with an impulse excitation from a pistol.

of the distance, as sound travels 344 metres (1129 feet) per second or approximately 1 foot per millisecond. The shortest path between the starting pistol and the listener is the direct path and therefore this is the first thing the listener hears. This component of the sound is called the direct sound and its propagation path and its associated time response is shown in Figure 6.2.

The direct component is important because it carries the information in the signal in an uncontaminated form. Therefore a high

Figure 6.2 The direct sound in a room.

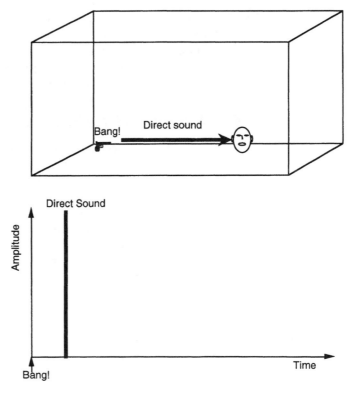

level of direct sound is required for a clear sound and good intelligibility of speech. The direct sound also behaves in the same way as sound in free space, because it has not yet interacted with any boundaries. This means that we can use the equation for the intensity of a free space wave some distance from the source to calculate the intensity of the direct sound. The intensity of the direct sound is therefore given, from Chapter 1, by:

$$I_{\text{direct sound}} = \frac{QW_{\text{Source}}}{4\pi r^2} \tag{6.1}$$

where $I_{\text{direct sound}}$ = the sound intensity (in W m^{-2})
Q = the directivity of the source (compared to a sphere)
W_{Source} = the power of the source (in W)
and r = the distance from the source (in m)

Equation 6.1 shows that the intensity of the direct sound reduces as the square of the distance from the source, in the same way as a sound in free space. This has important consequences for listening to sound in real spaces. Let us calculate the sound intensity of the direct sound from a loudspeaker.

Example 6.1 A loudspeaker radiates a sound intensity level of 102 dB at 1 m. What is the sound intensity level (I_{direct}) of the direct sound at a distance of 4 m from the loudspeaker?

The sound intensity of the direct sound at a given distance can be calculated, using Equation 1.18 from Chapter 1, as:

$$IL = 10 \log_{10}\left(\frac{W_{\text{Source}}}{W_{\text{ref}}}\right) - 20 \log_{10}(r) - 11 \text{ dB}$$

As we already know the intensity level at 1 m this equation becomes:

$$I_{\text{direct sound}} = I_{1\,\text{m}} - 20 \log_{10}(r)$$

which can be used to calculate the direct sound intensity as:

$$I_{\text{direct sound}} = 102 \text{ dB} - 20 \log_{10}(4) = 102 \text{ dB} - 12 \text{ dB} = 90 \text{ dB}$$

Example 6.1 shows that the effect of distance on the direct sound intensity can be quite severe.

6.1.2 Early reflections

A little time later the listener will then hear sounds which have been reflected off one or more surfaces (walls, floor, etc.), as

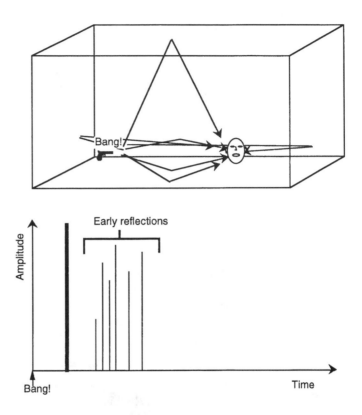

Figure 6.3 The early reflections in a room.

shown in Figure 6.3. These sounds are called early reflections and they are separated in both time and direction from the direct sound. These sounds will vary as the source or the listener move within the space. We use these changes to give us information about both the size of the space and the position of the source in the space. If any of these reflections are very delayed, total path length difference longer than about 30 milliseconds (33 feet), then they will be perceived as echoes. Early reflections can cause inter-ference effects, as discussed in Chapter 1, and these can both reduce the intelligibility of speech, and cause unwanted timbre changes in music, in the space. The intensity levels of the early reflections are affected by both the distance and the surface from which they are reflected. In general most surfaces absorb some of the sound energy and so the reflection is weakened by the reflection process. However it is possible to have surfaces which 'focus' the sound, as shown in Figure 6.4, and in these circum-stances the intensity level at the listener will be enhanced. It is important to note, however, that the total power in the sound will have been reduced by the interaction with the surface. This means that there will be less sound intensity at other positions in the room. Also any focusing structure must be large when

Figure 6.4 A focusing surface.

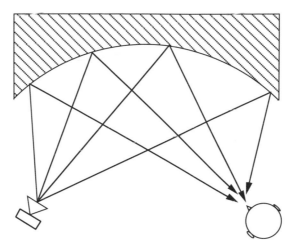

measured with respect to the sound wavelength, which tends to mean that these effects are more likely to happen for high-, rather than low-frequency components. In general therefore the level of direct reflections will be less than that predicted by the inverse square law due to surface absorption. Let us calculate the amplitude of an early reflection from a loudspeaker.

Example 6.2 A loudspeaker radiates a peak sound intensity of 102 dB at 1 m. What is the sound intensity level ($I_{\text{reflection}}$), and delay relative to the direct sound, of an early reflection when the speaker is 1.5 m away from a hard reflecting wall and the listener is at a distance of 4 m in front of the loudspeaker?

The geometry of this arrangement is shown in Figure 6.5 and we can calculate the extra path length due to the reflection by considering the 'image' of the loudspeaker, also shown in Figure 6.6, and by using Pythagoras' theorem. This gives the path length as 5 m.

Given the intensity level at 1 m, the intensity of the early reflection can be calculated because the reflected wave will also suffer from an inverse square law reduction in amplitude:

$$I_{\text{early reflection}} = I_{1\text{ m}} - 20 \log_{10}(\text{Path length}) \tag{6.2}$$

Which can be used to calculate the direct sound intensity as:

$$I_{\text{early reflection}} = 102 \text{ dB} - 20 \log_{10}(5) = 102 \text{ dB} - 14 \text{ dB} = 88 \text{ dB}$$

Comparing this with the earlier example we can see that the early reflection is 2 dB lower in intensity compared to the direct sound. The delay is simply calculated from the path length as:

$$\text{Delay}_{\text{early reflection}} = \frac{\text{Path length}}{344 \text{ ms}^{-1}} = \frac{5 \text{ m}}{344 \text{ ms}^{-1}} = 14.5 \text{ ms}$$

Similarly the delay of the direct sound is:

Figure 6.5 A geometry for calculating the intensity of an early reflection.

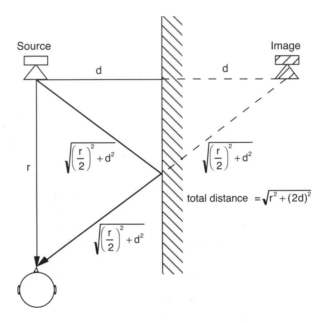

Figure 6.6 The maximum bounds for early reflections assuming no absorption or focusing.

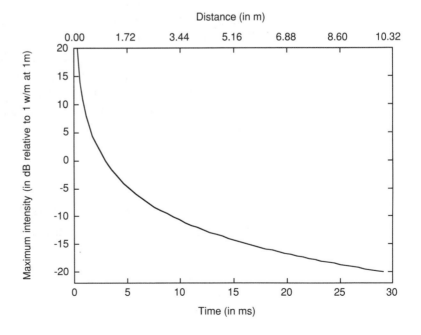

$$\text{Delay}_{\text{direct}} = \frac{r}{344 \text{ ms}^{-1}} = \frac{4 \text{ m}}{344 \text{ ms}^{-1}} = 11.6 \text{ ms}$$

So the early reflection arrives at the listener 14.5 ms – 11.6 ms = 2.9 ms after the direct sound.

Because there is a direct correspondence between delay, distance from the source and the reduction in intensity due to the inverse square law, we can plot all this on a common graph (see Figure 6.6), which shows the maximum bounds of the intensity level of reflections, providing there are no focusing effects.

6.1.3 The effect of absorption on early reflections

How does the absorption of sound affect the level of early reflections heard by the listener? The absorption coefficient of a material defines the amount of energy, or power, that is removed from the sound when it strikes it. In general the absorption coefficient of real materials will vary with frequency but for the moment we shall assume they do not. The amount of energy, or power, removed by a given area of absorbing material will depend on the energy, or power, per unit area striking it. As the sound intensity is a measure of the power per unit area this means that the intensity of the sound reflected is reduced in proportion to the absorption coefficient. That is:

$$\text{Intensity}_{\text{reflected}} = \text{Intensity}_{\text{incident}} \times (1 - \alpha) \tag{6.3}$$

where $\text{Intensity}_{\text{reflected}}$ = the sound intensity reflected after absorption (in W m^{-2})
$\text{Intensity}_{\text{incident}}$ = the sound intensity before absorption (in W m^{-2})
and α = the absorption coefficient

Because a multiplication of sound levels is equivalent to adding the decibels together, as shown in Chapter 1, Equation 6.3 can be expressed directly in terms of the decibels as:

$$I_{\text{absorbed}} = I_{\text{incident}} + 10 \log(1 - \alpha) \tag{6.4}$$

which can be combined with Equation 6.2 to give a means of calculating the intensity of an early reflection from an absorbing surface:

$$I_{\text{early reflection}} = I_{1 \text{ m}} - 20 \log_{10}(\text{Path length}) + 10 \log(1 - \alpha) \tag{6.5}$$

As an example consider the effect of an absorbing surface on the level of the early reflection level calculated earlier.

Example 6.3 A loudspeaker radiates a peak sound intensity of 102 dB at 1 m. What is the sound intensity level ($I_{\text{early reflection}}$) of an early reflection, when the speaker is 1.5 m away from a reflecting wall and the listener is at a distance of 4 m in front of the loudspeaker, and the wall has an absorption of 0.9, 0.69, 0.5?

As we already know the intensity level at 1m the intensity of the early reflection can be calculated using Equation 6.5 because the reflected wave also suffers from an inverse square law reduction in amplitude:

$$I_{\text{early reflection}} = I_{1\text{ m}} - 20 \log_{10}(\text{Path length}) + 10 \log(1 - \alpha)$$

The path length, from the earlier calculation, is 5 m so the sound intensity at the listener for the three different absorption coefficients is:

$$\begin{aligned} I_{\text{early reflection } (\alpha = 0.9)} &= 102 \text{ dB} - 20 \log_{10}(5\text{ m}) + 10 \log(1 - 0.9) \\ &= 102 \text{ dB} - 14 \text{ dB} - 10 \text{ dB} = 78 \text{ dB} \end{aligned}$$

$$\begin{aligned} I_{\text{early reflection } (\alpha = 0.69)} &= 88 \text{ dB} + 10 \log(1 - 0.69) = 88 \text{ dB} - 5 \text{ dB} \\ &= 83 \text{ dB} \end{aligned}$$

$$\begin{aligned} I_{\text{early reflection } (\alpha = 0.5)} &= 88 \text{ dB} + 10 \log(1 - 0.5) = 88 \text{ dB} - 3 \text{ dB} \\ &= 85 \text{ dB} \end{aligned}$$

6.1.4 The reverberant sound

At an even later time the sound has been reflected many times and is arriving at the listener from all directions, as shown in Figure 6.7. Because there are so many possible reflection paths, each individual reflection is very close in time to its neighbours and thus there is a dense set of reflections arriving at the listener. This part of the sound is called reverberation and is desirable as it adds richness to, and supports, musical sounds. Reverberation also helps integrate all the sounds from an instrument so that a listener hears a sound which incorporates all the instrument's sounds, including the directional parts. In fact we find rooms which have very little reverberation uncomfortable and generally do not like performing music in them; it is much more fun to sing in the bathroom compared to the living room. The time taken for reverberation to occur is a function of the size of the room and will be shorter for smaller rooms, due to the shorter time between reflections. In fact the time gap between the direct sound and reverberation is an important cue to the size of the space that the music is being performed in. Because some of the sound is

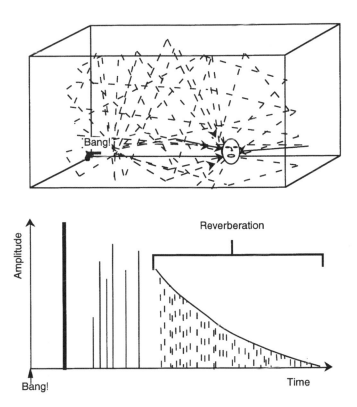

Figure 6.7 The reverberant sound in a room.

absorbed at each reflection it dies away eventually. The time that it takes for the sound to die away is called the reverberation time and is dependent on both the size of the space and the amount of sound absorbed at each reflection. In fact there are three aspects of the reverberant field that the space affects, see Figure 6.8.

- *The increase of the reverberant field level:* This is the initial portion of the reverberant field and is affected by the room size, which affects the time between reflections and therefore the time it takes the reverberant field to build up. The amount of absorption in the room also affects the time that it takes the sound to get to its steady state level. This is because, as shall be shown later, the steady state level is inversely proportional to the amount of absorption in the room. The sound level will take longer to reach a louder level than a smaller one, because the rate at which sound builds up depends on the time between reflections and the absorption.

- *The steady state level of the reverberant field:* If a steady tone, such as an organ note, is played in the space then after a period of time the reverberant sound will reach a constant level because at that point the sound power input balances

Figure 6.8 The time and amplitude evolution of the reverberant sound in a room.

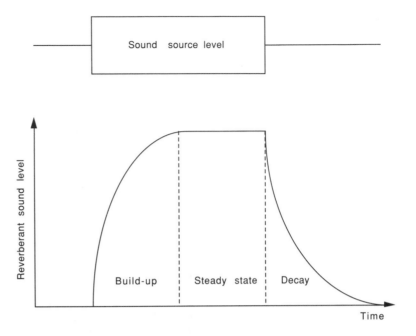

the power lost by absorption in the space. This means that the steady state level will be louder in rooms which have a small amount of absorption. Note that a transient sound in the space will not reach a steady state level.

- *The decay of the reverberant field level:* When a tone in the space stops, or after a transient, the reverberant sound level will not reduce immediately but will instead decay at a rate determined by the amount of sound energy that is absorbed at each reflection. Thus in spaces with a small amount of absorption the reverberant field will take longer to decay.

Bigger spaces tend to have longer reverberation times and well furnished spaces tend to have shorter reverberation times. Reverberation time can vary from about 0.2 of a second for a small well furnished living room to about 10 seconds for a large glass and stone cathedral.

6.1.5 The behaviour of the reverberant sound field

The reverberation part of the sound in a room behaves differently, compared to the direct sound and early reflections from the perspective of the listener. The direct sound and early reflections follow the inverse square law, with the addition of absorption effects in the case of early reflections, and so their amplitude varies with position. However the reverberant part of the sound remains constant with the position of the listener in the room.

Figure 6.9 The source of the steady state sound level of the reverberant field.

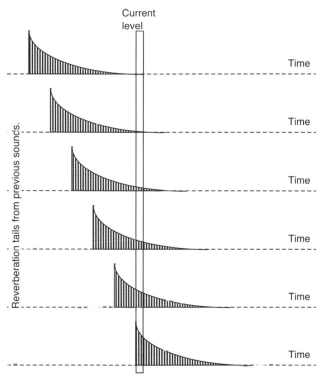

Current reverberation level is the sum of the previous reverberation tails

This is not due to the sound waves behaving differently from normal waves; instead it is due to the fact that the reverberant sound waves arrive at the listener from all directions. The result is that at any point in the room there are a large number of sound waves whose intensities are being added together. These sound waves have many different arrival times, directions and amplitudes because the sound waves are reflected back into the room, and so shuttle forwards, backwards and sideways around the room as they decay. The steady state sound level, at a given point in the room, therefore is an integrated sum of all the sound intensities in the reverberant part of the sound, as shown in Figure 6.9. Because of this behaviour the reverberant part of the sound in a room is often referred to as the reverberant field.

6.1.6 The balance of reverberant to direct sound

This behaviour of the reverberant field has two consequences. Firstly the balance between the direct and reverberant sounds will alter depending on the position of the listener relative to the source. This is due to the fact that the level of the reverberant field is independent of the position of the listener with respect

Figure 6.10 The composite effect of direct sound and reverberant field on the sound intensity, as a function of the distance from the source.

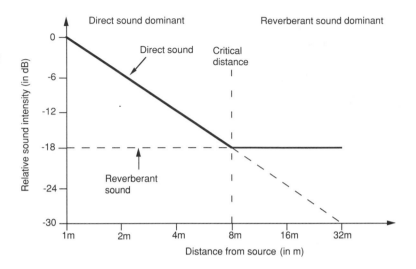

to the source, whereas the direct sound level is dependent on the distance between the listener and the sound source. These effects are summarised in Figure 6.10 which shows the relative levels of direct to reverberant field as a function of distance from the source. This figure shows that there is a distance from the source at which the reverberant field will begin to dominate the direct field from the source. The transition occurs when the two are equal and this point is known as the critical distance.

6.1.7 The level of the reverberant sound in the steady state

Secondly because, in the steady state, the reverberant sound at any time instant is the sum of all the energy in the reverberation tail the overall sound level is increased by reverberation. The level of the reverberation will depend on how fast the sound is absorbed in the room. A low level of absorption will result in sound that stays around in the room for longer and so will give a higher level of reverberant field. In fact, if the average level of absorption coefficient for the room is given by α, the power level in the reverberation sound in a room can be calculated using the following equation:

$$W_{\text{reverberant}} = W_{\text{Source}}\, 4\left(\frac{1 - \alpha}{S\alpha}\right) \tag{6.6}$$

where $W_{\text{reverberant}}$ = the reverberant sound power (in W)
S = the total surface area in the room (in m²)
W_{Source} = the power of the source (in W)
and α = the average absorption coefficient in the room

Figure 6.11 The leaky bucket model of reverberant field intensity level.

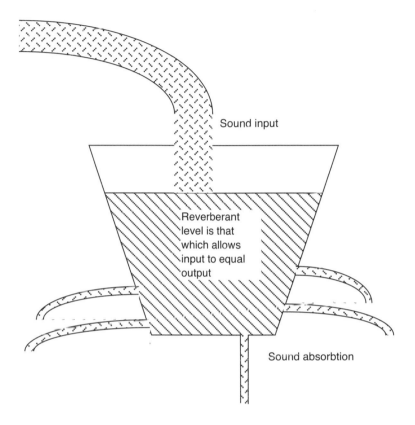

Sound input

Reverberant level is that which allows input to equal output

Sound absorbtion

Equation 6.6 is based on the fact that, at equilibrium, the rate of energy removal from the room will equal the energy put into its reverberant sound field. As the sound is absorbed when it hits the surface, it is absorbed at a rate which is proportional to the surface area times the average absorption, or $S\alpha$. This is similar to a leaky bucket being filled with water where the ultimate water level will be that at which the water runs out at the same rate as it flows in, see Figure 6.11. The amount of sound energy available for contribution to the reverberant field is also a function of the absorption because if there is a large amount of absorption then there will be less direct sound reflected off a surface to contribute to the reverberant field—remember that before the first reflection the sound is direct sound. The amount of sound energy available to contribute to the reverberant field is therefore proportional to the residual energy left after the first reflection, or $(1 - \alpha)$ because α is absorbed at the first surface. The combination of these two effects gives $(1 - \alpha)/S\alpha$ the term in Equation 6.6. The factor of four in Equation 6.6 arises from the fact that sound is approaching the surfaces in the room from all possible directions. An interesting result from Equation 6.6 is

259

that it appears that the level of the reverberant field depends only on the total absorbing surface area. In other words it is independent of the volume of the room. However in practice the surface area and volume are related because one encloses the other. In fact, because the surface area in a room becomes less as its volume decreases, the reverberant sound level becomes higher for a given average absorption coefficient in smaller rooms. Another way of visualising this is to realise that in a smaller room there is less volume for a given amount of sound energy to spread out in, like a pat of butter on a smaller piece of toast. Therefore the energy density, and thus the sound level, must be higher in smaller rooms.

The term $(1 - \alpha)/S\alpha$ in Equation 6.6 is often inverted to give a quantity known as the room constant, R, which is given by:

$$R = \frac{S\alpha}{(1 - \alpha)} \tag{6.7}$$

where R = the room constant (in m²)
and α = the average absorption coefficient in the room

Using the room constant Equation 6.6 simply becomes:

$$W_{\text{reverberant}} = W_{\text{Source}} \left(\frac{4}{R}\right) \tag{6.8}$$

In terms of the sound power level this can be expressed as:

$$SWL_{\text{reverberant}} = 10 \log_{10} \left(\frac{W_{\text{Source}}}{W_{\text{ref}}}\right) + 10 \log_{10} \left(\frac{4}{R}\right) \tag{6.9}$$

As α a is a number between 0 and 1 this also means that the level of the reverberant field will be greater in a room with a small surface area, compared to a larger room, for a given level of absorption coefficient. However one must be careful in taking this result to extremes. A long and very thin cylinder will have a large surface area, but Equation 6.6 will not predict the reverberation level correctly because in this case the sound will not visit all the surfaces with equal probability. This will have the effect of modifying the average absorption coefficient and so will alter the prediction of Equation 6.6. Therefore one must take note of an important assumption behind Equation 6.6 which is that *the reverberant sound visits all surfaces with equal probability and from all possible directions*. This is known as the diffuse field assumption. It can also be looked at as a definition of a diffuse field. In general the assumption of a diffuse field is reasonable and it is usually a design goal for most acoustics. However it is important to recognise that there are situations in which it breaks down, for example at low frequencies.

As an example consider the effect of different levels of absorption and surface area on the level of the reverberant field that might arise from the loudspeaker described earlier.

Example 6.4 A loudspeaker radiates a peak sound intensity of 102 dB at 1 m. What is the sound pressure level of the reverberant field if the surface area of the room is 75 m², and the average absorption coefficient is (a) 0.9, and (b) 0.2? What would be the effect of doubling the surface area in the room while keeping the average absorption the same?

From Equation 1.18 we can say:

$$SIL = 10 \log_{10} \left(\frac{W_{Source}}{W_{ref}} \right) - 20 \log_{10}(r) - 11 \text{ dB}$$

Thus the sound power level (SWL) radiated by the loudspeaker is:

$$SWL = 10 \log_{10} \left(\frac{W_{Source}}{W_{ref}} \right) = SIL + 11 \text{ dB} = 102 \text{ dB} + 11 \text{ dB} = 113 \text{ dB}$$

The power in the reverberant field is given by:

$$SWL_{reverberant} = 10 \log_{10} \left(\frac{W_{Source}}{W_{ref}} \right) + 10 \log_{10} \left(\frac{4}{R} \right)$$

The room constant 'R' for the two cases is:

$$R_{(\alpha = 0.9)} = \frac{S\alpha}{(1 - \alpha)} = \frac{75 \text{ m}^2 \times 0.9}{(1 - 0.9)} = 675 \text{ m}^2$$

$$R_{(\alpha = 0.2)} = \frac{S\alpha}{(1 - \alpha)} = \frac{75 \text{ m}^2 \times 0.2}{(1 - 0.2)} = 18.75 \text{ m}^2$$

The level of the reverberant field can therefore be calculated from:

$$SWL_{reverberant} = 10 \log_{10} \left(\frac{W_{Source}}{W_{ref}} \right) + 10 \log_{10} \left(\frac{4}{R} \right)$$

$$= 113 \text{ dB} + 10 \log_{10} \left(\frac{4}{R} \right)$$

which gives:

$$SWL_{reverberant \ (\alpha = 0.9)} = 113 \text{ dB} + 10 \log_{10} \left(\frac{4}{675} \right)$$

$$= 113 \text{ dB} - 22.3 \text{ dB} = 90.7 \text{ dB}$$

and:

$$SWL_{\text{reverberant } (\alpha = 0.2)} = 113 \text{ dB} + 10 \log_{10}\left(\frac{4}{18.75}\right)$$

$$= 113 \text{ dB} - 6.7 \text{ dB} = 106 \text{ dB}$$

The effect of doubling the surface area is to increase the room constant by the same proportion, so we can say that:

$$SWL_{\text{reverberant (S doubled)}} = 10 \log_{10}\left(\frac{W_{\text{Source}}}{W_{\text{ref}}}\right) + 10 \log_{10}\left(\frac{4}{2R}\right)$$

$$= 113 \text{ dB} + 10 \log_{10}\left(\frac{4}{R}\right) + 10 \log_{10}\left(\frac{1}{2}\right)$$

which gives:

$$SWL_{\text{reverberant (S doubled)}} = 113 \text{ dB} + 10 \log_{10}\left(\frac{4}{R}\right) - 3 \text{ dB}$$

Thus the effect of doubling the surface area is to reduce the level of the reverberant field by 3 dB in both cases.

Clearly the level of the reverberant field is strongly affected by the level of average absorption. The first example would be typical of an extremely 'dead' acoustic environment, as found in some studios, whereas the second is typical of an average living room. The amount of loudspeaker energy required to produce a given volume in the room is clearly much greater, about 15 dB, in the first room compared with the second. If there is a musician in the room then they will experience a 'lift' in output due to the reverberant field in the room. Because of this musicians feel uncomfortable playing in rooms with a low level of reverberant field and prefer performing in rooms which help them in producing more output. This is also one of the reasons we prefer singing in the bathroom.

6.1.8 Calculating the critical distance

The reverberant field is, in most cases, diffuse, and therefore visits all parts of the room with equal probability. Also at any point, and at any instant, we hear the total power in the reverberant field, as discussed earlier. Because of this it is possible to equate the power in the reverberant field to the sound pressure level. Thus we can say:

$$SPL_{\text{reverberant}} \approx SWL_{\text{reverberant}} = 10 \log_{10}\left(\frac{W_{\text{Source}}}{W_{\text{ref}}}\right) + 10 \log_{10}\left(\frac{4}{R}\right) \quad (6.10)$$

The distance at which the reverberant level equals the direct sound, the critical distance, can also be calculated using the above equations. At the critical distance the intensity due to the direct field and the power in the reverberant field at a given point are equal so we can equate Equation 6.1 and Equation 6.8 to give:

$$\frac{QW_{\text{Source}}}{4\pi r^2_{\text{critical distance}}} = W_{\text{Source}}\left(\frac{4}{R}\right)$$

Which can be rearranged to give:

$$r^2_{\text{critical distance}} = \left(\frac{R}{4}\right)\frac{Q}{4\pi}$$

Thus the critical distance is given by:

$$r_{\text{critical distance}} = \sqrt{\left(\frac{1}{16\pi}\right)}\sqrt{RQ} = 0.141\sqrt{RQ} \qquad (6.11)$$

Equation 6.11 shows that the critical distance is determined only by the room constant and the directivity of the sound source. Because the room constant is a function of the surface area of the room the critical distance will tend to increase with larger rooms. However many of us listen to music in our living rooms so let us calculate the critical distance for a hi-fi loudspeaker in a living room.

Example 6.5 What is the critical distance for a free standing, omnidirectional, loudspeaker radiating into a room whose surface area is 75 m², and whose average absorption coefficient is 0.2? What would be the effect of mounting the speaker into a wall?

The speaker is omnidirectional so the 'Q' is equal to 1. The room constant 'R' is the same as was found in the earlier example, 18.75 m². Substituting both these values into Equation 6.11 gives:

$$r_{\text{critical distance}} = 0.141\sqrt{RQ} = 0.141\sqrt{18.75 \times 1} = 0.61 \text{ m (61 cm)}$$

This is a very short distance! If the speaker is mounted in the wall the 'Q' increases to 2, because the speaker can only radiate into 2π steradians, so the critical distance increases to:

$$r_{\text{critical distance}} = 0.141\sqrt{RQ} = 0.141\sqrt{18.75 \times 2} = 0.86 \text{ m (86 cm)}$$

Which is still quite small!

As most people would be about 2 m away from their loudspeakers when they are listening to them this means that in a normal domestic setting the reverberant field is the most dominant source of sound energy from the hi-fi, and not the direct sound. Therefore the quality of the reverberant field is an important aspect of the performance of any system which reproduces recorded music in the home. There is also an effect on speech intelligibility in the space as the direct sound is the major component of the sound which provides this.

The level of the reverberant field is a function of the average absorption coefficient in the room. Most real materials, such as carpets, curtains, sofas and wood panelling have an absorption coefficient which changes with frequency. This means that the reverberant field level will also vary with frequency, in some cases quite strongly. Therefore in order to hear music, recorded or otherwise, with good fidelity, it is important to have a reverberant field which has an appropriate frequency response. As seen in the previous chapter, one of the cues for sound timbre is the spectral content of the sound which is being heard, and this means that when the reverberant field is dominant, as it is beyond the critical distance, it will determine the perceived timbre of the sound. This subject will be considered in more detail later in the chapter.

6.1.9 The effect of source directivity on the reverberant sound

There is an additional effect on the reverberation field, and that is the directivity of the source of sound in the room. Most hi-fi loudspeakers, and musical instruments, are omnidirectional at low frequencies but are not necessarily so at higher ones. As the level of the reverberant field is a function of both the average absorption and the directivity of the source, the variation in directivity of real musical sources will also have an effect on the reverberant sound field and hence the perception of the timbre of the sound. Consider the following example of a typical domestic hi-fi speaker in the living room considered earlier.

Example 6.6 A hi-fi loudspeaker, with a flat-on axis, direct field, response, has a 'Q' which varies from 1 to 25, and radiates a peak on axis sound intensity of 102 dB at 1 m. The surface area of the room is 75 m², and the average absorption coefficient is 0.2. Over what range does the sound pressure level of the reverberant field vary?

As the speaker has a flat-on axis response the intensity of the direct field given by Equation 6.1 should be constant.

That is:

$$I_{\text{directive source}} = \frac{QW_{\text{Source}}}{4\pi r^2} \tag{6.12}$$

where $I_{\text{directive source}}$ = the sound intensity (in W m^{-2})
Q = the directivity of the source (compared to a sphere)
W_{Source} = the power of the source (in W)
and r = the distance from the source (in m)

should be constant. Therefore the sound power radiated by the loudspeaker can be calculated by rearranging Equation 6.12 to give:

$$W_{\text{Source}} = \left(\frac{4\pi}{Q}\right) I_{\text{directive source}} \tag{6.13}$$

Equation 6.13 shows that in order to achieve a constant direct sound response the power radiated by the source must reduce as the 'Q' increases. The power in the reverberant field is given by:

$$W_{\text{reverberant}} = W_{\text{Source}}\left(\frac{4}{R}\right) \tag{6.14}$$

By combining Equations 6.13 and 6.14 the reverberant field due to the loudspeaker can be calculated as:

$$W_{\text{reverberant}} = I_{\text{directive source}}\left(\frac{4\pi}{Q}\right)\left(\frac{4}{R}\right)$$

which gives a level for the reverberant field as:

$$SWL_{\text{reverberant}} = 10\log_{10}\left(\frac{I_{\text{directvie source}}}{I_{\text{ref}}}\right) + 10\log_{10}(4\pi) - 10\log_{10}(Q)$$
$$+ 10\log_{10}\left(\frac{4}{R}\right)$$

The room constant 'R' is 18.75 m^2, as calculated in Example 6.4.

The level of the reverberant field can therefore be calculated as:

$$SWL_{\text{reverberant}} = 102\text{ dB} + 11\text{ dB} - 10\log_{10}(Q)$$
$$+ 10\log_{10}\left(\frac{4}{18.75\text{ m}^2}\right)$$

which gives:

$$SWL_{\text{reverberant }(Q=1)} = 102\text{ dB} + 11\text{ dB} - 10\log_{10}(1) - 6.7\text{ dB}$$
$$= 106.3\text{ dB}$$

for the level of the reverberant field when the 'Q' is equal to 1, and:

$$SWL_{\text{reverberant (Q = 25)}} = 102 \text{ dB} + 11 \text{ dB} - 10 \log_{10}(25) - 6.7 \text{ dB}$$

$$= 92.3 \text{ dB}$$

when the 'Q' is equal to 25.

Thus the reverberant field varies by $106.3 - 92.3 = 14$ dB over the frequency range.

The effect therefore of a directive source with constant on axis response is to reduce the reverberant field as the 'Q' gets higher. The subjective effect of this would be similar to reducing the high 'Q' regions via the use of a tone control which would not normally be acceptable as a sound quality. A typical reverberant response of a typical domestic hi-fi speaker is shown in Figure 6.12. Note that the reverberant response tends to drop in both the midrange and high frequencies. This is due to the bass and treble speakers becoming more directive at the high ends of their frequency range. The dip in reverberant energy will make the speaker less 'present' and may make sounds in this region harder to hear in the mix. The drop in reverberant field at the top end will make the speaker sound 'duller'. Some manufacturers try to compensate for these effects by allowing the on-axis response to rise in these regions, however this brings other problems. The reduction in reverberant field with increasing 'Q' is used to advantage in speech systems to raise the level of direct sound above the reverberant field and so improve the intelligibility.

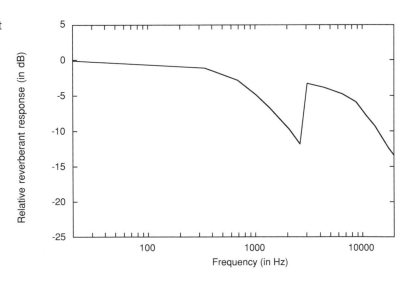

Figure 6.12 The reverberant response of a domestic two-way high-fidelity loudspeaker.

6.1.10 Reverberation time

Another aspect of the reverberant field is that sound energy which enters it at a particular time dies away. This is because each time the sound interacts with a surface in the room it loses some of its energy due to absorption. The time that it takes for sound at a given time to die away in a room is called the reverberation time. Reverberation time is an important aspect of sound behaviour in a room. If the sound dies away very quickly we perceive the room as being 'dead' and we find that listening to, or producing, music within such a space unrewarding. On the other hand when the sound dies away very slowly we perceive the room as being 'live'. A live room is preferred to a dead room when it comes to listening to, or producing, live music. On the other hand when listening to recorded music, which already has reverberation as part of the recording, a dead room is often preferred. However, as in many pleasurable aspects of life, reverberation must be taken in moderation. In fact the most appropriate length of reverberation time depends on the nature of the music being played. For example fast pieces of contrapuntal music, like that of Scarlatti or Mozart, require a shorter reverberation time compared with large romantic works, like that of Wagner or Berlioz, to be enjoyed at their best. The most extreme reverberation times are often found in cathedrals, ice rinks, and railway stations and these acoustics can convert many musical events to 'mush' yet to hear slow vocal polyphony, for example works by Palestrina, in a cathedral acoustic can be ravishing! This is because the composer has made use of the likely performance acoustic as part of the composition. Because of the importance of reverberation time in the perception of music in a room, and because of the differing requirements for speech and different types of music, much effort is focused on it. In fact a major step in room acoustics occurred when Wallace Clement Sabine enumerated a means of calculating, and so predicting, the reverberation time of a room in 1898. Much design work on auditoria in the first half of this century focused almost exclusively on this one parameter, with some successes and some spectacular failures. Nowadays other acoustical and psychoacoustical factors are also taken into consideration.

6.1.11 Calculating and predicting reverberation time

Clearly the length of time that it takes for sound to die is a function not only of the absorption of the surfaces in a room but is also a function of the length of time between interactions with the surfaces of the room. We can use these facts to derive an equation for the reverberation time in a room. The first thing to determine is the average length of time that a sound wave will travel between interactions with the surfaces of the room. This can be found from the mean free path of the room which is a measure of the average distances between surfaces, assuming all possible angles of incidence and position. For an approximately rectangular box the mean free path is given by the following equation:

$$MFP = \frac{4V}{S} \tag{6.15}$$

where MFP = the mean free path (in m)
  V = the volume (in m³)
and S = the surface area (in m²)

The time between surface interactions may be simply calculated from Equation 6.15 by dividing it by the speed of sound to give:

$$\tau = \frac{4V}{Sc} \tag{6.16}$$

where τ = the time between reflections (in s)
and c = the speed of sound (in ms⁻¹)

Equation 6.16 gives us the time between surface interactions and at each of these interactions α is the proportion of the energy absorbed, where α is the average absorption coefficient discussed earlier. If α of the energy absorbed at the surface then $(1 - \alpha)$ is the proportion of the energy reflected back to interact with further surfaces. At each surface a further proportion, α, of energy will be removed so the proportion of the original sound energy that is reflected will reduce exponentially. The combination of the time between reflections and the exponential decay of the sound energy, through progressive interactions with the surfaces of the room, can be used to derive an expression for the length of time that it would take for the initial energy to decay by a given ratio. See Appendix 3 for details.

There are an infinite number of possible ratios that could be used. However, the most commonly used ratio is that which corresponds to a decrease in sound energy of 60 dB, or 10^6. This gives an equation for the 60 dB reverberation time, known as T_{60}, which is, from Appendix 3:

$$T_{60} = \frac{-0.161 \; V}{S \ln(1 - \alpha)} \tag{6.17}$$

where T_{60} = the 60 dB reverberation time (in s)

Equation 6.17 is known as the Norris–Eyring reverberation formula, the negative sign in the top compensates for the negative sign that results from the natural logarithm resulting in a reverberation time which is positive. Note that it is possible to calculate the reverberation time for other ratios of decay and that the only difference between these and Equation 6.17 would be the value of the constant. The argument behind the derivation of reverberation time is a statistical one and so there are some important assumptions behind Equation 6.17. These assumptions are:

- that the sound visits all surfaces with equal probability, and at all possible angles of incidence. That is, the sound field is diffuse. This is required in order to invoke the concept of an average absorption coefficient for the room. Note that this is a desirable acoustic goal for subjective reasons as well; we prefer to listen to and perform music in rooms with a diffuse field.
- that the concept of a mean free path is valid. Again this is required in order to have an average absorption coefficient but in addition it means that the room's shape must not be too extreme. This means that this analysis is not valid for rooms which resemble long tunnels. However most real rooms are not too deviant and the mean free path equation is applicable.

6.1.12 The effect of room size on reverberation time

The result in Equation 6.17 also allows some broad generalisations to be made about the effect of the size of the room on the reverberation time, irrespective of the quantity of absorption present. Equation 6.17 shows that the reverberation time is a function of the surface area, which determines the total amount of absorption, and the volume, which determines the mean time between reflections in conjunction with the surface area. Consider the effect of altering the linear dimensions of the room on its volume and surface area. These clearly vary in the following way:

$$V \propto (\text{Linear dimension})^3$$

and

$$S \propto (\text{Linear dimension})^2$$

However, both the mean time between reflections, and hence the reverberation time, vary as:

$$\frac{V}{S} \propto \frac{(\text{Linear dimension})^3}{(\text{Linear dimension})^2} \propto \text{Linear dimension}$$

Hence as the room size increases the reverberation time increases proportionally, if the average absorption remains unaltered. In typical rooms the absorption is due to architectural features such as carpets, curtains, people, etc., and so tends to be a constant fraction of the surface area. The net result is that in general large rooms have a longer reverberation time than smaller ones and this is one of the cues we use to ascertain the size of a space, in addition to the initial time delay gap. Thus one often hears people referring to the sound of a 'big' or 'large', acoustic as opposed to a 'small' one when they are really referring to the reverberation time. Interestingly, now that it is possible to provide a long reverberation time in a small room, via electronic reverberation enhancement systems, with good quality, people have found that long reverberation times in a small room sound 'wrong' because the visual cues contradict the audio ones. That is, the listener, on the basis of the apparent size of the space and their experience, expects a shorter reverberation time than they are hearing. Apparently closing one's eyes restores the illusion by removing the distracting visual cue!

Let us use Equation 6.17 to calculate some reverberation times.

Example 6.7 What is the reverberation time of a room whose surface area is 75 m², whose volume is 42 m³, and whose average absorption coefficient is 0.9, 0.2? What would be the effect of doubling all the dimensions of the room while keeping the average absorption coefficients the same?

Using Equation 6.17 and substituting in the above values gives, for $\alpha = 0.9$:

$$T_{60} = \frac{-0.161V}{S \ln(1 - \alpha)} = \frac{-0.161 \times 42 \text{ m}^3}{75 \text{ m}^2 \times \ln(1 - 0.9)} = 0.042 \text{ s } (42 \times 10^{-3} \text{ s})$$

which is very small! For $\alpha = 0.2$ we get:

$$T_{60} = \frac{-0.161V}{S \ln(1 - \alpha)} = \frac{-0.161 \times 42 \text{ m}^3}{75 \text{ m}^2 \times \ln(1 - 0.2)} = 0.43 \text{ s}$$

which would correspond well with the typical T_{60} of a living room, which is in fact what it is.

If the room dimensions are doubled then the ratio of volume with respect to the surface area also doubles so the new reverberation times are given by:

$$\frac{V_{\text{doubled}}}{S_{\text{doubled}}} = (\text{Linear dimension})_{\text{doubled}} = 2$$

so the old reverberation times are increased by a factor of 2:

$$T_{60 \text{ doubled}} = T_{60} \times 2$$

which gives a reverberation time of:

$$T_{60 \text{ doubled}} = T_{60} \times 2 = 0.042 \times 2 = 0.084 \text{ s}$$

when $\alpha = 0.9$ and:

$$T_{60 \text{ doubled}} = T_{60} \times 2 = 0.43 \times 2 = 0.86 \text{ s}$$

when $\alpha = 0.2$.

6.1.13 The problem of short reverberation times

The very short reverberation times that occur when the absorption is high pose an interesting problem. Remember that one of the assumptions behind the derivation of the reverberation time calculation was that the sound energy visited all the surfaces in the room with equal probability. For our example room the mean time between reflections, using Equation 6.16 is given by:

$$\tau = \frac{4V}{Sc} = \frac{4 \times 42 \text{ m}^3}{75 \text{ m}^2 \times c} = \frac{2.24 \text{ m}}{344 \text{ ms}^{-1}} = 6.51 \text{ ms } (0.00651 \text{ s})$$

If the reverberation time calculated in Example 6.7, when $\alpha = 0.9$, is divided by the mean time between reflections then the average number of reflections that have occurred during the reverberation time can be calculated to be:

$$N_{\text{reflections}} = \frac{T_{60}}{\tau} = \frac{42 \times 10^{-3} \text{ s}}{6.51 \times 10^{-3} \text{ s}} = 6.45 \text{ reflections}$$

This is barely enough reflections to have hit each surface once! In this situation the reverberant field does not really exist; instead the decay of sound in the room is really a series of early reflections to which the concept of reverberant field or reverberation does not really apply. In order to have a reverberant field there must be much more than 6 reflections. A suitable number of reflections, in order to have a reverberant field, might be nearer 20, although this is clearly a hard boundary to accurately define. Many studios and control rooms have been treated so that they are very 'dead' and so do not support a reverberant field.

6.1.14 A simpler reverberation time equation

Although the Norris–Eyring reverberation formula is often used to calculate reverberation times there is a simpler formula known as the Sabine formula, named after its developer Wallace Clement Sabine, which is also often used. Although it was originally developed from considerations of average energy loss from a volume, a derivation which involves solving a simple differential equation, it is possible to derive it from the Norris–Eyring reverberation formula. This also gives a useful insight into the contexts in which the Sabine formula can be reasonably applied. Consider the Norris–Eyring reverberation formula below:

$$T_{60} = \frac{-0.161V}{S\ln(1-\alpha)}$$

The main difficulty in applying this formula is due to the need to take the natural logarithm of $(1-\alpha)$. However, the natural logarithm can be expanded as an infinite series to give:

$$T_{60} = \frac{-0.161V}{S\left(\alpha - \dfrac{\alpha^2}{2} - \dfrac{\alpha^3}{3} - ... - \dfrac{\alpha^n}{n} - ... - \dfrac{\alpha^\infty}{\infty}\right)} \tag{6.18}$$

Because $\alpha < 1$ the sequence always converges. However if $\alpha < 0.3$ then the error due to all the terms greater than $-\alpha$ is less than 5.7%. This means that Equation 6.18 can be approximated as:

$$T_{60(\alpha < 0.3)} \approx \frac{-0.161V}{S(-\alpha)} = \frac{0.161V}{S\alpha} \tag{6.19}$$

Equation 6.19 is known as the Sabine reverberation formula and, apart from being useful, was the first reverberation formula. It was developed on the basis of experimental measurements made by W.C. Sabine, thus initiating the whole science of architectural acoustics. Equation 6.19 is much easier to use and gives accurate enough results providing the absorption, α, is less than about 0.3. In many real rooms this is a reasonable assumption. However it becomes increasingly inaccurate as the average absorption increases and in the limit predicts a reverberation time when $\alpha = 1$, that is reverberation without walls!

6.1.15 Reverberation faults

As stated previously, the basic assumption behind these equations is that the reverberant field is statistically random, that is a diffuse field. There are however acoustic situations in which this is not the case. Figure 6.13 shows the decay of energy, in dB, as a function of time for an ideal diffuse field reverberation. In this case the decay is a smooth straight line representing an exponential decay of an equal number of dBs per second. Figure 6.14 on the other hand shows two situations in which the reverberant field is no longer diffuse. In the first situation all the absorption is only on two surfaces, for example an office with acoustic tiles on the ceiling, carpets on the floor, and nothing on the walls. Here

Figure 6.13 The ideal decay versus time curve for reverberation.

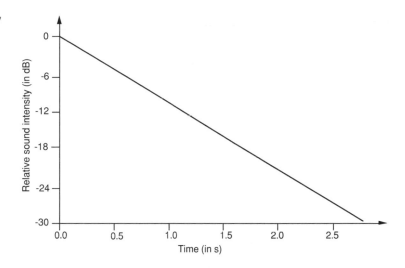

Figure 6.14 Two situations which give poor reverberation decay curves.

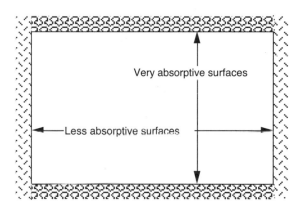

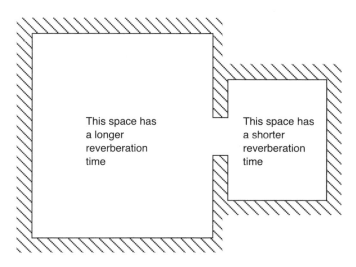

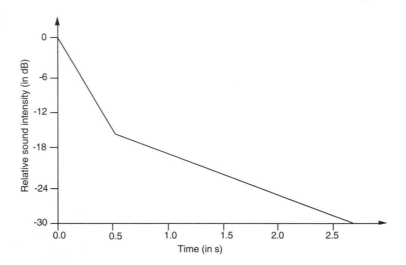

Figure 6.15 A double break reverberation decay curve.

the sound between the absorbing surfaces decays quickly whereas the sound between the walls decays much more slowly, due to the lower absorption. In the second case there are two connected spaces, such as the transept and nave in a church, or under the balconies in a concert hall. In this case the sound energy does not couple entirely between the two spaces and so they will decay at different rates which depend on the level of absorption in them. In both of these cases the result is a sound energy curve as a function of time which has two or more slopes, as shown in Figure 6.15. This curve arises because the faster decaying waves die away before the longer decaying ones and so allow them to dominate in the end. The second major acoustical defect in reverberant decay occurs when there are two precisely parallel and smooth surfaces, as shown in Figure 6.16. This results in a series of rapidly spaced echoes, onomatopoeiacally called flutter echoes, which

Figure 6.16 A situation which can cause flutter.

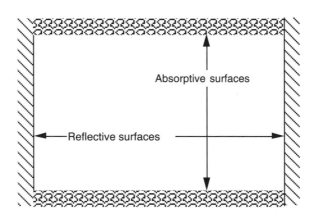

Figure 6.17 The decay versus time curve for flutter.

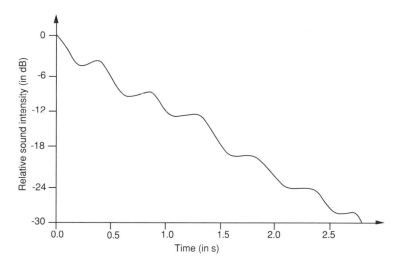

which result as the energy shuttles forwards and backwards between the two surfaces. These are most easily detected by clapping one's hands between the parallel surfaces to provide the packet of sound energy to excite the flutter echo. The decay of energy versus time in this situation is shown in Figure 6.17 and the presence of the flutter echo manifests itself as a series of peaks in the decay curve. Note that this behaviour is also often associated with the double-slope decay characteristic shown in Figure 6.15 because the energy shuttling between the parallel surfaces suffers less absorption compared with a diffuse sound.

6.1.16 Reverberation time variation with frequency

Equations 6.17 and 6.18 show that the reverberation time depends on the volume, surface area, and the average absorption coefficient in the room. However, the absorption coefficients of real materials are not constant with frequency. This means that, assuming that the room's volume and surface area are constant with frequency which is not an unreasonable assumption, the reverberation time in the room will also vary with frequency. This will subjectively alter the timbre of the sound in the room due to both the effect on the level of the reverberant field discussed earlier and the change in timbre as the sound in the room decays away. As an extreme example, if a particular frequency has a much slower rate of decay compared with other frequencies, then as the sound decays away this frequency will ultimately dominate and the room will 'ring' at that particular frequency. The sound power for steady state sounds will also have a strong peak at that frequency because of the effect on the reverberant field level.

Table 6.1 Typical absorption coefficients as a function of frequency for various materials

Material	Frequency					
	125 Hz	250 Hz	500 Hz	1 kHz	2 kHz	4 kHz
Plaster on lath	0.14	0.10	0.06	0.05	0.04	0.03
Carpet on concrete	0.02	0.06	0.14	0.37	0.60	0.65
Floor (wood joist)	0.15	0.11	0.10	0.07	0.06	0.07
Painted plaster	0.01	0.01	0.02	0.02	0.02	0.02
Walls (½ inch plasterboard)	0.29	0.10	0.05	0.04	0.07	0.09
Windows (float glass)	0.35	0.25	0.18	0.12	0.07	0.04
Wood panelling	0.30	0.25	0.20	0.17	0.15	0.10
Curtains (cotton draped to half area)	0.07	0.31	0.49	0.81	0.66	0.54
Air absorption (per m³ @ 20°C and 30% RH)	–	–	–	–	0.012	0.038

Table 6.1 shows some typical absorption coefficients for some typical materials which are used in rooms as a function of frequency. Note that they are measured over octave bands. One could argue that third octave band measurements would be more appropriate psychacoustically, as the octave measurement will tend to blur variations within the octave which might be perceptually noticeable. In many cases, because the absorption coefficient varies smoothly with frequency, octave measurements are sufficient. However, especially when considering resonant structures, more resolution would be helpful. Note also that there are often no measurements of the absorption coefficients below 125 Hz, this is due to both the difficulty in making such measurements and the fact that below 125 Hz other factors in the room become more important, as we shall see later.

In order to take account of the frequency variation of the absorption coefficients we must modify the equations which calculate the reverberation time as follows:

$$T_{60} = \frac{-0.161V}{S \ln(1 - \alpha(f))}$$

where $\alpha(f)$ = frequency dependent absorption coefficient

for the Norris–Eyring reverberation time equation and:

$$T_{60(\alpha < 0.3)} = \frac{0.161V}{S\alpha(f)}$$

for the Sabine reverberation time equation.

6.1.17 Reverberation time calculation with mixed surfaces

In real rooms we must also allow for the presence of a variety of different materials, as well as accounting for their variation of absorption as a function of frequency. This is complicated by the fact that there will be different areas of material, with different absorption coefficients, and these will have to be combined in a way that accurately reflects their relative contribution. For example, a large area of a material with a low value of absorption coefficient may well have more influence than a small area of material with more absorption. In the Sabine equation this is easily done by multiplying the absorption coefficient of the material by its total area and then adding up the contributions from all the surfaces in the room. These resulted in a figure which Sabine called the 'equivalent open window area' as he assumed, and experimentally verified, that the absorption coefficient of an open window was equal to one. The denominator in the Sabine reverberation equation, Equation 6.19, is also equivalent to the open window area of the room, but has been calculated using the average absorption coefficient in the room. It is therefore easy to incorporate the effects of different materials by simply calculating the total open window area for different materials, using the method described above, and substituting it for $S\alpha$ in Equation 6.19. This gives a modified equation which allows for a variety of frequency-dependent materials in the room as:

$$T_{60(\alpha < 0.3)} = \frac{0.161V}{\displaystyle\sum_{\text{All surfaces } S_i} S_i \alpha_i(f)} \tag{6.20}$$

where $\alpha_i(f)$ = absorption coefficient for a given material
and S_i = its area

For the Norris–Eyring reverberation time equation the situation is a little more complicated because the equation does not use the open window area directly. However the Norris–Eyring reverberation time equation can be rewritten in a modified form, as shown in Appendix 4, which allows for the variation in material absorption due to both nature and frequency, as:

$$T_{60} = \frac{-0.161V}{\displaystyle\sum_{\text{All surfaces } S_i} S_i \ln(1 - \alpha_i(f))} \tag{6.21}$$

Equation 6.21 is also known as the Millington–Sette equation. Although Equation 6.21 can be used irrespective of the absorption level it is still more complicated than the Sabine equation and, if the average absorption coefficient is less than 0.3 it can be approximated very effectively by it, as discussed previously. Thus in many contexts the Sabine equation, Equation 6.20, is preferred.

Table 6.2 Absorption and reverberation time calculations for an untreated living room

Surface (material)	Area (m²)	125 Hz	250 Hz	500 Hz	1 kHz	2 kHz	4 kHz
Ceiling (plaster on lath)	16.8	2.35	1.68	1.01	0.84	0.67	0.50
Floor (carpet on concrete)	16.8	0.34	1.01	2.35	6.22	10.08	10.92
Walls (painted plaster)	35.4	0.35	0.35	0.71	0.71	0.71	0.71
Windows (float glass)	6.0	2.10	1.50	1.08	0.72	0.42	0.24
Total open window area		5.14	4.54	5.15	8.48	11.88	12.37
Room volume (m³)	42						
Reverberation time (s)		1.32	1.49	1.31	0.80	0.57	0.55

Equation 6.20 is readily used in conjunction with tables of absorption coefficients to calculate the reverberation time and can be easily programmed into a spreadsheet. As an example consider the reverberation time calculation for a living room outlined in Example 6.8.

Example 6.8 What is the 60 dB reverberation time (T_{60}) of a living room as a function of frequency whose surface area is 75 m² and whose volume is 42 m³. The floor is carpet on concrete, the ceiling is plaster on lath, and both have an area of 16.8 m². There are 6 m² of windows and the rest of the surfaces are painted plaster on brick, ignore the effect of the door.

Figure 6.18 The reverberation time for the untreated room as a function of frequency.

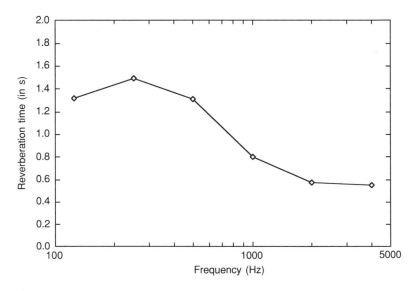

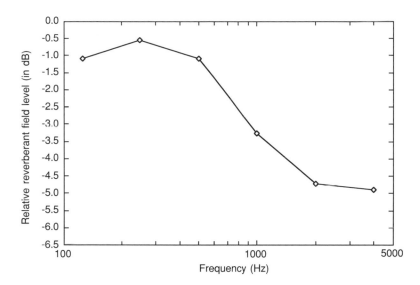

Figure 6.19 The reverberation field level for the untreated room as a function of frequency.

Using the data in Table 6.1 set up a spreadsheet or table, as shown in Table 6.2 and calculate the equivalent open window area for each surface as a function of frequency. Having done that add up the individual surface contributions for each frequency band and apply the Equation 6.20 to the result in order to calculate the reverberation time.

From the results shown in Table 6.2, which are also plotted in Figure 6.18, one can see that the reverberation varies from 1.49 seconds at low frequencies to 0.55 seconds at high frequencies. This is a normal result for such a structure and would tend to sound a bit 'woolly' or 'boomy'. The relative level of reverberant field for this room, is also shown in Figure 6.19 and this shows approximately a 5 dB increase in the reverberant field at low frequencies.

6.1.18 Reverberation time design

The results of Example 6.8 beg the question: 'How can we improve the evenness of the reverberation time?' The answer is to either add, or remove, additional absorbing materials into the room in order to achieve the desired reverberation characteristic. Here the concept of an open window area budget is useful. The idea is that, given the volume of the room, and the desired reverberation time, the necessary open window area required is calculated. The open window area already present in the room is then examined and, depending on whether the room is over or under budget, appropriate materials are added or removed.

Consider Example 6.9 which tries to improve the reverberation of the previous room.

Example 6.9 Which single material could be added to the room in Example 6.8 which would result in an improved reverberation time, and what amount would be required to effect the improvement?

A material which has a high absorption at low frequencies, such as wood panelling, needs to be added to the room. If the absorption budget is set as being equivalent to the open window area at 4 kHz then we must achieve an open window area of 12.5 m over the whole frequency range. The worst frequency in the previous example is 250 Hz, which only has 4.5 m of open window area at that frequency. This means that any additional absorber must add 12.5 – 4.5 = 8 m of open window area at that frequency. The absorption of wood panelling, from Table 6.1, at 250 Hz is 0.25. Therefore the amount of wood panelling required is:

$$\text{Area}_{\text{Wood panelling}} = \frac{\text{Required open window area}}{\text{Absorption coefficient}} = \frac{8 \text{ m}}{0.25} = 32 \text{ m}$$

Table 6.3, Figure 6.20 and Figure 6.21 show the effect of applying the treatment which dramatically improves the reverberation time characteristics. The reverberation time now only varies from 0.59 to 0.41 s which is a much smaller variation than before. The peak-to-peak variation in the level of the reverberant field has also been reduced to less than 2 dB.

Table 6.3 Absorption and reverberation time calculations for a treated living room

Surface (material)	Area (m²)	125 Hz	250 Hz	500 Hz	1 kHz	2 kHz	4 kHz
Ceiling (plaster on lath)	16.8	2.35	1.68	1.01	0.84	0.67	0.50
Floor (carpet on concrete)	16.8	0.34	1.01	2.35	6.22	10.08	10.92
Walls (painted plaster)	35.4	0.35	0.35	0.71	0.71	0.71	0.71
Windows (float glass)	6.0	2.10	1.50	1.08	0.72	0.42	0.24
Wood panelling	32.0	9.60	8.00	6.40	5.44	4.80	3.20
Total open window area		14.74	12.54	11.55	13.92	16.68	15.57
Room volume (m³)	42.0						
Reverberation time (s)		0.46	0.54	0.59	0.49	0.41	0.43

Figure 6.20 The reverberation time for the treated room as a function of frequency.

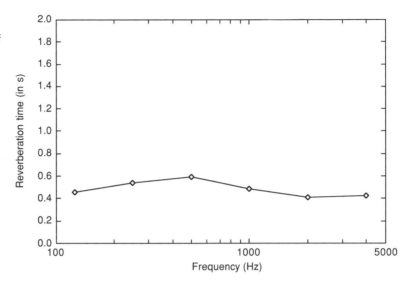

Figure 6.21 The reverberation field level for the treated room as a function of frequency.

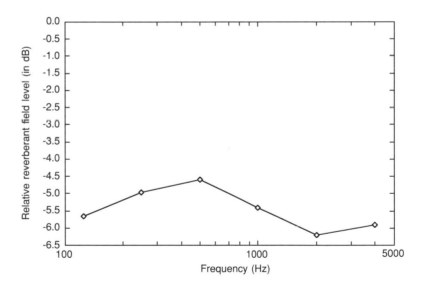

However the overall reverberation time has gone down, especially at the lowest frequencies, because of the effect of the wood panelling at frequencies other than the one being concentrated on. Thus in practice an iterative approach to deciding on the most suitable treatment for a room is often required. Another point to consider is that the treatment proposed only just fits in the room, and sometimes it proves impossible to achieve a desired reverberation characteristic due to physical limitations.

6.1.19 Ideal reverberation time characteristics

What is an ideal reverberation characteristic? We have seen that the decay should be a smooth exponential of a constant number of decibels of decay per unit time. We also know that different sorts of music require different reverberation times. In many cases the answer is, 'it depends on the situation'. However there are a few general rules which seem to be broadly accepted. Firstly, there is a range of reverberation times which are a function of the type of music being played; music with a high degree of articulation needs a drier acoustic than music which is slower and more harmonic. Secondly, as the performance space gets larger the reverberation time required for all types of music becomes longer. This result is summarised in Figure 6.22 which shows the 'ideal' reverberation time as a function of both music and room volume. Thirdly, in general, listeners prefer a rise in reverberation time in the bass (125 Hz) of about 40% relative to the midrange (1 kHz) value as shown in Figure 6.23. This rise in bass reverberation adds 'warmth' and it also helps increase the sound level of bass instruments, which often have weak fundamentals, by raising the level of the reverberant field at low frequencies. However, when recording instruments, or when listening to recorded music, this bass lift due to the reverberant field may be undesirable and therefore a flat reverberation characteristic preferred.

There are many other aspects of reverberation, too numerous to mention here, which must be considered when designing acoustic spaces. However, there are four aspects that are worthy

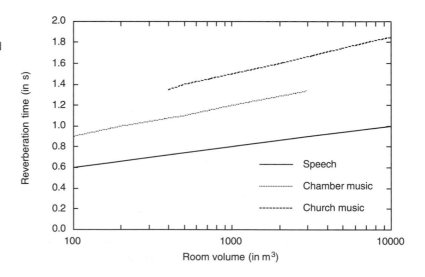

Figure 6.22 Ideal reverberation times as a function of room volume and musical style.

Figure 6.23 The ideal reverberation time versus frequency curves.

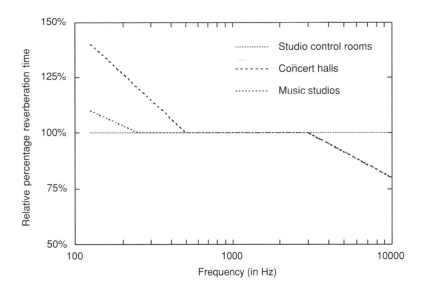

of mention as they have proved to be the downfall of more than one acoustic designer, or manufacturer of reverberation units.

6.1.20 Early decay time

The first aspect is that the measure of reverberation time as being the time it takes the sound to fall by 60 dB is not particularly relevant psychoacoustically; it is also very difficult to measure in situ. This is due to the presence of background noise, either unwanted or the music being played, which often results in less than 60 dB of energy decay before the decay sound becomes less than the residual noise in the environment. Even in the quieter environment of a Victorian town in the days before road traffic, Sabine had to do measurements, using his ears, at night to avoid the results being affected by the level of background noise. Because we rarely hear a full reverberant decay, our ears and brains have adapted, quite logically, to focus on what can be heard. Thus we are more sensitive to the effects of the first 20 to 30 dB of the reverberant decay curve. In principle, providing we have an even exponential decay curve, the 60 dB reverberation is directly proportional to the earlier curves and so this should not cause any problems. However if the curve is of the double-slope form shown in Figure 6.15 then this simple relationship is broken. The net result is that, although the T_{60} reverberation time may be an appropriate value, because of the faster early decay to below 30 dB we perceive the reverberation as being shorter than it really is. The psychoacoustic effect of this is that the space sounds 'drier' than one would expect from a simple measurement of T_{60}. Modern acoustic designers therefore worry much more about the early decay time (EDT) than they used to when designing concert halls.

6.1.21 Lateral reflections

The second factor which has been found to be important for the listener is the presence of dense diffuse reflections from the side walls of a concert hall, called lateral reflections, as shown in Figure 6.24. The effect of these are to envelop or bathe the listener in sound and this has been found to be necessary for the listener to experience maximum enjoyment from the sound. It is important that these reflections be diffuse, as specular reflections will result in disturbing comb filter effects, as discussed in Chapter 1, and distracting images of the sound sources in unwanted and unusual directions. Providing diffuse reflections is thus important and this has been recognised for some time. Traditionally, the use of plaster mouldings, niches and other decorative surface irregularities have been used to provide diffusion in an ad hoc manner. More recently diffusion structures based on patterns of wells whose depths are formally defined by an appropriate mathematical sequence have been proposed and used. However it is not just the provision of diffusion on the side walls that must be considered. The traditional concert hall is called a shoe-box hall, because of its shape, as shown in Figure 6.25, and this naturally provides a large number of lateral reflections to the audience. This shape, combined with the Victorian penchant for florid plaster decoration, resulted in some excellent sounding spaces. Unfortunately shoe-box halls are harder to make a profit with because they cannot seat as many people as some other structures. Another popular structure, which has a different behaviour as regards to lateral reflec-

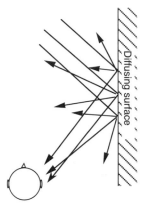

Figure 6.24 Lateral reflections in a concert hall.

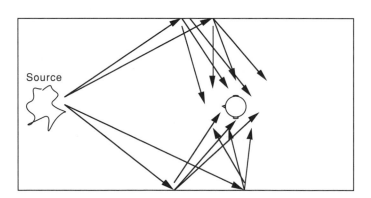

Figure 6.25 Lateral reflections in a shoe-box concert hall.

Figure 6.26 Lateral reflections in a fan-shaped concert hall.

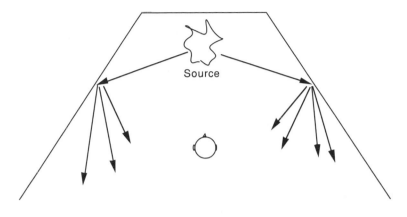

Figure 6.27 Lateral reflections from ceiling diffusion in a concert hall.

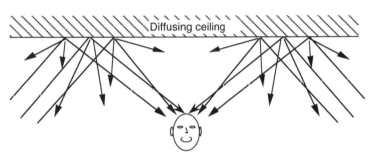

tions, is the fan-shaped hall shown in Figure 6.26. This structure has the advantage of being able to seat more people with good sightlines but unfortunately it directs the lateral reflections away from the audience and those few that do arrive are very weak at the wider part of the fan. The situation can be improved via the use of explicit diffusion structures on the walls, ceilings, and mid-air as floating shapes, as shown in Figure 6.27. However it has been found that the pseudo-lateral diffuse reflections from the ceiling are not quite comparable in effect to reflections from the side walls, and so the provision of a good listening environment within the realities of economics is still a challenge.

6.1.22 Early reflections and performer support

A third factor, which is often ignored, is the acoustics that the performers experience. Pop groups have known about this for years and take elaborate precautions to provide each performer on stage with their own individual balance of acoustic sounds via a technique known as foldback. In fact some performers now receive their foldback directly into their ears via a technique known as 'in-ear monitoring' and in many large gigs the equipment providing foldback to the performer can rival, or even

Figure 6.28 Early reflections to provide acoustic foldback for the performer.

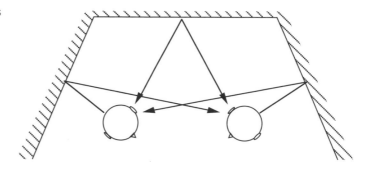

Figure 6.29 The effect of diffusion on the acoustic foldback for the performer.

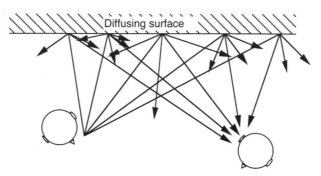

Diffusing surface

exceed, that which provides the sound for the audience. The classical musician, however, only has the acoustics of the hall to provide them with 'foldback'. Thus the musicians on the stage must rely on reflections from the nearby surfaces to provide them with the necessary sounds to enable them to hear themselves and each other. There are two requirements for the sound reaching the performer on stage. Firstly, it must be at a sufficient level and arrive soon enough to be useful. To begin with it is important that the surfaces surrounding the performers direct some sound back to them. Note that there is a conflict between this and providing a maximum amount of sound to the audience so some compromise must be reached. The usual compromise is to make use of the sound which radiates behind the performers and direct it out to the audience via the performers, as shown in Figure 6.28. This has the twofold advantage of providing the performers with acoustic foldback and redirecting sound energy that might have been lost back to the audience. Ideally the sound that is redirected back to the performers should be diffuse as this will blend the sounds of the different instruments together for all the performers, whereas specular reflectors can have hot and cold spots for a given instrument on the stage, as shown in Figure 6.29. An important aspect of acoustic foldback, however, is the time that it takes to arrive back at the performers. Ideally it should arrive immediately, and some does via the floor and direct sound from the instrument. However, the majority will have to travel to a reflecting or diffusing surface and back to the performers. There is evidence to show that, in order to maintain good ensemble comfortably, the musicians should receive the sound from other musicians within about 20 ms of the sound being produced. This means that ideally there should be a reflecting or diffusing surface within 10 ms (3.44 m or 11.5 ft) of the performer; the time is divided by 2 to allow for going to the reflecting surface and back. In practice some of the surfaces may have to be further away when large orchestral forces are being mustered, although the staging used can assist the provision of acoustic foldback. Sometimes, however, the orchestra enclosure is so large that the reflections arrive later than this. If they arrive later than about 50 ms the musicians will perceive them as echoes and ignore them. On the other hand if these reflections arrive at the boundary between perceiving it as part of the sound or an echo of a previous sound it can cause severe disruption of the performers' perception of it. The net effect of these 'late early reflections' is to damage the performers' ability to hear other instruments close to them and this further reduces their ability to maintain ensemble. In one prestigious hall, the reason musicians used to complain that they couldn't hear each other and so hated playing there was traced to the problem of late early reflections. As a postscript it is interesting to note that the orchestra enclosure in shoe-box halls often did the right things. However, in modern multipurpose facilities it is often a challenge to provide the necessary acoustic foldback while allowing space for scenery and machinery, etc.

6.1.23 The effect of air absorption

The fourth aspect of reverberation, which caught early reverberation unit designers by surprise, is an observation. The observation is that, as well as suffering many reflections, the sound energy in a reverberant decay will have travelled through a lot of air. In fact the distance that the sound will have travelled will be directly proportional to the reverberation time, so a one second reverberation time implies that the sound will have travelled 344 m by the end of the decay. Although for low frequencies air absorbs a minimal amount of sound energy, at high frequencies this is not the case. In particular humidity, smoke particles and other impurities will absorb high-frequency energy and so reduce the level of high frequencies in the sound. This is one of the reasons that people sound duller when they are speaking at a distance. In terms of reverberation time, and also the level of the reverberant field, the effect of this extra absorption is to reduce the reverberation time, and the level of the reverberant field, at high frequencies. Fortunately this effect only becomes dominant at higher frequencies, above 2 kHz. Unfortunately it is dependent on the level of humidity and smoke in the venue and so the high-frequency reverberation time, and the reverberant field level, will change as the audience stays in the space. Note this is an additional dynamic effect over and above the static absorption simply due to the presence of a clothed person in a space and is due to the fact that people exhale water vapour and perspire. Clearly then the degree of change will be a function of both the physical exertions of the audience and the quality of the ventilation system! As the effect of air absorption is determined by the distance the sound has travelled, rather than its interaction with a surface, it is difficult to incorporate the effect into the reverberation time equations discussed earlier. An approximation that seems to work is to convert the effect of the air absorption into an equivalent absorption area by scaling an air absorption coefficient by the volume of the space. This is reasonable because as the volume of the room increases the more air that the wave travels through and the longer the distance that it travels. This coefficient is shown at the bottom of Table 6.1 and from it one can see that for small rooms the effect can be ignored because until the volume becomes greater than 40 m^3 the equivalent absorbing area is less than 1 m^2. The effect does become significant if one is designing artificial reverberation units because, if it is not allowed for, the result will be an overbright reverberation, which sounds unnatural.

In this section the concept of reverberation time and reverberant field has been discussed. The assumption behind the equations has been that the sound field is diffuse. However, if this is not the case then the equations are invalid. Although at mid and high audio frequencies a diffuse field can be possible, either by accident or design, at low frequencies this is not the case due to the effect of the room's boundaries causing standing waves.

6.2 Room modes and standing waves

When a room is excited by an impulse, the sound energy is reflected from its surfaces. At each reflection some of the sound is absorbed and therefore the sound energy decays exponentially. Ideally the sound should be reflected from each surface with equal probability, forming a diffuse field. This results in a

Figure 6.30 Cyclic reflection
paths in a room.

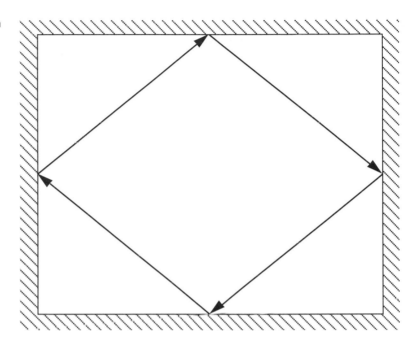

single exponential decay with a time constant proportional to the
average absorption in the room. However in practice not all the
energy is reflected in a random fashion. Instead some energy is
reflected in cyclic paths, as shown in Figure 6.30. If the length
of the path is a precise number of half wavelengths then they
will form standing waves in the room. These standing waves
have pressure and velocity distributions which are spatially
static and so behave differently to the rest of the sound in the
room in the following ways:

- They do not visit each surface with equal probability.
  Instead a subset of the surfaces are involved.
- They do not strike these surfaces with random incidence.
  Instead a particular angle of incidence is involved in the
  reflection of the standing wave.
- They require a coherent return of energy back to an original
  surface, a cyclic path. This is of necessity strongly frequency-
  dependent and so these paths only exist for discrete frequencies which are determined by the room geometry.

Another name for these standing waves in a room are room
modes and the frequencies at which they occur are known as
modal frequencies. Because the modes are spatially static there
will be a strong variation of sound pressure level as one moves
around the room, which is undesirable. There are three basic
types of room mode which are outlined in Sections 6.2.1 to 6.2.3.

Figure 6.31 Axial modal
paths.

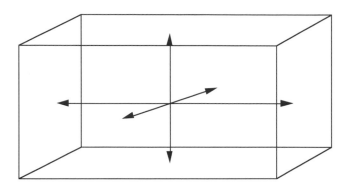

Figure 6.31 Axial modal
paths.

6.2.1 Axial modes

These modes occur between two opposing surfaces, as shown in Figure 6.31, and so are a function of the linear dimensions of the room. The frequencies of an axial mode are given by the following equation:

$$f_{x(axial)} = \frac{c}{2}\left(\frac{x}{L}\right)$$

where $f_{x(axial)}$ = the axial mode frequencies (in Hz)
x = the number of half wavelengths that fit between the surfaces (1, 2, ..., ∞)
L = the distance between the reflecting surfaces (in m)
and c = the speed of sound (in ms⁻¹)

This equation shows that there are an infinite number of possible modal frequencies at which an integer number of wavelengths fit into the room with lowest modal frequency occurring when just one half wavelength fits into the space between the reflecting surfaces.

Figure 6.32 Tangential
modal paths.

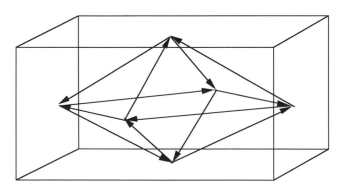

6.2.2 Tangential modes

These modes occur between four surfaces, as shown in Figure 6.32, and so are a function of two of the dimensions of the room. The frequencies of the tangential modes are given by the following equation:

$$f_{xy(tangential)} = \sqrt{\frac{c}{2}\left(\frac{x}{L}\right)^2 + \left(\frac{y}{W}\right)^2}$$

where $f_{xy(tangential)}$ = the tangential modal frequencies (in Hz)
x = the number of half wavelengths between one set of two surfaces (1, 2, ..., ∞)
y = the number of half wavelengths between the other set of surfaces (1, 2, ..., ∞)
and L, W = the distance between the reflecting surfaces (in m)

There is also an infinite number of tangential modes, but they must fit an integral number of half wavelengths in two dimensions. This has the interesting consequence that the lowest modal frequencies are higher than the axial modes, despite the fact that the apparent path length is greater. The reason is that the standing waves must fit between the opposing surfaces, that is on the sides rather than the hypotenuse of the triangular path, and as the propagating wave travels down the hypotenuse, the effective wavelength, or phase velocity, on the sides of the room is

Figure 6.33 The phase velocity of tangential modes.

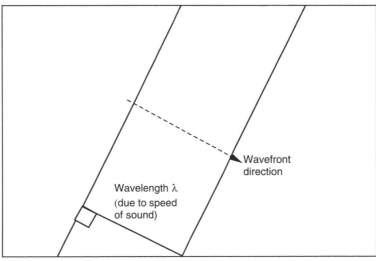

Wavefront direction

Wavelength λ (due to speed of sound)

Phase wavelength λ (due to apparently higher phase velocity)

Figure 6.34 An oblique modal path.

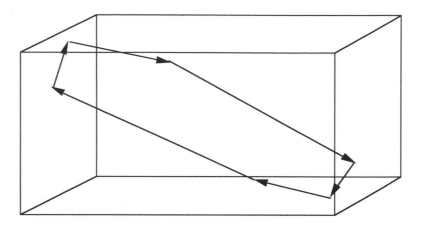

larger, as shown in Figure 6.33. The lowest modal frequency for a tangential mode occurs when precisely one half wavelength, at the phase velocity, fits into each dimension.

6.2.3 Oblique modes

These modes occur between all six surfaces, as shown in Figure 6.34, and so are a function of all three dimensions of the room. The frequencies of the oblique modes are given by the following equation:

$$f_{xyz(oblique)} = \frac{c}{2} \sqrt{\left(\frac{x}{L}\right)^2 + \left(\frac{y}{W}\right)^2 + \left(\frac{z}{H}\right)^2}$$

where $f_{xyz(oblique)}$ = the oblique modal frequencies (in Hz)
x, y, z = the number of half wavelengths between the surfaces (1, 2, ..., ∞)
and L, W, H = the distance between the reflecting surfaces (in m)

The lowest frequencies of these modes are also higher than the lowest axial modes, for the reasons discussed earlier.

6.2.4 A universal modal frequency equation

The combination of these three types of modes form a dense set of possible standing wave frequencies in the room and they can be combined into one equation by simply allowing x, y, and z in the oblique mode equation to range from 0, 1, 2 to infinity, giving the following equation which will give the frequencies of all possible modes in the room:

Table 6.4 Some favourable room dimensions

	Height	Width	Length
A	1.00	1.14	1.39
B	1.00	1.28	1.54
C	1.00	1.60	2.33

$$f_{xyz} = \frac{c}{2} \sqrt{\left(\frac{x}{L}\right)^2 + \left(\frac{y}{W}\right)^2 + \left(\frac{z}{H}\right)^2} \qquad (6.22)$$

where x, y, z = the number of half wavelengths between the surfaces $(0, 1, 2, ..., \infty)$

The above equation also shows that if any of the dimensions are integer multiples of each other then some of the modal frequencies will be the same and this can cause problems. It is therefore better to choose non-commensurate ratios for the wall dimensions to ensure that the modes are spread out as much as possible. Much work has been done on ideal room ratios and one set of favourable room dimensions is shown in Table 6.4. However, these dimensions are not necessarily the only optimum ones for all room sizes. It is also important to realise that room modes are inherent in any structure which encloses the sound sources. This means that changing the shape of the room, for example by angling the walls, does not remove the resonances, it merely changes their frequencies from values which are easily calculated to ones that are not.

6.2.5 The Bonello criteria

In general the number of resonances within a given frequency bandwidth increases with frequency. In fact it can be shown that they increase proportional to the square of the frequency, and in large well-behaved acoustical spaces, that sound good, this increase in mode density with frequency is smooth. This is the rationale behind a method for assessing the modal behaviour in a room known as the Bonello criteria. These criteria try to ascertain how significant the modal behaviour of a room is in perceptual terms. It does this by dividing the audio frequency spectrum into third octave bands, as an approximation of critical bands, and then counting the number of modes per band. If the number of modes per third octave band increase monotonically then there is a good chance that we will perceive the room as having a 'smooth' frequency response despite the resonances. If the number of resonances per third octave drops as the frequency rises then there will be a perceptually noticeable peak in the frequency response. Coincident modes are also another way of creating a perceptually noticeable frequency response peak and the Bonello criteria does further stipulate that there should be no modal coincidence within a third octave band unless there are at least three additional non-coincident resonances to balance the two that are coincident. As an example of the calculation of mode frequencies let us calculate some for a typical living room.

Example 6.10 Calculate the lowest frequency mode in a room which measures 3.5 m × 5 m × 2.5 m. At what frequency would a tangential mode with one half wavelength along the 3.5 m dimension and three half wavelengths along the 5 m dimension occur, at what frequency would the (2 2 2) oblique mode occur, and at what frequency is the first coincident mode?

Using Equation 6.21 calculate the modes as follows. The lowest frequency mode is the first axial mode along the longest dimension of the room which is the (0 1 0) or axial mode in this example so the lowest modal frequency in the room is:

$$f_{010} = \frac{c}{2} \sqrt{\left(\frac{0}{3.5 \text{ m}}\right)^2 + \left(\frac{1}{5 \text{ m}}\right)^2 + \left(\frac{0}{2.5 \text{ m}}\right)^2} = \frac{344 \text{ ms}^{-1}}{2} \sqrt{\left(\frac{1}{5 \text{ m}}\right)^2}$$

$$= 34.4 \text{ Hz}$$

The mode with one half wavelength along the 3.5 m dimension and three half wavelengths along the 5 m dimension is the (1 3 0) or tangential mode so its frequency is:

$$f_{130} = \frac{c}{2} \sqrt{\left(\frac{1}{3.5 \text{ m}}\right)^2 + \left(\frac{3}{5 \text{ m}}\right)^2 + \left(\frac{0}{2.5 \text{ m}}\right)^2}$$

$$= 172 \text{ ms}^{-1} \sqrt{0.082 + 0.36} = 114.4 \text{ Hz}$$

The frequency of the (2 2 2) or oblique mode is:

$$f_{222} = \frac{c}{2} \sqrt{\left(\frac{2}{3.5 \text{ m}}\right)^2 + \left(\frac{2}{5 \text{ m}}\right)^2 + \left(\frac{2}{2.5 \text{ m}}\right)^2}$$

$$= 172 \text{ ms}^{-1} \sqrt{0.327 + 0.16 + 0.64} = 182.6 \text{ Hz}$$

The dimensions of 2.5 m and 5 m are related by a factor of 2 so the second axial mode along the 5 m dimension will be at the same frequency as the first axial mode along the 2.5 m dimension. That is:

$$f_{020} = f_{001}$$

The (0 2 0) mode has a frequency of:

$$f_{020} = \frac{c}{2} \sqrt{\left(\frac{0}{3.5\ \text{m}}\right)^2 + \left(\frac{2}{5\ \text{m}}\right)^2 + \left(\frac{0}{2.5\ \text{m}}\right)^2}$$

$$= 172\ \text{ms}^{-1} \sqrt{\left(\frac{2}{5\ \text{m}}\right)^2} = 68.8\ \text{Hz}$$

and the (0 0 1) mode has a frequency of:

$$f_{001} = \frac{c}{2} \sqrt{\left(\frac{0}{3.5\ \text{m}}\right)^2 + \left(\frac{0}{5\ \text{m}}\right)^2 + \left(\frac{1}{2.5\ \text{m}}\right)^2}$$

$$= 172\ \text{ms}^{-1} \sqrt{\left(\frac{1}{2.5\ \text{m}}\right)^2} = 68.8\ \text{Hz}$$

which are both at the same frequency and are therefore coincident.

6.2.6 The behaviour of modes

As has been already discussed, modes behave differently to diffuse sound and this has the following consequences:

- The standing wave is not absorbed as strongly as sound which visits all surfaces. This is due to both the reduction in the number of surfaces visited and the change in absorption due to non-random incidence.
- This reduction in absorption is strongly frequency-dependent and results in less absorption and therefore a longer decay time at the frequencies at which standing waves occur.
- The decay of sound energy in the room is no longer a single exponential decay with a time constant proportional to the average absorption in the room. Instead there are several decay times. The shortest one tends to be due to the diffuse sound field whereas the longer ones tend to be due to the modes in the room. This results in excess energy at those frequencies with the attendant degradation of the sound in the room.

How does the energy in a mode decay as a function of time, how can it be related to the reverberation, and what is the effect of absorption in a mode on the frequency response?

6.2.7 The decay time of axial modes

The decay of sound energy in modes, is in many respects, identical to the decay of sound energy which is analysed in

Appendix 3. The main difference is that the absorption coefficient is sometimes smaller, because the modal sound wave does not have random incidence, it will also be specific to the surfaces involved instead of being an average value for the whole room. In addition the time between reflections will be dependent on the length of the modal path rather than the mean free path. This means that the decay time for a mode is likely to be different to the diffuse sound.

For example the length of an axial mode path is determined by the distance between the two reflecting surfaces that support the mode, which will be one of the room's dimensions. Thus for an axial mode the energy left after a given time period is given by modifying Equation A3.5 in Appendix 3 using the distance between the surfaces instead of the mean free path to give:

$$\text{Modal energy}_{\text{After t seconds}} = \text{Modal energy}_{\text{Initial}}\,(1 - \alpha_{\text{mode}})^{t(c/L_{\text{mode}})}$$

where L_{mode} = distance between the surfaces in the mode (in m)

and α_{mode} = absorption per reflection in the modal structure

The above equation can be manipulated to give a 60 dB decay time, analogous to Equation 6.17, which is:

$$T_{60\,(\text{Modal})} = \left(\frac{L_{\text{mode}}}{c}\right)\frac{\ln(10^{-6})}{\ln(1 - \alpha_{\text{mode}})} = \left(\frac{L_{\text{mode}}}{\ln(1 - \alpha_{\text{mode}})}\right)\frac{(-13.82)}{344\ \text{ms}^{-1}}$$

$$= \frac{-0.04 L_{\text{mode}}}{\ln(1 - \alpha_{\text{mode}})}$$

where $T_{60\,(\text{Modal})}$ = the 60 dB decay time of the mode (in s)

This expresses a similar result to Equation 6.17 except for the difference caused by the differing length of the modal structure with respect to the mean free path. If the length of the modal structure is longer than the mean free path then, assuming similar levels of absorption, the decay time for the mode will be longer than the diffuse field whereas if the length is smaller then the modal decay will be shorter than the diffuse field. The length between reflections is both a function of the surfaces that support the mode and the type of mode—axial, tangential, or oblique—that occurs. For axial modes the mode length, L_{mode}, is simply the relevant room dimension.

6.2.8 The decay time of other mode types

For the other types of mode the situation is more complicated, as shown in Figure 6.35 for a tangential mode. However, one

Figure 6.35 The path length for a tangential mode.

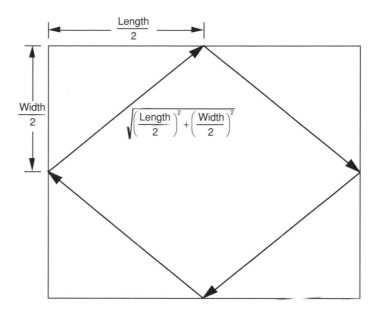

could argue that the path length for this type of mode is given by the length of the hypotenuse of the triangle formed by half the length and half the width of the four walls that support the mode. That is, the modal length is given by:

$$L_{\text{mode (tangential)}} = \sqrt{\left(\frac{\text{Length}}{2}\right)^2 + \left(\frac{\text{Width}}{2}\right)^2}$$

$$= \frac{1}{2}\sqrt{(\text{Length}^2 + \text{Width}^2)}$$

This equation shows that the distance between reflections for a tangential mode is essentially the diagonal dimension between the four surfaces that support the mode divided by two to allow for the fact that the wave suffers two reflections along this path. The length so derived can then be used as the modal length in the equation for the modal decay. A similar argument can be applied to the oblique modes, which visit all six surfaces. Because of this the modal supporting structure is a cuboid and the diagonal between opposing corners must be used. In addition the wave will suffer three reflections along this path. This gives a path length for the oblique mode as:

$$L_{\text{mode (oblique)}} = \sqrt{\left(\frac{\text{Length}}{3}\right)^2 + \left(\frac{\text{Width}}{3}\right)^2 + \left(\frac{\text{Height}}{3}\right)^2}$$

$$= \frac{1}{3}\sqrt{(\text{Length}^2 + \text{Width}^2 + \text{Height}^2)}$$

Figure 6.36 The bandwidth of modes for a given value of absorption.

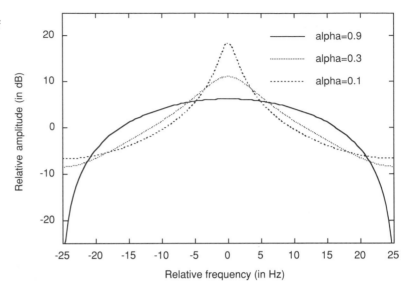

The absorption does more than cause the mode to decay; it reduces the total energy stored in the mode, in a similar manner to the effect of absorption on the reverberant field, and also causes the mode to have a finite bandwidth which is proportional to the amount of absorption, as shown in Figure 6.36. The absorption also reduces the peak to minimum variation in the standing wave pattern, and so reduces the spatial variation of the sound pressure, as shown in Figure 6.37. The bandwidth of a mode can be calculated from the 60 dB decay time using the following equation:

Figure 6.37 The spatial variation in the amplitude of modes for a given value of absorption.

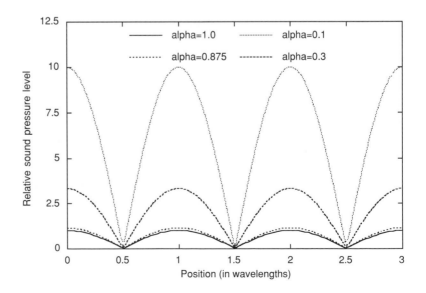

Figure 6.38 Typical variation in the amplitudes and bandwidths for the different mode types, assuming an even distribution of absorption.

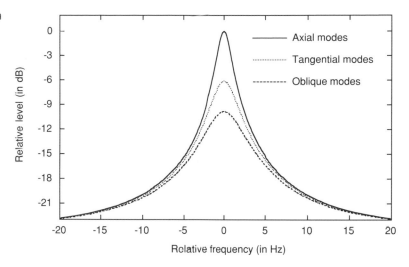

$$Bw_{\mathrm{mode}} = \frac{2.2}{T_{60 \, (\mathrm{Modal})}}$$

where Bw_{mode} = the –3 dB bandwidth of the mode (in Hz)

Because it is not always possible to calculate the true modal decay time this equation can be dangerously approximated using the reverberation time of the room as:

$$Bw_{\mathrm{mode}} \approx \frac{2.2}{T_{60}}$$

This assumption is dangerous because the mode will not be diffuse whereas the reverberation time calculation assumes a diffuse sound field. In general the bandwidths and intensity levels of a mode are proportional to the number of reflections required to support them. This means that axial modes tend to be the strongest followed by tangential and then oblique modes in order of strength, as shown in Figure 6.38. However, this is not always the case as a tangential mode in a room with four reflecting surfaces could be stronger than the axial mode between the other two absorbing surfaces.

Example 6.11 Calculate the approximate modal bandwidth in a room which has a reverberation time (T_{60}) of 0.44 seconds. What would be the modal bandwidth of axial modes along the 5 m dimension of the room if the absorption coefficients on the opposing walls were equal to the average room absorption coefficient of 0.2?

The approximate modal bandwidth can be calculated as:

$$Bw_{\text{mode}} \approx \frac{2.2}{T_{60}} = \frac{2.2}{0.44 \text{ s}} = 5 \text{ Hz}$$

To answer the second question one must calculate the modal decay time, $T_{60 \text{ (Modal)}}$, which is given by:

$$T_{60 \text{ (Modal)}} = \frac{-0.04L}{\ln(1 - \alpha_{\text{mode}})} = \frac{-0.04 \times 5 \text{ m}}{\ln(1 - 0.2)} = 0.9 \text{ s}$$

This value of decay can be used to calculate the actual modal bandwidth as:

$$Bw_{\text{mode}} = \frac{2.2}{T_{60 \text{ (Modal)}}} = \frac{2.2}{0.9 \text{ s}} = 2.4 \text{ Hz}$$

Clearly care must be taken when calculating modal bandwidths due to the fact that the diffuse field assumptions no longer apply. In the case above, even though an even distribution of absorption was assumed, the decay time and bandwidth were radically different to that predicted by the diffuse field assumptions simply because the path travelled by the sound wave was longer than the mean free path. In practice the absorption coefficient is likely to be different as well, making prediction even more difficult. Note that, if the absorption remains constant with frequency, the bandwidths of a mode are independent of their frequency—they are simply a function of the modal decay time.

6.2.9 Critical frequency

Because all rooms have modes in their lower frequency ranges there will always be a frequency below which the modal effects dominate and the room can no longer be treated as diffuse. Even anechoic rooms have lower frequency limits to their operation. One of the effects of room modes is to cause variations in the frequency response of the room, via its effect on the reverberant field. The frequency response due to modal behaviour will also be room position dependent, due to the spatial variation of standing waves. An important consequence of this is that the room no longer supports a diffuse field in the modal region and so the reverberation time concept is invalid in this frequency region. Instead an approach based on modal decay should be used. But at what frequency does the transition occur, can it be even calculated? Consider the typical frequency response of a room, shown in Figure 6.39. In it, three different frequency regions can be identified.

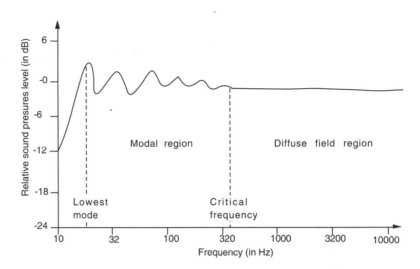

Figure 6.39 The frequency response of a typical room.

- *The cut-off region:* the region below the lowest resonance, sometimes called the room cut-off region. In this region the room is smaller than a half wavelength in all dimensions. This does not mean that the room does not support sound propagation, in fact it behaves more like the air in a bicycle pump when the end is blocked. This means that the environment 'loads' any sources of sound in the room differently (such as loudspeakers or musical instruments), and often the effect of this loading is to reduce the ability of the source to radiate sound into the room and so results in reduced sound levels at these frequencies. The low frequency cut-off can be calculated simply from:

$$f_{\text{cut-off}} = \frac{c}{2 \times \text{Longest dimension}}$$

$$= \frac{344 \text{ ms}^{-1}}{2 \times \text{Longest dimension (m)}}$$

- *The modal region:* the next region is the modal region in which the modal behaviour of the room dominates its acoustic performance. In this region the analysis based on the assumption of a diffuse field is doomed to fail.
- *The diffuse field region:* the final region is the region in which a diffuse field can exist and therefore the concept of reverberation time is valid. In general this region of the frequency range is the one that will sound the best, providing the reverberation characteristics are good, because the effects of room modes are minimal and so the listener experiences an even reverberant sound level throughout the room.

301

The transition boundary between the region of modal behaviour and the region of diffuse behaviour is known as the critical frequency. As is usual in these situations, although the critical frequency is a single frequency it is not a sharp boundary, it represents some defined point in a transition region between the two regions.

6.2.10 Acoustically 'large' and 'small' rooms

The concept of critical frequency allows us to define the difference between rooms which are 'large' and 'small' in acoustical terms. In an acoustically large room the critical frequency is below the lowest frequency of the sound that will be generated in the room whereas an in an acoustically small room the critical frequency will occur within the frequency range of the sounds being produced in it. Examples of acoustically large rooms would be concert halls, cathedrals and large recording studios. Most of us listen to and produce music in acoustically small rooms such as bedrooms, bathrooms, living rooms, etc., and there is an increasing trend—due to the effect of computer recording and editing technology and because it's cheaper—to perform more and more music and sound production tasks in small rooms.

6.2.11 Calculating the critical frequency

How can the critical frequency be calculated? There are two main approaches. The first is to recognise that when the wavelength of the sound approaches the mean free path of the room then the likelihood of modal behaviour increases, because a sound wave is 'in touch' with all the walls in the room. This approach can be used to set an approximate lower frequency bound on the critical frequency below which it is likely to be difficult to prevent modal effects from dominating the acoustics without extreme measures being taken. This approach gives the following expression for calculating the critical frequency, which assumes that modal behaviour dominates once the mean free path is equal to one and a half wavelengths:

$$f_{critical} = \left(\frac{3}{2}\right)\frac{c}{MFP} = \left(\frac{3}{2}\right)\frac{344 \text{ ms}^{-1}}{MFP}$$

This expression is useful for making a rapid assessment of the likelihood of achieving a particular critical frequency in a given room. However, the real critical frequency may well be higher because a room can have significant modal behaviour at high frequencies, if the absorption is low. Because of this the accepted definition of critical frequency is based on the mode bandwidth,

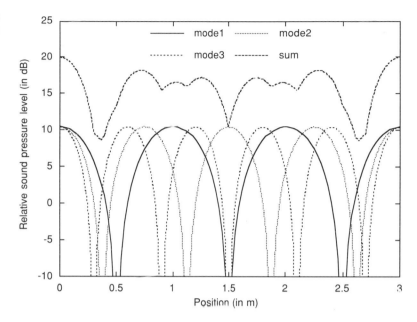

Figure 6.40 The composite effect of adjacent modes on the spatial variation in a room.

although this can result in a chicken and egg situation at the initial design stages, hence the earlier equation. The rationale for this is as follows. The main consequence of modal behaviour is the frequency and spatial variation caused by it. This means that if a given frequency excites only one mode, then this variation will be very strong. However, if a given frequency excites more than one mode, both the spatial and frequency variation will be reduced. Figure 6.40 shows the effect of adding three adjacent modes together and it shows that once more than three adjacent modes are added together the variation is considerably reduced. The way to excite adjacent modes with a single frequency is to increase their bandwidth until the three bandwidths associated with the three modes overlap a given frequency point, as shown in Figure 6.41. The critical frequency is defined as when the modal overlap equals three, so at least three modes are excited by a given frequency, and is given by:

$$f_{critical} = 2102 \sqrt{\left(\frac{T_{60}}{V}\right)}$$

This equation shows that the critical frequency is inversely proportional to the square root of the room volume and is proportional to the square root of the reverberation time, which is also proportional to the cube root of the volume, if the absorption remains constant, as discussed earlier. The net result of this is that, as expected, the critical frequencies for larger rooms are

303

Figure 6.41 The way to guarantee that at least three modes are excited by a given frequency.

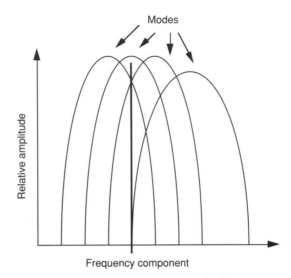

Modes

Relative amplitude

Frequency component

generally lower than those of smaller ones. Thus big rooms are acoustically 'large' as well.

As an example let us calculate the critical frequency of our typical living room:

Example 6.12 What is the critical frequency of a room whose surface area is 75 m², whose volume is 42 m³? What is the critical frequency of the same room if the average absorption coefficient is 0.2?

Using the first equation calculate the lowest bound on the critical frequency using the mean free path as:

$$f_{critical} = \left(\frac{3}{2}\right) \frac{c}{MFP} = \left(\frac{3}{2}\right) \frac{c}{\frac{4V}{S}} = 1.5 \times \frac{344 \text{ ms}^{-1}}{\frac{4 \times 42 \text{ m}^3}{75 \text{ m}^2}}$$

$$= 1.5 \times \frac{344 \text{ ms}^{-1}}{2.24 \text{ m}} = 230 \text{ Hz}$$

Using the second equation calculate the critical frequency using the reverberation time. First calculate the reverberation time as:

$$T_{60} = \frac{-0.161V}{S \ln(1 - \alpha)} = \frac{-0.161 \times 42 \text{ m}^3}{75 \text{ m}^2 \times \ln(1 - 0.2)} = 0.43 \text{ s}$$

Then, using the second equation, calculate the critical frequency as:

$$f_{critical} \approx 2102 \sqrt{\left(\frac{T_{60}}{V}\right)} = 2102 \sqrt{\left(\frac{0.43 \text{ s}}{42 \text{ m}^3}\right)} = 213 \text{ Hz}$$

The second equation predicts a slightly lower critical frequency compared to the first one. However, the agreement is surprisingly good. Although the modal overlap has been calculated using a reverberation time, and hence a diffuse field assumption, this is probably just valid at this frequency which represents the boundary between the two regions. The critical frequency results show that, for this room, frequencies below 213 Hz must be analysed using modal decay time rather than reverberation time.

6.3 Absorption materials

Absorption materials are clearly important in their effects on the acoustics, and this section briefly looks at the factors which affect the performance of these materials and their effects on an acoustic space.

There are two basic forms of absorption materials—porous absorbers and resonant absorbers—which behave differently because their mechanisms of absorption are different.

6.3.1 Porous absorbers

Porous absorbers, such as carpets, curtains and other soft materials, work due to frictional losses caused by the interaction of the velocity component of the sound wave with the surface of the absorbing material. In Chapter 1 we saw that the velocity component arose because the air molecules had to move between the compression and rarefaction states. A given pressure variation will require a greater pressure gradient, and hence higher peak velocities, as the wavelength gets smaller with rising frequency. Because the pressure gradient of a sound wave increases with frequency, the friction due to interaction with a surface will also increase with frequency and therefore the absorption of these types of materials also rises with frequency. Clearly the larger the surface area available for interaction, the higher the friction and therefore the absorption. This means that porous materials which consist of a large number of fibres per unit volume, such as high-density rockwool or fibreglass, plush carpets, etc., will tend to have a high level absorption. This also explains why curtains which are draped to a fraction of their cloth area absorb more strongly than ones which are flat. A typical absorption curve for a porous absorber is shown in Figure 6.42. Because porous absorbers interact with the velocity component of the sound wave, they are affected by the space between them and the wall and their thickness. This is due to the fact that at the surface of a hard surface, such as a wall,

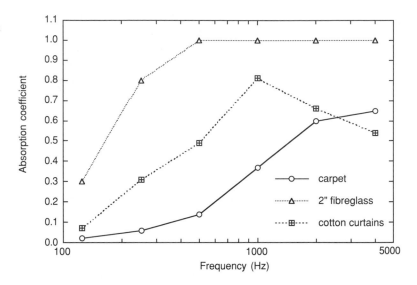

Figure 6.42 Typical absorption curves for porous absorbers.

the velocity component is zero whereas at a quarter of a wavelength away from the wall the velocity component will be at a maximum, as shown in Chapter 1, and so a porous material will absorb more strongly at frequencies whose quarter wavelength is less than either the spacing of the material from the wall, or the thickness of the material if it is bonded directly to the surface. This effect is shown in Figure 6.43. Although in principle there could be a variation in the absorption coefficient as the frequency increases above the quarter wavelength point, due to the inherent variation of the velocity component as a function

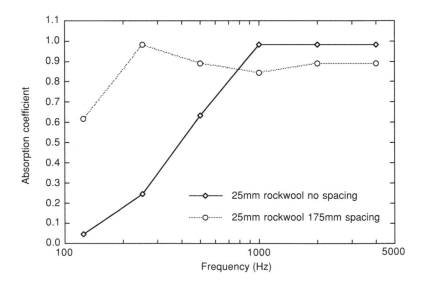

Figure 6.43 The effect of spacing a porous absorber away from a hard surface.

Figure 6.44 Typical construction of a panel absorber.

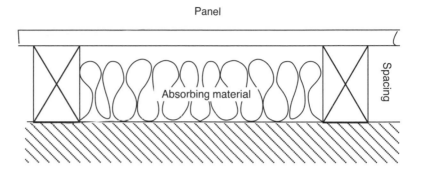

of wavelength at a fixed distance from a surface, in practice this does not occur unless the material is quite thin.

6.3.2 Resonant absorbers

Resonant absorbers such as wood panelling work because the incident sound energy causes vibrations in the absorber and these are converted to frictional losses within the absorbing structure itself. This makes them sensitive to the pressure component of the sound wave and so they work well when attached to walls. The typical construction of a panel absorber is shown in Figure 6.44. In the case of wood panels it is the internal frictional losses in the wood, and in the perforated absorber, discussed later, it is due to the enhancement of velocity that happens in the perforations at resonance. Because the absorbers are resonant their absorption increases at low frequencies, as

Figure 6.45 Typical absorption curves for panel absorbers.

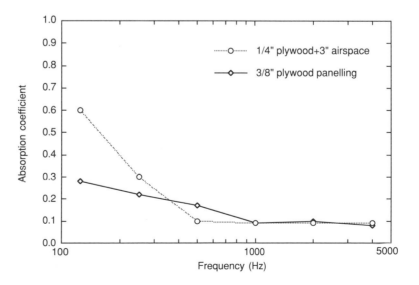

shown in Figure 6.45. The resonant characteristics of these absorbers enables them to be tuned to low frequencies and so allows them to have absorption characteristics which complement those of porous absorbers. The peak absorption frequency of a resonant absorber is a function of the space behind the absorber and the effective mass of the front panel. To use an analogy with a spring and weight, the rear cavity acts like a spring whose stiffness is inversely proportional to the depth of the cavity and the effective mass per unit area of the front panel determines the size of the weight. As the spring gets less stiff and the effective mass becomes greater the resonant frequency drops. Thus deeper rear cavities result in lower resonances for both types. For the panel absorbers the mass per unit area of the panel is directly related to the effective mass, so heavier front panels result in a lower resonant frequency. The resonance frequency of panel absorbers can be calculated using the following equation.

$$f_{resonance} = \frac{60}{\sqrt{Md}}$$

where M = panel's mass per unit area (in kg m^{-2})
and d = depth of the airspace (in m)

However, this equation must be applied with some caution because it assumes that the panel has no stiffness. This assumption is valid for thin panels but becomes less applicable as the panel becomes thicker and thus more stiff.

6.3.3 Helmholtz absorbers

Another form of resonant absorber is based on the use of the resonance that occurs when air is trapped in a tube above an air space. This type of resonance is called a Helmholtz resonance and is the resonance that occurs in a beer bottle when you blow across it. The cavity acts like a spring and the air in the tube above the cavity acts like the mass. The construction of this type of absorber consists of a perforated panel above an airspace, as shown in Figure 6.46.

For the perforated panels the effective mass is a function of both the depth of the perforations and their effective area as a percentage of the total area. Their effective mass increases as the depth increases and the percentage hole area reduces. Typical absorption curves for this type of absorber are shown in Figure 6.47. This type of absorber is often used to add extra absorption at high-bass and low-midrange frequencies.

Figure 6.46 Typical construction of a Helmholtz resonant absorber.

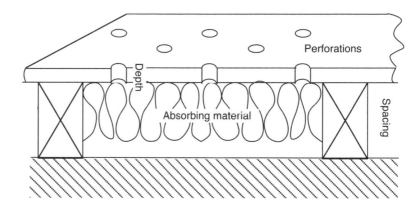

Figure 6.47 Typical absorption curves for Helmholtz resonant absorbers.

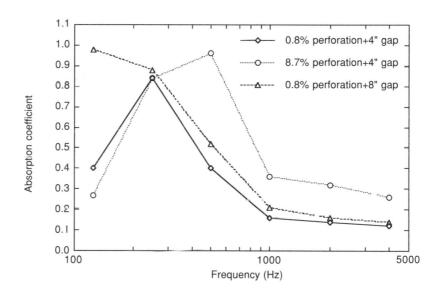

6.3.4 Wideband absorbers

It is possible to combine the effects of porous and resonant absorbers to form wideband absorbers. A typical construction is shown in Figure 6.48 and its performance is shown in Figure 6.49. As with all absorbers using rockwool or fibreglass one must take precautions to prevent the egress of irritating fibres from the absorber into the space being treated.

An alternative means of achieving wideband absorption is to use a large depth of porous absorber, for example one metre, and this can provide effective absorption with a flat frequency response, but at the cost of considerable depth.

Figure 6.48 Construction of a wideband absorber.

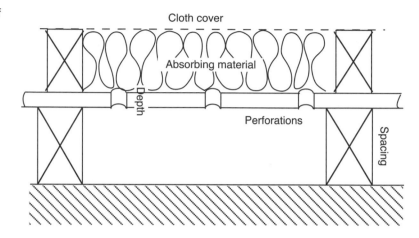

Figure 6.49 Typical absorption curves for wideband absorbers.

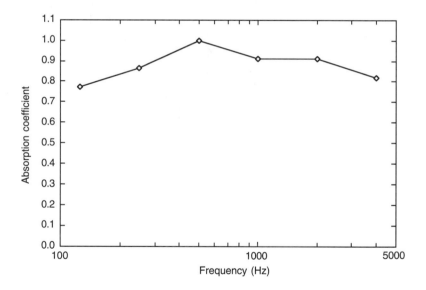

6.3.5 Summary

With these basic types of absorption structures it is possible to achieve a high degree of control over the absorption coefficient in a room as a function of frequency. In many cases much of the required absorption can be achieved by using materials which fit naturally in the room. For example much baroque music was performed in the halls of mansions which had a balanced acoustic due to the extensive use of wood panelling in their decoration. This panelling acted as an effective low-frequency absorber and in conjunction with the flags, drapes and tapestries which also decorated these spaces provided the necessary acoustic absorption.

6.4 Diffusion materials

As well as absorption it is essential that the sound be diffused when it strikes a surface. Ideally we want the acoustic equivalent of a matte surface. Unfortunately most surfaces, including large areas of absorbing material, act like acoustic mirrors, with varying shade of darkness. In order to have a matt surface one needs a 'bumpy wall' and many things can be used to provide this. Unfortunately the bumps need to be at least an eighth, and preferably a quarter, of a wavelength in size to be effective. This results in the requirement for very large objects at low frequencies, 1.25–2.5 m at 34 Hz, and very small objects at higher frequencies, 1.25–2.5 cm at 3.4 kHz. If the objects are too small, that is, less than one eighth of a wavelength, they will not diffuse properly, if they are too big, that is, greater than about a half a wavelength, they will behave as acoustic mirrors in their own right and so will not diffuse effectively. Clearly effective diffusion is a difficult thing to achieve in an ad hoc manner. Curved and angled structures can help at mid and high frequencies, and at very high frequencies, greater than about 4 kHz, the natural rough textures of materials such as brick and rough cut stone are effective. Because of the need to achieve well defined diffusion characteristics, diffusion structures based on patterns of wells whose depths are formally defined by an appropriate mathematical sequence have been proposed and used (Schroeder 1975 and D'Antonio 1984). The design of these structures is quite involved and the reader is directed to the references if they want more information. However, a brief description of how they work is as follows.

6.4.1 How diffusers work

Consider a hard surface consisting of bumps of height d. Also consider an acoustic wavefront approaching it from a normal direction. The way this wavefront is reflected will depend on the height of the bumps relative to its wavelength. Let us consider three cases:

- In the case of $d << \lambda$ the surface will behave like a flat surface and specularly reflect the wavefront.
- In the case of $d = \frac{1}{4}$ the wavefronts which are reflected from the front of the bumps are reflected $\frac{1}{2}$ earlier than those from the surface. This means that in the normal direction the wavefronts cancel and so no sound pressure is propagated in this direction. However, there has been no energy loss in the system so the wavefront must be reflected in some direction. In fact as one moves away from the normal direction

the relative path lengths between the bump and the surface become less and the amplitude of the wavefront increases as one moves off the normal direction. This is the basic principle behind diffusion using hard reflectors. That is, the diffusing surface modifies the phase of the wavefronts so that the reflected wave must propagate in directions other than the specular direction.

- In the case of $d = ½$ the wavefronts from the bumps and surface are delayed by λ and so arrive back in phase. Thus the bumps disappear and the surface behaves as if it were flat. That is, it behaves like a specular reflector.

So, one has a problem, a regular sequence of bumps will diffuse but only at frequencies at which it is an odd multiple of ¼. Note also that these frequencies will depend on the angle of incidence of the incoming wavefront.

What is required is a pattern of bumps which alter the phases of the incident in such a way that two objectives are satisfied:

1 The sound is scattered in some 'optimum' manner.
2 The scattering is optimum over a range of frequencies.

These objectives can be satisfied by several different sequences, but they share two common properties:

- The Fourier transform of the sequence is constant except for the d.c. component which may be the same or lower. This satisfies objective (1) because it can be shown that reflection surfaces with such a property scatter energy equally in all directions. The effect of a reduced d.c. component is to further reduce the amount of energy which is reflected in the specular direction.
- The second desirable property of these sequences is that the Fourier transform is unaffected if the wavelength of the incident sound varies. This has the effect of changing the scale of the sequence but again one can show that the resulting sequence still has the same properties as the original sequence.

Both the above properties arise because the sequences work by perturbing the wavefronts over a full cycle of the waveform. Such sequences are called phase reflection gratings because they perturb the phase of the wavefront.

To make this a little clearer, let us consider two sequences which are used for diffusers.

1 *Quadratic residue sequences*   well depth = n^2 mod p where p is a prime number. If $p = 5$ this gives a set of well depths of

0, 1, 4, 4, 1, 0, 1 ....etc.

so the sequence repeats with a period of 5.

2 *Primitive root sequences* well depth = a^n mod p where p is a prime and a is a suitable constant called a primitive root. For a = 2 and p = 5 we get the sequence

1, 2, 4, 3, 1, 2 ......etc.

Here we have a sequence which has a period of 4 (5–1).

At the lowest design frequency for these examples a well of depth 5 would correspond to ½. At higher frequencies the sequences still have the same properties and thus scatter sound effectively. However, when the frequency gets high enough so that ½ becomes equal to the minimum difference in depths (1) then the surface again becomes equivalent to a flat surface.

The typical construction of these structures is shown in Figure 6.50 and their performance is shown in Figure 6.51.

Figure 6.50 Typical construction of a quadratic residue diffuser.

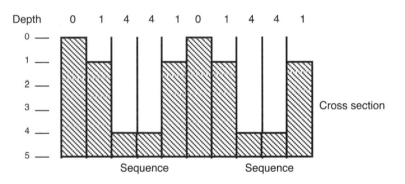

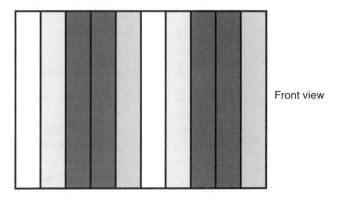

Figure 6.51 Typical performance of a quadratic residue diffuser compared to a flat plate.

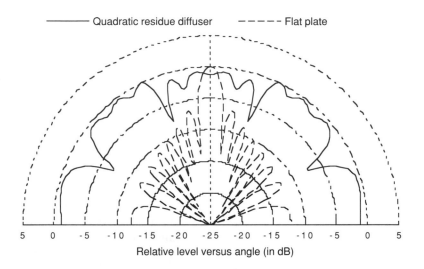

──── Quadratic residue diffuser ‒‒‒‒ Flat plate

5 0 -5 -10 -15 -20 -25 -20 -15 -10 -5 0 5

Relative level versus angle (in dB)

6.4.2 Discussion

As we have seen, these sequences achieve their performance by spreading the phase of the reflected wavefront over at least one cycle of the incident wavefront. In order to do this, their maximum depth must be ½ at the lowest design frequency. This means that to achieve diffusion a reasonable depth is required. For example, to have effective diffusion down to 500 Hz a depth of 34 cm (13.5 inches) is required. To get down to 250 Hz one would need to double this depth. However, as we have seen, a simple bump of ¼ can provide diffusion, albeit somewhat frequency dependently. This is half the depth of the above sequences and represents the ultimate limit for a diffusing object.

It is possible to have sequences which achieve the phase scatter required for good diffusion using a depth closer to ¼ at the lowest frequency (4) and so allow better performance diffusers in restricted spaces. However even ¼ at low frequencies is often too large to be useful. What one really requires is a diffuser which is effective without using any depth!

6.4.3 Amplitude reflection gratings

It is not just physically observable bumps on the wall that can cause diffusion of the sound. In fact any change in the reflecting characteristics of the surface will cause diffusion. The change from an absorbing region on a wall to a reflecting one is an example of a change that will cause the sound to scatter. Thus it is always better to distribute the absorption in small random amounts around a room rather than concentrate it in one particular area. As well as encouraging diffusion this strategy will avoid the possibility that some modes might shuttle between surfaces with minimum absorption. There are also mathematically based procedures for the optimum placement of absorbing materials to encourage diffusion and more details may be found in Angus (1995). What is required is an amplitude weighting, that is, a pattern of absorbers, which gives a flat Fourier transform.

The most obvious sequences to consider are binary, that is they contain the only levels 0 and 1 where 1 represents reflection from a hard surface and 0 represents absorption from some form of absorbing material. Clearly not all acoustic absorbers are 100% absorbing but this can be simply allowed for by using (1-absorption) instead of zero in the sequence. The net effect of less than 100% absorption would be to increase the level of the specular component. Of the many possible binary sequences M-sequences would seem to be a good starting point as they have desirable Fourier properties. There are many other bi-level sequences which have flat Fourier transforms but M-sequences are well documented.

Thus amplitude reflection gratings consist of a surface treatment which consists of strips of absorbing material whose width is less than ½ at the highest frequency of use laid out in a pattern in which strips of absorber represent zero and strips of reflecting wall represent 1 (see Figure 6.52). Note that because we are not depending on depth we do not have a low-frequency limit

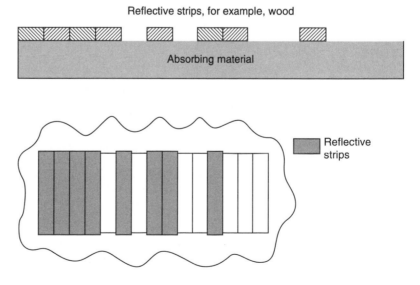

Figure 6.52 Simple implementation of a length 15 one-dimensional Binary Amplitude Diffuser.

Reflective strips, for example, wood

Absorbing material

Reflective strips

Figure 6.53 An implementation of a length 1023 two-dimensional Binary Amplitude Diffuser (white dots are holes over absorber).

to the range of diffusion only a high-frequency limit which is a function of the width of the strips. A two-dimensional example of an amplitude reflection grating is shown in Figure 6.53.

Amplitude gratings provide some diffusion but cannot be as good at diffusing as phase reflection gratings but, because of their size, they are useful at low frequencies. It also is possible to develop curved diffusion structures, although there are no simple mathematical recipes for them. For further details see Cox (1996). Other structures are possible and the reader is referred to the references for more information.

6.5 Sound isolation

No discussion of the quality of sound in a room would be complete without a brief discussion of how to keep unwanted sound from entering a room, or how to keep the wanted sound in, so as not to disturb the pleasure of people inside or outside it.

The first thing to note is that just because a material is a good absorber of sound doesn't mean that it is a good isolator of

Figure 6.54 Sound transmission versus sound absorption in a material.

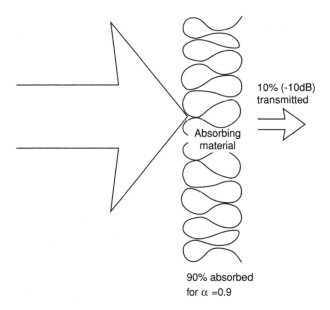

10% (-10dB) transmitted

Absorbing material

90% absorbed for $\alpha = 0.9$

sound. In fact most absorbing materials are terrible at sound isolation. This is because, in the sound isolation case, we are interested in the amount of sound that travels through a structure rather than the amount that is absorbed by it, as shown in Figure 6.54. A poor value of sound isolation would be around 20 dB which corresponds to only one hundredth of the sound being transmitted. A good absorber with an absorption coefficient of 0.9 would let one tenth of the sound through which corresponds to a sound isolation of only 10 dB! As we are more interested in sound isolations of 40 dB as a minimum, absorption is clearly not the answer!

6.5.1 Ways of achieving sound isolation

There are only two ways to achieve sound isolation, using either stiffness or mass. Figure 6.55 shows the attenuation of a partition as a function of frequency and from it one can see that stiffness is effective at low frequencies due to the fact that the sound wave must push against the stiffness of the partition. This is known as the stiffness controlled isolation region. As the frequency rises the partition needs to move less distance to re-radiate a given level of sound and so it gets less effective until at the resonant frequency of the partition its level of attenuation is at its lowest value. This is due to the fact that at resonance the partition can be moved easily by the incident sound wave and so re-radiates the sound effectively. As the frequency rises above the partition's resonant frequency, the mass-controlled

Figure 6.55 Sound transmission as a function of frequency for a partition.

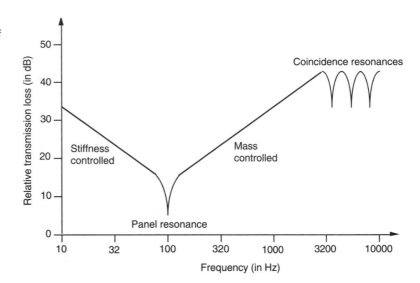

region of isolation is entered. In this region, the fact that the sound must accelerate a heavy mass provides the isolation. Because more force is required to move the partition at higher frequencies, the attenuation rises as the frequency rises. At even higher frequencies there are resonances in which both the thickness of the partition, and the way sound propagates within it, interact with the incident sound to form coincident resonances that reduce the attenuation of the partition. Most practical partitions operate in the mass-controlled region of the isolation curve with coincident resonances limiting the performance at higher frequencies. Figure 6.56 shows the attenuation

Figure 6.56 Sound transmission versus frequency for typical partitions.

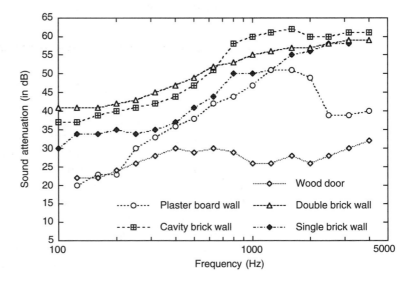

of a variety of single partitions as a function of frequency. In particular note that the plaster board wall has a significant coincidence resonance. The performance of a single partition increases by 3 dB every time its mass is doubled but the coincidence resonances also move lower in frequency as well. These coincidence resonances limit the ultimate performance of single partitions. In addition the cost, and size, of single partitions get unreasonable for large attenuations.

6.5.2 Independent partitions

The solution is to have two or more partitions which are independent of each other. If the two partitions are truly independent then the total attenuation, or effective sound isolation, is the product of the attenuations of individual partitions, that is the dB attenuation is the sum of the dB attenuations of the individual partitions. In practice the partitions are not independent and the isolation is improved dramatically, but not as much as would be predicted by simply summing the dB attenuations. Coincidence resonances also reduce the effectiveness of a partition and it is important to ensure that the two partitions have different resonances. This is most easily assured by having them made with either a different thickness, or a different material. As an amusing example Figure 6.57 shows the measured results, from Inman (1994), for single and double glazing made with similar and different thicknesses of glass and spacing.

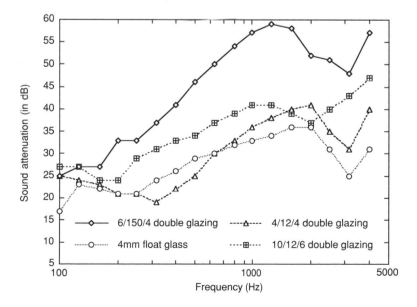

Figure 6.57 Sound transmission versus frequency for single and double glazing.

Because of the effect of the coincidence resonances the double glazed unit with 4 mm glass is actually worse than a single pane of 4 mm glass! As the other two curves show if the glass is dissimilar the result is much improved, and is further improved if the spacing is increased so as to reduce the coupling between the individual partitions. Often absorbing material is placed in the cavity between the two partitions to reduce the effect of coincidence resonances but it is important to ensure that the absorbing materials do not make contact with both partitions or else flanking may occur.

6.5.3 Flanking paths

Flanking paths, which are the main limitation to sound isolating structures, arise when there are other paths that the sound can travel through in order to get round, that is flank, the sound isolating structure, as shown in Figure 6.58. Typical paths for flanking are the building structure, heating pipes, and most commonly ventilation systems or air leaks. The effect of the building structure can be reduced by building a 'floating room', as shown in Figure 6.59, which removes the effect of the building structure by floating the room on springs away from it. In practice ensuring that no part of the building is touching the floating room by any means (plumbing pipes and electrical wiring conduits are popular offenders in this respect), is extremely difficult. The effect of ventilation systems and air leaks are also a major source of flanking in many cases. In fact in the domestic situation the sound isolation is almost entirely dominated by air leaks and draught paths, and it is the removal of these that allow double glazing salesmen to advertise a dramatic improvement in sound isolation, despite having two 4 mm panes of glass in the double glazing.

Figure 6.58 Flanking paths in a structure.

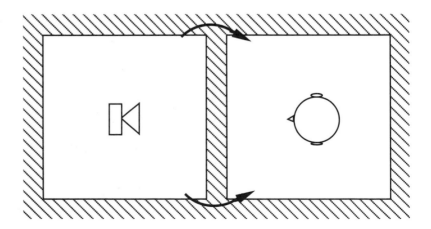

Figure 6.59 Floating room construction.

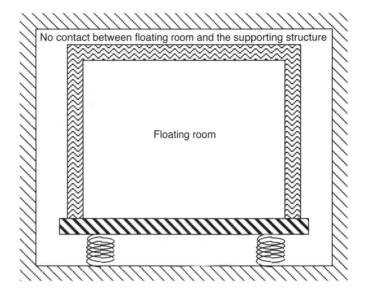

No contact between floating room and the supporting structure

Floating room

So in order to have good sound isolation one needs good partitions and an air-tight, draught-free, structure. Achieving this in practice while still allowing the occupants to breathe is a challenge.

6.6 Energy–time considerations

The main advances in acoustical design for listening to music have arisen from the realisation that, as well as reverberation time, the time evolution of the first part of the sound energy build up in the room must be considered. There are now acoustic measurement systems that can measure the energy time curve of a room directly, thus allowing a designer to see what is happening, rather than relying on a pair of 'golden ears'. An idealised energy–time curve for a room is shown in Figure 6.60 and it has three major features:

- A gap between the direct sound and first reflections. This happens naturally in most spaces and gives a cue as to the size of the space. The gap should not be too long, less than 30 ms, or the early reflections will be perceived as echoes. Some delay, however, is desirable as it gives some space for the direct sound and so improves the clarity of the sound, but a shorter gap does add 'intimacy' to the space.
- The presence of high level diffuse early reflections which come to the listener predominately from the side, that is lateral early reflections. This adds spaciousness and is easier

Figure 6.60 An idealised energy–time curve.

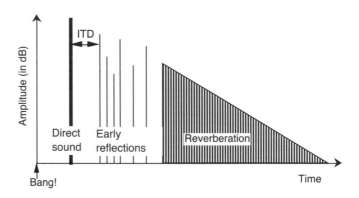

to achieve over the whole audience in a shoe-box hall rather than a fan-shaped one. The first early reflections should ideally arrive at the listener within 20 ms of the direct sound. The frequency response of these early reflections should ideally be flat and this, in conjunction with the need for a high level of lateral reflections, implies that the side walls of a hall should be diffuse reflecting surfaces with minimal absorption.

• A smoothly decaying diffuse reverberant field which has no obvious defects, no modal behaviour, and whose time of decay is appropriate to the style of music being performed. This is hard to achieve in practice so a compromise is necessary in most cases. For performing acoustic music a gentle bass rise in the reverberant field is desirable to add 'warmth' to the sound but in studios this is less desirable.

6.6.1 Reflection-free zones

These conditions apply to the design of concert hall and, to a lesser extent, the design of the part of the studio that the musicians play in. However for the home listener, or sound engineer in the control room of a studio, the ideal would be an acoustic which allows them to 'listen through' the system to the original acoustic that the sound was recorded in. Unfortunately the room in which the recorded sound is being listened to is usually much smaller than the original space and this has the effect shown in Figure 6.61. Here the first reflection the listener hears is due to the wall in the listening room and not the acoustic space of the sound that has been recorded. Because of the precedence effect this reflection dominates and the replayed sound is perceived as coming from a space the size of the listening room which is clearly undesirable. What is required is a means of making the sound from the loudspeakers appear as if it is

Figure 6.61 The effect of a shorter initial time delay gap in the listening room.

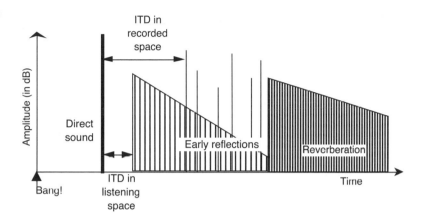

Figure 6.62 Maximising the initial time delay gap by suppressing early reflections.

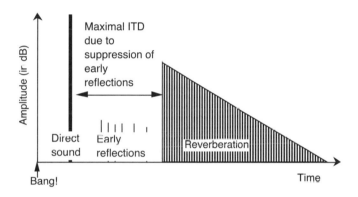

coming from a larger space by suppressing the early reflections from the nearby walls, as shown in Figure 6.62. One way of achieving this is to use absorption, as shown in Figure 6.63. The effect can also be achieved by using angled or shaped walls. This is known as the reflection-free zone technique because it relies on the suppression of early reflections in a particular area of the room to achieve a larger initial time delay gap. This effect can only be achieved over a limited volume of the room, unless the room is made anechoic which is undesirable. The idea is that by absorbing, or reflecting away, the first reflections from all walls except the furthest one away from the speakers, the initial time delay gap is maximised. If this gap is larger than the initial time delay gap in the original recording space, the listener will hear the original space, and not the listening room. However this must be achieved while satisfying the need for even diffuse reverberation and so the rear wall in such situations must have some explicit form of diffusion structure on it to assure this. The initial time delay gap at the listening position should be as large as possible, but is clearly limited by the time it takes the sound

Figure 6.63 Achieving a reflection-free zone using absorption.

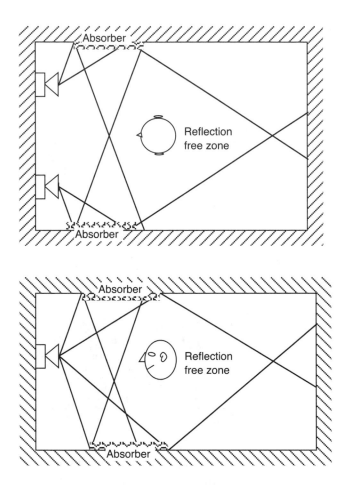

to get to the rear wall and back to the listener. Ideally this gap should be about 20 ms but it should not be much greater or it will be perceived as an echo. In most practical rooms this requirement is automatically satisfied and initial time delay gaps in the range of 8 ms to 20 ms are achieved.

6.6.2 Absorption level required for reflection-free zones

In order to achieve a reflection-free zone it is necessary to suppress early reflections, but by how much? Figure 6.64 shows a graph of the average level that an early reflection has to be in order to disturb the direction of a stereo image and from this we can see that the level of the reflections must be less than about 15 dB to be subjectively inaudible. Allowing for some reduction due to the inverse square law, this implies that there must be about 10 dB, or $\alpha = 0.9$ of absorption on the surfaces contributing to the first reflections. In a domestic setting it is possible to

Figure 6.64 The degree of reflection suppression required to assure a reflection-free zone (data from Toole, 1990).

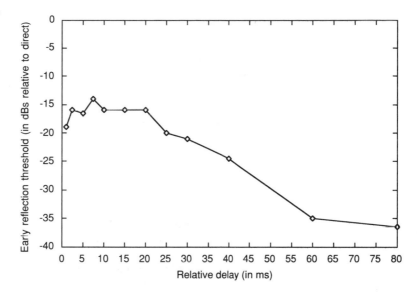

get close using carpets and curtains, and bookcases can form effective diffusers, although persuading the other occupants of the house that carpets or curtains on the ceiling is chic can be difficult! In a studio more extreme treatments can be used. However it is important to realise that the overall acoustic must still be good and comfortable, that is not anechoic, and that, due to the wavelength range of audible sound, this technique is only applicable at mid to high frequencies where small patches of treatment are significant with respect to the wavelength.

In this chapter we have examined how the space in which the sound is reproduced affects the way we hear and have analysed various situations and examined many techniques for achieving a good acoustic environment for hearing music.

References

Angus, J.A.S. (1995). Sound diffusers using reactive absorption gratings. Audio Engineering Society 98th Convention, 25–28 February, Paris, preprint #3953.

Angus, J.A.S. and McManmon, C.I. (1998) Orthogonal sequence modulated phase reflection gratings for wide-band diffusion. *Journal of the Audio Engineering Society*, **46**, (12), December.

Cox, T.J. (1996). Designing curved diffusers for performance spaces. *Journal of the Audio Engineering Society*, **44**, (5), 354.

D'Antonio, P. and Cox, T. (1998). Two decades of diffuser design and development, Part 1: Applications and design. *Journal of the Audio Engineering Society*, **46**, (11), November.

D'Antonio, P. and Konnert, J.H. (1984). The reflection phase grating diffusor: design theory and application. *Journal of the Audio Engineering Society*, **32**, (4), April, 228–238.

Holman, T. (1999) *5.1 Surround Sound*. Boston: Focal Press.

Inman, C. (1994) A practical guide to the selection of glazing for acoustic performance in buildings. *Acoustic Bulletin*, **19**, (5), September–October, 19–24.

Newell, P. (1995) *Studio Monitoring Design*. Oxford: Focal Press.

Newell, P. (2000) *Project Studios: A more professional approach*. Oxford: Focal Press.

Schroeder, M.R. (1975). Diffuse sound reflection by maximum length sequences. *Journal of the Acoustical Society of America*, **57**, (1), January, 149–150.

Toole, F.E. (1990). Loudspeakers and rooms for stereophonic sound reproduction. In *The Proceedings of the Audio Engineering Society 8th International Conference, The Sound of Audio*, 3–6 May, Washington DC, 71–91.

Walker, R. (1993). A new approach to the design of control room acoustics for stereophony. *Audio Engineering Society Convention*, preprint #3543, 94.

Walker, R. (1996). Optimum dimension ratios for small rooms. *Audio Engineering Society Convention*, preprint #4191, 100.

Walker, R. (1998). A controlled-reflection listening room for multi-channel sound. *Audio Engineering Society Convention*, preprint #4645, 104.

7 Processing sound electronically

In the previous chapters we have examined the basic elements of acoustics and psychoacoustics in relation to their application to music. In particular cases we have highlighted areas of application to studio, theatre and musical instrument design. However one important area of modern music making and production is the electronic manipulation of sound. In this final chapter we shall look at electronic sound processing from an acoustic and psychoacoustic point of view.

7.1 Filtering

The simplest form of electronic sound processing is to filter the signal in order to remove unwanted components. For example, often low-frequency noises, such as ventilation and traffic rumble, need to be removed from the signal picked up by the microphone. This would be accomplished by a high-pass filter and mixing desks often provide some form of high pass filtering for this reason. High frequencies also often need to be removed to either ameliorate the effects of noise and distortion or to remove the high-frequency components that would cause alias distortion in digital systems. This is achieved through the use of a low-pass filter. A third type of filter is the notch filter which is often used to remove tonal interference from signals. Figure 7.1 shows the effect of these different types of filter on the spectrum of a typical music signal.

Figure 7.1 Types of filter and their effect.

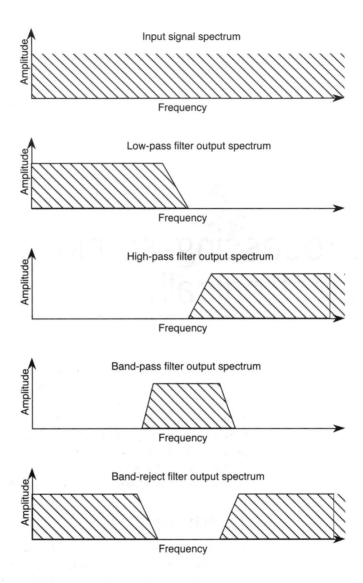

In these cases the ideal would be to filter the signal in a way that minimised the subjective effect on the desired signal. Ideally the timbre of the sound being processed should not change after filtering but in practice there will be some effects. What are these effects and how can they be minimised in light of acoustic and psychoacoustic knowledge?

The first way of minimising the effect is to recognise that many musical instruments do not cover the whole of the audible frequency range. Few instruments have a fundamental frequency which extends to the lowest frequency in the audible

Figure 7.2 The
specifications of a real filter.

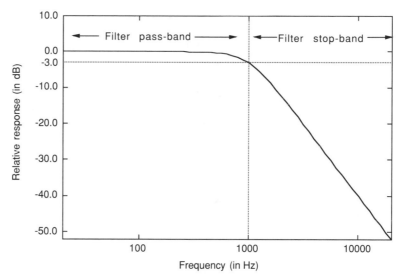

range and many of them do not produce harmonics or components which extend to the upper frequencies of the audible range. Therefore, in theory, one can filter these instruments such that only the frequencies present are passed with no audible effect. In practice this is not easily achieved for two reasons.

- *Filter shape:* Real filters do not suddenly stop passing signal components at a given frequency. Instead there is a transition from passing the signal components to attenuating them, as shown in Figure 7.2. The cut-off frequency of a filter is usually expressed as the point at which it is attenuating the signal by 3 dB relative to the pass-band, see Figure 7.2. Thus if a filter's cut-off is set to a given frequency there will be a region within the pass-band which affects the amplitude of the frequency components of the signal. This region can extend as far as an octave away from the cut-off point. However, as the order of the filter increases, both the slope of the attenuation as a function of frequency, and the sharpness of the cut-off, increase. Therefore, in practice, the filter's cut-off frequency must be set beyond the pass-band that one would expect from a simple consideration of the frequency range of the instruments.
- *Time domain effects:* Filters also have a response in the time domain. Any form of filtering which reduces the bandwidth of the signal will also spread it over a longer period of time. In most practical filter circuits these time domain effects are most pronounced near the cut-off frequency and become worse as the cut-off becomes sharper. Again, as in the case

of filter shape, these effects can extend well into the pass-band of the filter. Note that even the notch filter has a time response which gets longer as the notch bandwidth reduces. Interestingly, particular methods of digital filtering are particularly bad in this respect because they result in time domain artefacts which precede the main signal in their output. These artefacts are easily unmasked and so become subjectively disturbing. Again the effect of the filter's time response is to require that the filter cut-off be set beyond the value that would be expected from a simple consideration of the frequency range of the instruments.

Because of these effects it is difficult to design filters which achieve the required filtering effect without subjectively altering the timbre of the signal.

The second way of minimising the subjective effects is to recognise that the ear uses the spectral shape as a cue to timbre. Therefore the effect of removing some frequency components by filtering can be partially compensated by enhancing the amplitudes of the frequency components nearby. Note that this is a limited effect and cannot be carried too far. Figure 7.3 shows how a filter shape might be modified to provide some compensation. Here a small amount, between 1 dB and 2 dB, of boost has been added to the region just before cut-off in order to enhance the amplitude of the frequencies near to those that have been removed.

Figure 7.3 Partially compensating for filtered components with a small boost at the band-edge.

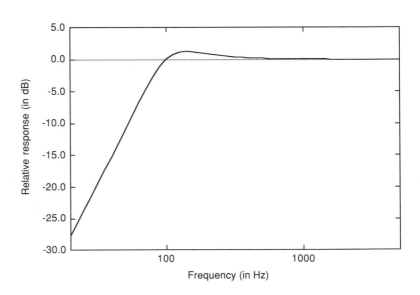

7.2 Equalisation and tone controls

A related and important area of signal processing to filtering is equalisation. Unlike filtering, equalisation is not concerned with removing frequency components but with selectively boosting or cutting/reducing them to achieve a desired effect. The process of equalisation can be modelled as a process of adding or subtracting a filtered version of the signal from the signal, as shown in Figure 7.4. Adding the filtered version gives a boost to the frequencies selected by the filter whereas subtracting the filtered output reduces the frequency component amplitudes in the filter's frequency range. The filter can be a simple high- or low-pass filter, which results in a treble or bass tone control, or it can be a band-pass filter to give a bell-shaped response curve. The cut-off frequencies of the filters may be either fixed or variable depending on the implementation. In addition the bandwidths of the band-pass filters and, less commonly, the slopes of the high- and low-pass filters can be varied. An equaliser in which all the filter's parameters can be varied is called a parametric equaliser. However in practice many implementations, especially those in mixing desks, only use a subset of the possible controls for both economy and simplicity of use. Typically in these cases only the cut-off frequencies of the band-pass filters, and in some cases the low- and high-pass filters, are variable. There is an alternative version of the equaliser structure which uses a bank of closely spaced fixed-frequency band-pass filters to cover the audio frequency range. This approach results in a device known as the graphic equaliser with typical band widths of the individual filters ranging from ⅓ octave to 1 octave. For parametric equalisers the bandwidths can become quite small.

Because a filter is required in an equaliser they also have the same time domain effects that filters have, as discussed earlier. This is particularly noticeable when narrow-bandwidth equalisation is used, as the associated filter can 'ring' for a considerable length of time in both boost and cut modes.

Figure 7.4 Block diagram of tone control function.

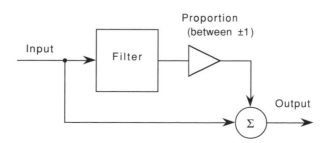

Equalisers are used in three main contexts which each have different acoustic and psychoacoustic rationales, as outlined in Sections 7.2.1 to 7.2.3.

7.2.1 Correcting frequency response faults due to the recording process

This was one of the original functions of an equaliser in the early days of recording which to some extent is no longer required because of the improvement in both electroacoustic and electronic technology. However in many cases there are effects due to the acoustic environment and the placement of microphones which need correction. There are three common acoustic contexts which often require equalisation.

- *Close miking with a directional microphone:* The acoustic bass response of a directional microphone increases as it is moved closer to an acoustic source due to the proximity effect. This has the effect of making the recorded sound bass heavy and some vocalists often deliberately use this effect to improve their vocal sound. This effect can be compensated for by applying some bass-cut to the microphone signal and this often has the additional benefit of further reducing low-frequency environmental noises. Note that some microphones have this equalisation built-in but that in general a variable equaliser is required to fully compensate for the effect.
- *Compensating for the directional characteristics of a microphone:* Most practical microphones do not have an even response at all angles as a function of frequency. In general they become more directional as the frequency increases. As most microphones are designed to give a specified on-axis frequency response, in order to capture the direct sound accurately, this results in a response to the reverberant sound which falls as the frequency rises. For recording contexts in which the direct sound dominates, for example close miking, this effect is not important. However in recordings in which the reverberant field dominates, for example classical music recording, the effect is significant. It can be compensated for by applying some high frequency boost to the microphone signal.
- *Compensating for the frequency characteristics of the reverberant field:* In many performance spaces the reverberant field does not have a flat frequency response and therefore subjectively colours the perceived sound if distant miking is used. Typically the bass response of the reverberant field rises

more than is ideal, resulting in a bass-heavy recording. Again the use of some bass-cut can help to reduce this effect.

All the above uses of equalisation compensate for limitations imposed by the acoustics of the recording context. To make intelligent use of it in these contexts requires some idea of the likely effects of the acoustics of the space at a particular microphone location, especially in terms of the direct to reverberant sound balance.

7.2.2 Timbre modification of sound sources

A major role for equalisers is the modification of the timbre of both acoustically and electronically generated sounds for artistic purposes. In this context the ability to boost or cut selected frequency ranges is used to modify the sounds spectrum to achieve a desired effect on its timbre. For example boosting selected high-frequency components can add 'sparkle' to an instrument's sound whereas adding a boost at low frequencies can add 'weight' or 'punch'. Equalisers achieve these effects through spectral modification only; they do not modify the envelope or dynamics of a music signal. Any alteration of the timbre is purely due to the modification, by the equaliser, of the long-term spectrum of the music signal. There is also a limit to how far these modifications can be carried before the result sounds odd, although in some cases this may be the desired effect.

When using equalisers to modify the timbre of a musical sound it is important to be careful to avoid 'psychoacoustic fatigue'. This arises because the ear and brain adapt to sounds. This has the effect of dulling the effect of a given timbre modification over a period of time. Therefore one puts in yet more boost to which the listener adapts, and so on. The only remedy for this condition is to take a break from listening to that particular sound for a while and then listen to it again later. Note that this effect can happen at normal listening levels and so is different to the temporary threshold shifts that happen at excessive sound levels.

7.2.3 Altering the balance of sounds in mixes

The other major role is to alter the balance of sounds in mixes, in particular the placing of sound 'up-front' or 'back' in the mix. This is because the ability of the equaliser to modify particular frequency ranges can be used to make a particular sound become more or less masked by the sounds around it. This is

Figure 7.5 Spectrum of a masked soloist in the mix.

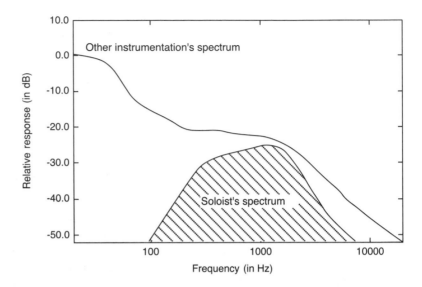

similar to the way the singer's formant is used to allow a singer to be heard above the orchestra as mentioned in Chapter 4. For example suppose one has a vocal line which is being buried by all the other instrumentation going on. The spectrum of such a situation is shown in Figure 7.5 and from this it is clear that the frequency components of the instruments are masking those of the vocals. By selectively reducing the frequency components of the instruments at around 1.5 kHz, while simultaneously boosting the components in the vocal line over the same frequency range, the frequency components of the vocal line can become

Figure 7.6 Spectrum after the use of equalisation to unmask the soloist.

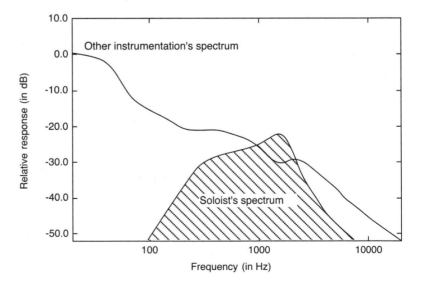

unmasked, as shown in Figure 7.6. This has the subjective effect of bringing the vocal line out from the other instruments. Similarly, performing the process in reverse would further reduce the audibility of the vocal line in the mix. To achieve this effect successfully requires the presence of frequency components of the desired sound within the frequency range of the equalisers' boost and cut region. Thus different instruments require different boost and cut frequencies for this effect. Again it is important to apply the equalisation gently in order to avoid substantial changes in the timbre of the sound sources.

Equalisers therefore have a broad application in the processing of sound. However, despite their utility, they must be used with caution, firstly to avoid extremes of sound character, unless that is desired, and secondly, to avoid unwanted interactions between different equaliser frequency ranges. As a simple example consider the effect of adding treble, bass and mid-range boost to a given signal. Because of the inevitable interaction between the equaliser frequency responses, the net effect is to have the same spectrum as the initial one after equalisation. All that has happened is that the gain is higher. Note this can happen if a particular frequency range is boosted and then, because the result is a little excessive, other frequency ranges are adjusted to compensate.

7.3 Artificial reverberation

Another major form of signal processing for music is artificial reverberation. This used to be mainly provided through the use of electro-mechanical devices such as metal springs or plates, although real rooms were sometimes used. Nowadays this effect is provided using digital signal processing techniques, which convert the audio signal into a numerical representation and then use computer operations and components to effect the processing. The details of this processing are beyond the scope of this book. However, the concepts behind the processing can be explained without reference to the detail; suffice to say that digital techniques allow for easy storage of audio signals, and therefore the delay, and precise manipulation of them.

7.3.1 Analogue reverberation techniques

These electromechanical devices were mechanically excited by the signal which needed reverberation and then their output was used for the artificial reverberation signal, as shown in Figure 7.7. Because the metal had very little absorption, the sound

Figure 7.7 Plate
reverberation device.

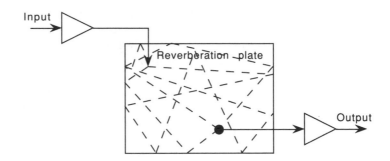

coupled into it would bounce around for some considerable time
before it decayed away, thus simulating an artificial reverberation source. However, the system was not without its problems;
the reverberation produced was not the same as natural reverberation, the system was sensitive to external noise and vibration, and, although the best examples were very good, the
reverberation produced could sound harsh and metallic.

To understand the acoustic and psychoacoustic reasons for this
one needs to re-examine the ideal reverberation curve for a
room, which was discussed earlier in Chapter 6 and is shown in
Figure 7.8. Note that it has three important aspects to its time
evolution:

- an initial time delay gap,
- early reflections, and
- a smooth exponential decay, or reverberant tail, of dense
  diffuse reflections.

The reverberant tail also should have a decay which is slightly
faster at high frequencies, to allow for the effect of air absorption. It should also not have any dominant or distinguishable
resonances.

Figure 7.8 Ideal
reverberation characteristic.

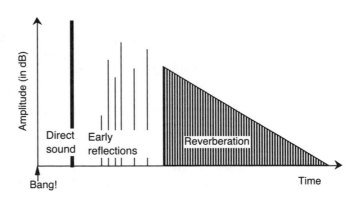

A spring or plate device, being made of metal, has a different speed of sound to air. It is also much smaller than the average hall. Designers used modes of excitation of the plates and springs which had the lowest speed of sound, and used materials to assist that goal. These effects meant that the effective acoustic size of the plate or spring was small and this had several consequences:

- There was virtually no initial time delay gap as it was determined by both the speed of sound in the material and the spacing between the send and receive transducers, which was limited.
- There were virtually no early reflections for both the previous reasons and the limited number of boundaries in the plate or spring. This resulted in a rapid transition to a reverberant tail of dense reflections, although the density tended to be less than that of a real room due to the reduced number of boundaries.
- The resonances tended to be further apart which tended to give a 'rougher' or 'ringing' quality to the reverberation.

In addition, because the absorption characteristics of the plate or spring were different to air, the increase in decay with frequency tended to be less. This made the reverberation sound 'brighter' than natural reverberation.

7.3.2 Digital reverberation techniques

Modern reverberation devices use digital techniques to achieve a simulation of reverberation. Digital techniques remove many of the limitations of the electro-mechanical devices described earlier. Because storage is inexpensive any length of delay can be realised and the main design effort is in producing algorithms to provide the necessary patterns of reflections. Modern reverberation algorithms tend to consist of three distinct sections which require different algorithms, although they may be realised in one piece of hardware. A block diagram of the typical algorithm structure is shown in Figure 7.9 and its three blocks perform the following operations:

- *Initial time delay:* A simple digital delay line is used here to provide a delay to simulate the initial time delay, as shown in Figure 7.9. The delay is usually variable to allow different spaces to be simulated, and different effects to be achieved. For example a delay of 10 to 20 ms gives some space between the direct sound and reverberation whereas a delay of greater than 50 ms starts being perceived as a distinct echo.

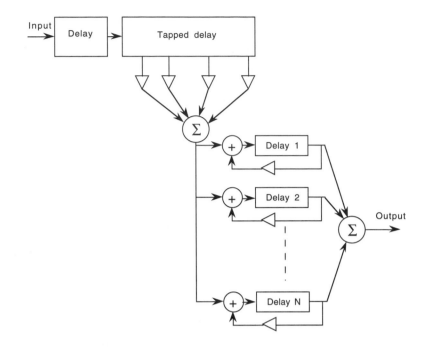

Figure 7.9 A digital reverberation device.

- *Early reflection simulation:* A delay line with taps and multipliers is used here to provide a means of simulating the early reflections, as shown in Figure 7.9. The position of the taps and the corresponding multipliers are adjustable. The position of the taps sets the time at which the early reflection happens and the multiplier determines the strength of the early reflection. As the multipliers and taps require computation there is usually a limit to the number that can be calculated which is dependent on the processing power available. Some of the more sophisticated reverberation devices will even allow for the possibility of frequency-dependent early reflections by using extra multipliers and taps or filters.
- *Reverberant tail simulation:* Here several delay lines of different lengths are used. However, unlike the early reflection simulation, the outputs of these delay lines are fed back to the inputs, thus forming recirculating delay lines. The multiplier in the feedback path must be less than one so that an exponential decay results. The closer the multiplier is to one the longer the decay and conversely as the multiplier becomes smaller the decay becomes shorter. If instead of a simple multiplier, a filter is used with a response which is lower at high frequencies compared with low frequencies, then the decay will be shorter at high frequencies than low

Figure 7.10 The frequency response of a recirculating delay line.

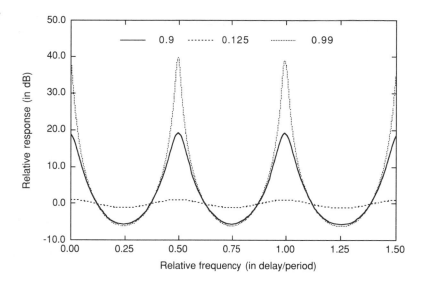

frequencies. This technique is used to simulate the effect of air absorption on reverberation time. The main difference between 'bright' and 'dark' room simulations is the length of the high-frequency reverberation which is longer for bright rooms than dark ones which therefore require more high-frequency attenuation in the feedback path.

The main difficulty with the reverberation simulation is providing enough delay lines of different lengths to provide sufficient density of reflections to assure an even decay response. This is because each delay line simulates only one possible recirculating signal path in the room whereas in fact there are many. This also affects the frequency response of the reverberation because each recirculating signal path in the room also represents a modal structure as mentioned earlier, in Chapter 6. The frequency response of a recirculating delay line, which is shown in Figure 7.10, consists of a set of equally spaced resonances whose spacing is inversely proportional to the length of the delay line. The bandwidth is determined by the amount of signal fed back and becomes more narrow as the amount fed back approaches one. In fact each recirculating delay line effectively simulates a set of resonances which lie on a radius on a plot of the modal frequencies, described in Chapter 6, and shown in Figure 7.11. Figure 7.11 shows the difficulty of adequately simulating all the possible paths in a room as there would need to be a separate resonator for each possible radius. Therefore this part of a digital reverberation device suffers from similar problems to that of the earlier

Figure 7.11 The locus of a recirculating delay line with respect to possible modal frequencies.

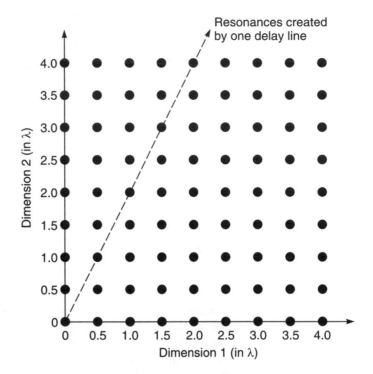

plate and spring devices in that they both have difficulties simulating the reflection and modal density of real rooms. Clearly the number of recirculating delay lines which can be used is a function of the processing power available, and therefore the cost, and this is one of the areas that less expensive devices have to compromise on in order to meet their cost targets. However, there are a variety of ways of improving the density of the reflections over and above the simple techniques outlined. Many of these are proprietary; however, they include techniques such as cross-coupling the individual recirculating delay lines, modifying the feedback multipliers continuously with time, and others.

So far the reverberation devices considered process only monophonic signals because they process only one input and produce only one output. Figure 7.12 shows some possible arrangements for providing stereo reverberation. Figure 7.12a shows one way of obtaining stereo reverberation from a monophonic signal. In this arrangement the signal is fed to two different reverberation circuits of the type outlined previously and these then provide the necessary left and right channel reverberation outputs. However, it is important that the reverberation units do not have an identical set of reverberation

Figure 7.12 (a) Stereo reverberation from a monophonic input. (b) Stereo reverberation from a stereo input. (c) A better way of obtaining stereo reverberation from a stereo input.

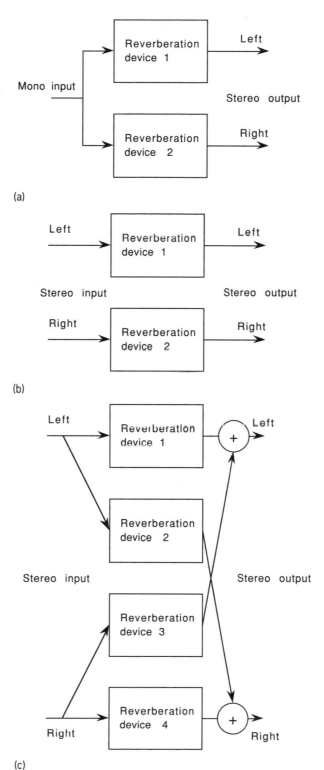

(a)

(b)

(c)

coefficients, otherwise they will produce an identical reverberation in each output channel. Instead the following similarities and differences are required:

*   The initial time delay and early reflection patterns should be different for both reverberation devices because the initial time delay and early reflection patterns would be different for a source in a real room.
*   The reverberation tails should have the same decay rate for both devices. However, the reverberation tails should be uncorrelated with respect to each other because in an ideal diffuse field there is no correlation between sound arriving from different directions.

Note that in principle the early reflection and initial time delay should change as the source is panned across the stereo sound stage because this would also happen in a real room. A simpler approach is to merely alter the inputs to the two reverberation devices. However, this will give an incorrect illusion because, although the amplitudes of the early reflections will alter as a source is panned, this method will result in the reverberation tail amplitudes also varying as the source is panned. In the extreme case of a source being panned full left or right, there would be no reverberation in the other channel. This is not how the reverberation tail would behave in a real room. Instead it would be independent of the location of the source, as discussed in Chapter 6.

The approach shown in Figure 7.12a may be appropriate for single sound sources but it is not appropriate for stereo sources which may contain more than one source of sound panned to different stereo positions. One possible but simplistic solution is to use two reverberation devices separately, as shown in Figure 7.12b. The two reverberation devices would have similar characteristics to those required for mono to stereo reverberation outlined earlier. However, this arrangement also suffers from the fact that the reverberation tail amplitude will depend on the panned position of the sources. A better arrangement is shown in Figure 7.12c. In this arrangement there are two extra reverberation devices which provide cross-coupled reverberation between the two stereo channels, in addition to the two providing reverberation within the channels. The requirements for the initial time delay, early reflection patterns, and reverberation tails of these additional reverberation devices are the same as the ones described earlier for the mono to stereo reverberation arrangement. The advantage of this arrangement is that the amplitude of the reverberant tail is independent of the panned position of the source, as in a real acoustic environment. This is because as the sound is panned, the cross-coupled path feeds the necessary reverberation into the other channel. In addition, because the effect of panning is to change the relative amplitudes of a sound source in the left and right channels, the relative amplitudes of the different early reflection patterns and initial time delay varies as the source is moved within the stereo sound stage. Although the early reflection pattern does not vary in exactly the same way as it would in a real acoustic—for example the time position of the early reflections will not vary as they would in a real acoustic—it nevertheless provides a useful variation which assists in a more realistic simulation of real reverberation. The technique of Figure 7.12c can also be used to provide mono to stereo reverberation and in this context will allow an easy way of varying the early reflection pattern as a source is moved in the stereo picture.

7.4 Chorus, artificial double tracking (ADT), phasing and flanging effects

Chorus, ADT, phasing and flanging effects are related to artificial reverberation in that they use similar technology and algorithms to it. ADT (artificial double tracking), phasing and flanging effects all require the use of delay lines.

- *ADT:* This uses a delay line to simulate the effect of someone singing the same line twice. A typical arrangement is shown in Figure 7.13. The delay needs to be sufficient to ensure that the delayed version is perceived as a separate event, that is greater than 30 ms or so. However, even then it can be difficult to distinguish the second voice so in practice additional processing such as delay modulation and or pitch shifting are applied to make the illusion more effective.
- *Phasing:* This is an effect which uses the comb filtering effect created by adding the output of a delay line to the undelayed signal, as shown in Figure 7.14a. The equivalent

Figure 7.13 A block diagram for automatic double tracking (ADT).

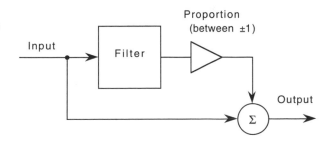

Figure 7.14 Phasing. (a)
Block diagram. (b) Frequency
response.

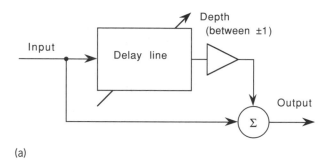

(a)

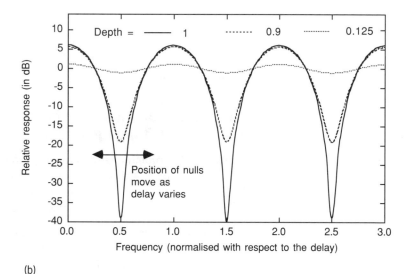

(b)

frequency response is shown in Figure 7.14b. By modulating
the delay of the delay line, at a low frequency, a pleasant
'swooshing' effect is achieved. This is caused by the
movement of notches through the spectrum of the sound
being processed. Interestingly, if the notches are static, and
the sound is steady in pitch, the effect of the notches are
considerably less perceptible.

• *Flanging:* This effect is similar to phasing except that a recirculating delay line is used instead of a simple delay as
shown in Figure 7.15a. This has the effect of creating a set
of regularly spaced resonances, as shown in Figure 7.15b.
Again the delay is modulated which causes the resonance
peaks to move and so provide a time varying effect. Also,
like phasing, the effect is enhanced through the movement
of the resonance peaks.

Figure 7.15 Flanging. (a)
Block diagram. (b) Frequency
response.

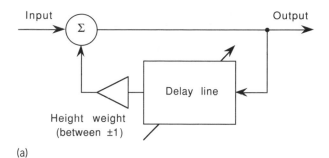

(a)

(b)

These effects had their origin in tape technology and were achieved through the use of simultaneous replay of two tape copies of the same sound track. In fact the term flanging comes from the method of causing the modulation of the effect which was achieved by subtly altering the relative delay of the two tapes by physically manipulating the tape flanges of the tape recorders which were replaying the sound tracks.

Another interesting factor associated with these effects is that they all work much better if there is some time variability in either the sound source or the processing device. This is due to the fact that we are much more sensitive to sounds which change. This effect is also noticeable in reverberation devices. If a reverberation effect is fed a sound which is constant in both frequency and amplitude then after a short period of time the reverberation effect can no longer be easily heard. This is because the resulting output, although different, is static. If some vibrato is imposed on the sound signal the reverberation will become much more noticeable because the vibrato will alter the pitch in a time-varying fashion. This means that at any given time the listener will be hearing the different delays due to reverberation, because they will be associated with different time varying pitches. This is yet another reason why vibrato seems to be such a prevalent musical device in both the playing of real acoustic instruments, sound synthesis and singing.

*   *Chorus:* Chorus effects try to achieve the illusion of a group or chorus of sounds from a single source. The effect was popular in the 1970s in a variety of synthesisers which tried to create the illusion of a string or brass section of an orchestra. In order to do this a chorus device must somehow simulate the effect of different versions of the same source. Simply replicating them with delay is not sufficient. In a real ensemble of the same instrument, including voice, each instrument would have slightly different pitches and pitch variation. They would also have slightly different timbres and amplitude characteristics, and finally they would have variations in the precise timing of notes.

    One way of achieving this is shown in Figure 7.16. In this technique the signal is fed to several delay lines simultaneously and the delay of each line is modulated. The modulation may be low-frequency random noise or sine waves. The important aspect is that they should be different for each delay line by either varying the frequency, delay or the source of the noise. The effect of varying the delay of a delay line is

Figure 7.16 A diagram for chorus effects.

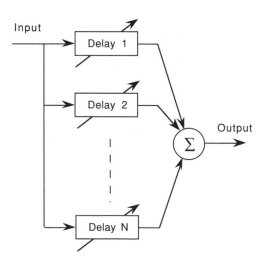

to vary the frequency of its output because the output will be read out at a slightly different rate to the input. Thus the effect of having several delay lines being modulated independently is to give a variation in both frequency and timing that can give the illusion of more than one instrument sounding.

There are other ways of achieving chorus effects using time-varying resonant filters, pitch shifting, etc., but they all attempt to break up some of the correlation between the original sound source and the replicated copies in some fashion or other.

This completes the sections on signal processing which involve delays and reverberation. In some of these applications mention was made of the processing technique of pitch shifting as a useful adjunct to reverberation and delay effects and this is the subject of the next section.

7.5 Pitch processing and time modification

Pitch processing is the generic term used to describe signal processing which operates on the pitch of the signal and, as we shall see, time modification is a related effect. In modern music production, especially on re-mixes and club-style music, pitch and time modification are essential tools of the trade. Most pitch processing involves modifying the existing pitch of the signal in some fashion or other. This may mean simply shifting the pitch up or down by a small amount, imposing an external pitch modulation envelope, or even providing harmonies from a single source. In many cases pitch modification may also involve some associated time modification of the signal, or may be required to remove the effects of time modification of a sound source. Note that the basic principles of sampling-type synthesisers are a form of pitch processing because they provide a variety of pitches from a single or small set of prototype waveforms. We shall see that some of the problems associated with sampling synthesisers are identical to those that arise from modifying the pitch of any signal.

The simplest way of modifying pitch is to store the signal to be shifted in a circular piece of buffer memory, so called because the data is wrapped round when the end of the memory is reached thus forming a circle. This data is then read out at a different rate, either higher or lower, depending on whether the pitch is to be shifted upwards or downwards.

When the pitch is shifted upwards the readout rate is higher than the input rate and the readout pointer will catch up with

Figure 7.17 Pitch shifting up in frequency.

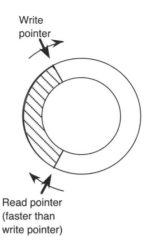

Write pointer

Read pointer
(faster than
write pointer)

Output signal

| Repeated section | Repeated section | Repeated section | Repeated section |

and overtake the input data pointer. Because the buffer is circular, the data that is stored in front of the input pointer is simply the data that was previously written into the buffer. Thus the effect of the output pointer catching up and overtaking the input pointer is to repeat sections of the signal, as shown in Figure 7.17. The amount of signal data repeated depends on the buffer length, and the rate at which a repeat occurs depends on the amount of frequency shift required; a higher repeat rate is required for a larger frequency shift.

When the pitch is shifted downwards the readout rate is lower than the input rate and the input data pointer will catch up with and overtake the readout data pointer. Because the buffer is circular, the data that is stored in front of the readout data pointer is simply data that was previously written into the buffer. Thus the effect of the input data pointer catching up and overtaking the readout pointer is to delete sections of the signal, as shown in Figure 7.18. The amount of data deleted depends on the buffer length and the rate at which a deletion occurs depends on the amount of frequency shift required; more deletions are required for a larger frequency shift.

This technique, although simple has several problems:

• *Waveform discontinuities:* The repeats or deletions will introduce discontinuities into the waveform. As these will be at

Figure 7.18 Pitch shifting
down in frequency.

Write
pointer

Read pointer
(slower than
write pointer)

Output signal

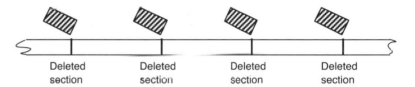

| Deleted | Deleted | Deleted | Deleted |
| section | section | section | section |

a constant rate, which is determined by the amount of pitch shifting required, the effect will be to produce an additional 'buzz' in the output signal at the repeat/deletion rate. A variety of strategies have been used to reduce this effect. One way is to make the length of the repeats/deletions as long as possible, by having a large size, as this will reduce the number of repeats/deletions per unit time for a given pitch shift. However, there is a limit to how long the buffer can be set to before both the delay and the effects of the repeats/deletions become noticeable. From considerations of masking this would set the maximum buffer size at about 30 ms. In addition there are algorithms to minimise the discontinuities such as fading across the join rather than an abrupt cut, repeating or deleting only between zero crossings of the waveform in the same direction, and repeating or deleting a length of signal which is related to the pitch period of the signal. These techniques may be used singly but are often used in combination. Of these techniques, only pitch related repeats or deletions are likely to be inaudible because in this case the additional harmonics generated by process will be pitch related. This means that they will contribute only to the existing harmonic spectrum and thus will be easily masked by the existing frequency components. In all the other techniques the additional harmonics generated by process will not be pitch related and so are much more noticeable. The effect of cross-fading, and using zero crossings of the waveform in the same direction, is to reduce both the amplitude and frequency extent of the additional frequency components and so increase the likelihood that they will be masked by the existing spectral content in the waveform. Unfortunately pitch synchronous repeats and deletions are only possible with monophonic sources, because the task of finding a repeat or deletion length which is a multiple of all the pitches in a chord, especially in equal temperament, is extremely difficult. Thus practical pitch shifters have to adopt the other strategies in order to cover those situations. This also has the advantage of effectively covering up any mistakes in the pitch detector.

- *Formant shifting:* The algorithm achieves pitch shifting by altering the output data rate. This is equivalent to speeding up or slowing down a tape recording and the function of the repeats or deletions is to equalise the time back to that of normal speed play back. Therefore, like tape recorder pitch shifting, the effect of using this algorithm is to shift the spectrum up or down in proportion to the degree of pitch shifting, see Figure 7.19. In extreme cases this can result in

Figure 7.19 The effect of pitch shifting on the formants of a signal.

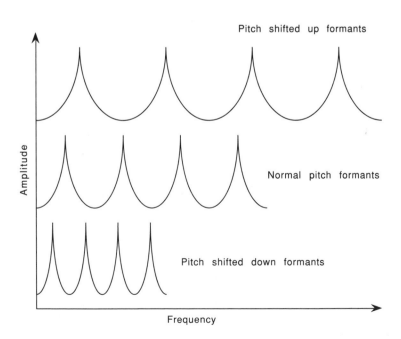

Pitch shifted up formants

Normal pitch formants

Pitch shifted down formants

Amplitude

Frequency

the 'chipmunk' and other effects. Note this effect also occurs in sampling synthesisers for the same reason; they also use sample rate modification to alter the pitch of a sample. Because of this effect the range of pitch shifting that can be used, before the output begins to sound strange, is limited but nonetheless it can be useful for a variety of effects such as chorus, ADT, thickening the texture of sounds, and even correcting small errors of intonation in a recorded performance. In samplers this limitation is eased with the use of multi-samples across the pitch range of the instrument, thus limiting the range of pitch shifts required for any given sample. Unfortunately for recorded or live sound sources this is not an option. However, one application for pitch shifting is to correct the effect of replaying recorded music at a different speed. This operation is required in the broadcast industry where it allows one to modify the running time of a recorded programme to precisely fit the time available for broadcast. It also is heavily used in the dance and remix music industry because it allows a DJ or producer to match the tempo of different pieces of music and they may also want some pitch shifting in addition to match the key as well. In this context the formant shifting ability of this type of pitch shifting algorithm is precisely what is required and thus this technique finds broad application in areas of music production which require time modification.

There are other ways of modifying the pitch of a sound signal which do not cause the formant shifting of the simple method. One of these is to separate the sound into its separate source and filter components, following the acoustic model discussed in Chapter 5. Once this is done the previous algorithm can be used to pitch shift the source component independently of the filter component. The resulting shifted source component can then be recombined with the source component to give a pitch-shifted output with no formant shift. This is because the formants are contained in the filter component which is not shifted with this technique. The filter component can be extracted through the use of an adaptive inverse filter and a block diagram of the complete shifting process is shown in Figure 7.20. This method has not been popular in the past because it requires a large amount of processing to realise the inverse filters. However, with modern technology, the amount of processing required is now more feasible.

Another way of achieving the pitch shifting without altering the formant structure, which is easier to implement, is also based on

Figure 7.20 Pitch shifting using adaptive inverse filtering.

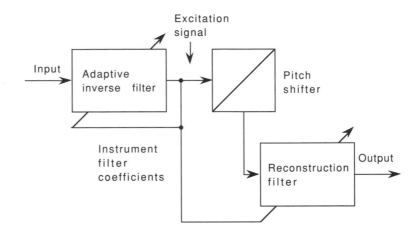

the source–filter acoustical model of musical instruments. The method is based on the fact that the only thing that alters when the pitch changes is the rate at which the excitation is applied to the formant filter, as shown in Figure 7.20. Furthermore, if the excitation is an impulse then the filtering function can become a simple readout of waveform values. These waveform values correspond to the actual waveform values in the input signal over a pitch period, as shown in Figure 7.21. Therefore the pitch shifting algorithm becomes one of pitch detection of the input rate followed by extraction of the most recent waveform within a pitch period. This waveform is then read out whenever the output pitch generates an excitation pulse. The process is shown in Figure 7.22, and the resulting waveforms in Figure 7.23, for both up and down shifts.

Note that the amount of waveform that can be extracted is a function of the pitch and in general will be only part of the total

Figure 7.21 Idealised waveform from a single instrument.

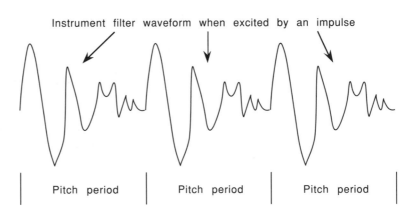

Figure 7.22 Waveform-based pitch shifter.

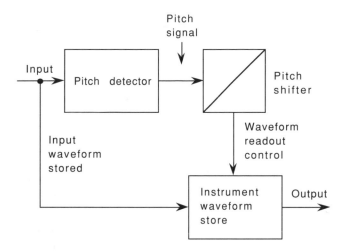

waveform that would result from exciting the instrument's resonances with an impulse. This results in a bell-like quality in the shifted output due to the inevitable truncation of the filter's impulse response that happens in practice. When this truncated waveform is shifted downward in pitch there will be a gap

Figure 7.23 Waveforms output by a waveform-based pitch shifter.

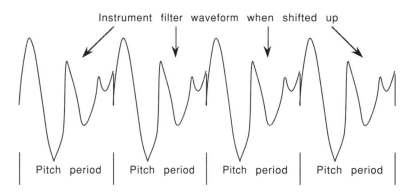

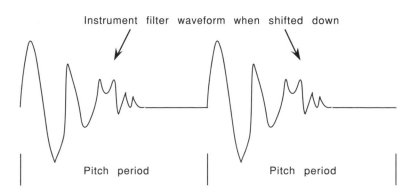

between each pitch period, see Figure 7.23. This imparts a 'buzzy' quality to the pitch shifted sound because it effectively broadens the bandwidths of the formants. When the pitch is shifted upwards the waveforms for each pitch period will overlap. One way of dealing with this is to simply stop reading out the waveform when the next pitch period occurs, which is computationally efficient, and this further enhances the bell-like quality. The alternative is to add the overlapped samples together. This requires some additional processing and the amount available will limit the maximum upward shift amount. For example if a maximum of four samples can be added together then the maximum upward shift amount is a factor of four, or two octaves. Because of the need to pitch detect, the technique is limited to monophonic sound sources, as discussed earlier. However, despite these limitations the technique proves an effective technique for shifting pitch by large amounts with reasonable quality and could form the basis for an effective alternative sampling synthesiser technique.

7.6 Sound morphing and vocoding

Another form of signal processing which makes use of both the source–filter acoustic model and our perception is sound morphing and vocoding. In this form of processing the input sound is first separated into its source and excitation components. This may be accomplished through the use of an adaptive inverse filter or a parallel bank of filters, as shown in Figure 7.24. Ideally the filter bandwidths should be similar to a critical band, that is approximately one third of an octave. However, in practice they are often wider in order to reduce the number of filters required.

Figure 7.24 Separating a sound into source and filter components.

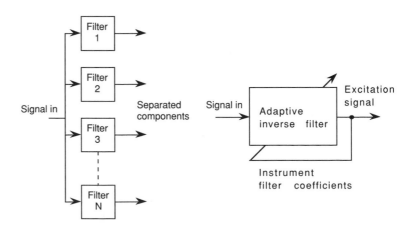

Figure 7.25 A vocoder block diagram.

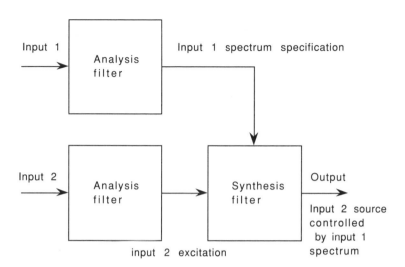

Once the sound has been separated into source and filter components, a number of interesting forms of processing become possible. In particular a source from a different sound can be applied to an instrument's filter function, thus giving a combination of the two. Thus it becomes possible to have the seashore whisper or a piano talk. This type of operation is often known as vocoding. A related operation is that of sound morphing, in which different sources and filter functions are combined to provide sounds not normally available. In one case the source from a soprano singer was combined with the filter functions from a counter tenor to give an overall sound that was similar to a castrato singer, of which no modern examples exist. Morphing may also be carried out dynamically to effect a smooth transition between two different sounds, for example transforming the 'zz' sound in bees into the sound of a swarm of bees.

Sound morphing requires the ability to separate different possible sources into their separate components and then some means of combining them; a block diagram of a typical system is shown in Figure 7.25. Vocoding and sound morphing require very similar hardware and algorithms, the difference being that generally vocoding uses the all of the other sound source as the excitation signal rather than separating it into separate components. It is the usage of the complete alternative signal that can give illusions like the wind whispering, see Figure 7.26. Sound morphing on the other hand requires more control over the individual components to realise its effects.

Both these effects rely on the source filter model to realise the processing and so generally require monophonic sources to

Figure 7.26 Using a vocoder to impose one spectrum on a different source.

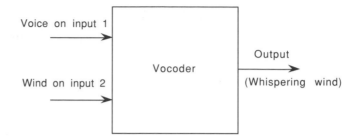

work effectively, although vocoding can easily use a polyphonic signal as the alternative source. Both methods of processing also tend to use filter banks with bandwidths that approximate the ear's critical bandwidths in order to make effective use of the way we hear.

7.7 Spatial processing

Spatial processing is an important aspect of modern music production and a variety of systems are currently in use.

7.7.1 Stereo

The simplest such system is stereophonic reproduction which uses mainly intensity cues to give the illusion of direction to sound coming from two speakers. This is due both to the fact that at high frequencies, above 1 kHz, the direction of a sound is determined by the relative amplitudes at each ear, because of head shading, and to the fact that at low frequencies there is some trading between delay and amplitude. The net result is that amplitude stereo can give a convincing directional image of a set of sound sources, as described in Section 2.6.6.

7.7.2 Dolby Stereo™

A related form of spatial processing, Dolby Stereo™, is used in film and video presentations. Here the objective is to provide not only clear dialogue but also stereo music, and sound effects, as well as a sense of ambience. The typical speaker layout is shown in Figure 7.27. Here in addition to the conventional stereo speakers there are some additional ones to provide the additional requirements. These are as follows:

- *Centre dialogue speaker:* The dialogue is replayed via a central speaker because this has been found to give better speech intelligibility over a stereo presentation. Interestingly, the

Figure 7.27 Typical speaker arrangement for Dolby Stereo™.

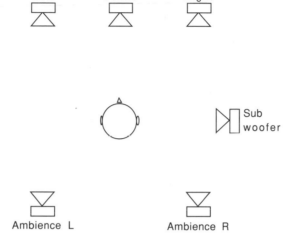

Left Centre Right

Sub woofer

Ambience L   Ambience R

fact that the speech is not in stereo is not noticeable because the visual cue dominates so that we hear the sound coming from the person speaking on the screen even if their sound is coming from a different direction.

- *Ambience speakers:* The ambient sounds and sound effects are diffused via rear-mounted speakers. However they are not supposed to provide directional effects and so are deliberately designed, and fed signals, which minimise their correlation with each other and the front speakers. The effect of this is to fool the hearing system into perceiving the sound as all around with no specific direction.

- *Sub woofer speaker:* This speaker is present to enhance the low-frequency content of the sound. This is required because many of the sound effects used in film and video, such as explosions and punches, have substantial low-frequency and infrasonic content. Thus a specialised speaker is needed to reproduce these sounds properly.

All these signals can be encoded into, and derived from, a two-channel format but better performance is obtained from a multi-channel format.

7.7.3 Ambisonics

Another form of spatial sound presentation is called ambisonics. This is a three-dimensional sound reproduction system that tries to simulate the sound field at a given point in the room. It does this by recognising that at a given point there are: (i) the component of pressure at the point, (ii) velocity in the left to right direction, (iii) velocity in the front to back direction, and (iv) velocity

Figure 7.28 Ambisonic microphone signals. Note: all microphones are coincident so the polar patterns overlap.

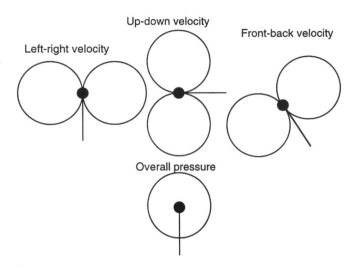

in the down to up direction. These components can be measured through the use of an omnidirectional, pressure microphone and three figure-of-eight, pressure gradient or velocity microphones which are oriented in the three orthogonal directions, as shown in Figure 7.28. This gives four channels of information which represent the first order sound field at the recording point. This soundfield can be reproduced in a room via an array of speakers. The minimum recommended is eight and a typical arrangement is shown in Figure 7.29. Unlike stereo, however, no particular speaker is assigned to a particular direction, such as left, right, etc. Instead all the speakers work in concert to provide

Figure 7.29 Ambisonic speaker layout.

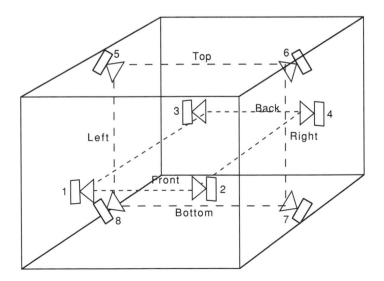

the necessary sound field. That is, all the speakers work together to provide a pressure component and, by feeding appropriate levels and phases of the other signals to the loudspeakers all co-operate to synthesise the necessary velocity components. As an example the front-to-back component would be synthesised by feeding equal levels of the front-to-back signal to speakers 1 and 2 and the same level but in anti-phase to speakers 3 and 4. Speakers 5 to 8 would receive no signal because they fall on the null of the figure-of-eight polar pattern for front-to-back sounds. Note that these weightings are only true for the speaker arrangement shown in Figure 7.29 and if other dimension ratios are used then the weightings require change, because they are determined by the shape of the polar patterns of the different components. The system can be simplified by simply removing channels, thus a two-dimensional sound field requires only three channels, and the system becomes stereo if only two channels are used. In principle it is possible to extend the system to sample acceleration components, in addition to the velocity components, to provide greater accuracy but this is difficult to achieve in practice.

7.7.4 Binaural stereo

For recorded sound that is to be presented on headphones binaural stereo reproduction is more appropriate as described in Section 2.6.6. This can result in amazingly realistic recordings, especially if your own ears are used. However, if other people's ears are used the effect is less realistic, because a given person learns the responses of their own head and ears. In fact it seems that reasonably simple approximations to a head, which just handle the effect of head shading and inter-aural delay, provide the most benefit, probably because they feed less confusing directional cues to the listener. The other problem with binaural presentation is that the stereo image does not change as the listener moves their head. This is not a natural effect and is one of the reasons that binaural signals, especially those from normal intensity stereo sources, appear within the head. In order to avoid this effect it would be necessary to detect the head movement and adjust the signals to the two ears to take account of the change in head direction. This type of processing is required in virtual reality systems where sound sources need to remain associated with their visual cue. One way of achieving this is to measure the effective filter response to the two ears for different directions, as shown in Figure 7.30, and these are known as head related transfer functions (HRTFs). These filter functions can then be used directly to 'pan' a monaural source

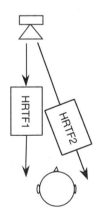

Figure 7.30 Binaural stereo head related transfer functions (HRTFs).

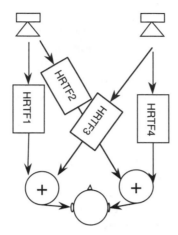

Figure 7.31 Converting stereo signals to binaural signals.

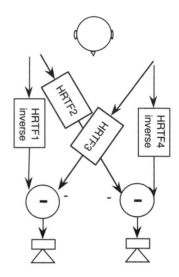

Figure 7.32 Converting binaural signals to stereo signals.

to a particular direction in the binaural stereo perspective. Because it is difficult to interpolate between HRTFs for different angles these functions have to be stored as a large table of data in both azimuth and elevation. If a recorded binaural signal is to be processed then the difference between the HRTF for zero angle, and the actual angle of the head must be used to correct for head movement.

Normal stereo signals do not reproduce correctly on headphones and binaural recordings do not reproduce correctly on normal stereo equipment. This is because in the former, the effect of the head is removed from the listening environment whereas in the latter the effect of head is applied twice, once in the recording and once in the listening. It is possible to convert the signals from one format to the other by making use of the head-related transfer functions.

To convert normal stereo signals into signals suitable for binaural presentation it is necessary to filter the left and right channels with the HRTFs associated with the positions of left and right speakers, as shown in Figure 7.31. Note that four filters are involved because one must model not only the transfer function to each ear but also the cross-coupling of the loudspeakers between the two ears.

To convert binaural signals into a form suitable for normal stereo presentation it is necessary to cancel out the HRTFs associated with the positions of left and right speakers, as shown in Figure 7.32. This involves using two filters which are the inverse of the HRTF to each ear and two other filters which model the cross-coupling between the two ears and subtract the effect. Clearly this is only going to be possible for a particular speaker layout and a forward facing listener with no head movement. However, this technique is used in some proprietary spatial enhancement systems.

7.8 Loudness processing

Sometimes the level of an audio signal needs to be manipulated, usually to modify its dynamic range in order to make it more suitable for a particular musical context. The two main devices used for this are called compressor/limiters or expander/noise-gates. Both of these functions involve detecting the amplitude of the signal and then using this amplitude to control some form of gain control device, usually a voltage controlled amplifier, although a simple multiplication can be used in the digital domain.

The difference between compressor/limiters or expander/gates are as follows:

- *Compressor/limiters:* These devices reduce the dynamic range of a signal so that a large dynamic range becomes smaller. The reduction can either be achieved at an even rate over the whole dynamic range or, more commonly, above a certain signal level or threshold. The parameters that can be controlled are the threshold level and the amount of dynamic range compression, usually called the compression slope, where 1:1 corresponds to no compression, 2:1 corresponds to some compression, and 20:1 corresponds to extreme compression. Extreme compression results in an output level which does not appreciably alter as the input signal level changes and so is often known as limiting. Typical compression curves are shown in Figure 7.33.

  The effect of compression is to make the signal seem louder because it increases the average signal energy, which is an important psychoacoustic cue to loudness. This is often what it is used for, other than controlling the peaks of dynamic range in a signal in order to prevent distortion.

  It is often important that the compressor does not cause any audible artefact, such as sudden changes in gain, modulation of the noise floor of the signal, and signal distortion. This requires careful setting of the time constants associated with the compressor. Typically the user has

Figure 7.33 Output level versus input level for a compressor.

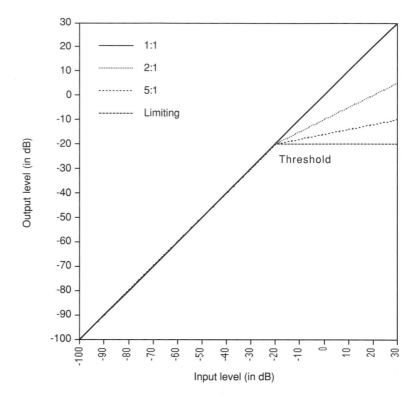

control of the rate at which the compressor acts, called the attack time, and the rate at which the effect of the compressor action reduces after the signal has reduced in amplitude, this is known as the release time. A possible rule of thumb is to set the release time to a time related to the reverberation time of the signal as this would tend to control the time that it would take the signal to change amplitude.

- *Expander/gates:* These devices increase the dynamic range of a signal so that a small dynamic range becomes larger. The increase can either be achieved at an even rate over the whole dynamic range or, more commonly, below a certain signal level or threshold. The parameters that can be controlled are the threshold level and the amount of dynamic range compression, usually called the expansion slope, where 1:1 corresponds to no expansion, 1:2 corresponds to some expansion, and 1:20 corresponds to extreme expansion. Extreme expansion results in an output level which disappears as the input signal level goes below a threshold and so is often known as gating. Often there is also a means of limiting the amount of gain reduction, this is known as a range control. Typical expansion curves are shown in Figure 7.34.

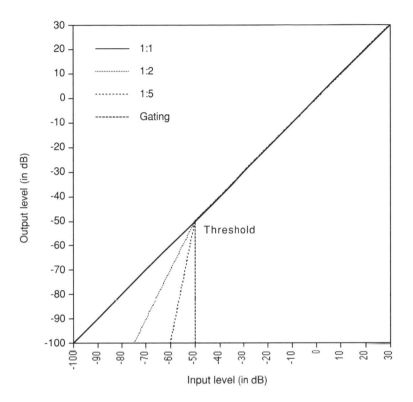

Figure 7.34 Output level versus input level for an expander.

Expanders can make the signal seem quieter because they decrease the average signal energy. However their main use is to reduce the effect of noise on a recorded signal because they can be used to automatically fade down a signal when it drops below the level at which other noises become dominant. This has the effect of reducing the subjective effect of the noise, because it is masked when the signal is present and is faded down when the signal disappears.

In order for an expander to perform its function it must also have its attack and release controls set properly. The attack constant, which determines how quickly the gain increases when the signal rises above the threshold, should be short. For some sounds, for example percussion, it should be as fast as possible. Unfortunately this can cause an audible click when sounds with slower attack times are put through it, so in general this time is controllable.

It should be set as fast as possible for sounds with rapid attack transients but should be set for a slightly slower time if the attack time is slower. As in the compressors, an appropriate time constant to set the release time, which determines the rate at which the expander reduces the gain when the signal drops below the threshold, is on the order of the reverberation time.

Both of these forms of processing are available in forms in which the threshold is not a sharp 'knee' but instead is a soft transition over a region of sound levels. Such soft knee processing can often result in less audible artefacts in the output signal.

Although loudness processing is often used to control the dynamic range and noise in a signal it can also be used creatively to modify the amplitude envelopes of sound signals and so generate new sounds. The techniques can also be used to create stereo effects such as auto-panning in which a sound source is dynamically moved in the stereo sound stage.

7.9 Summary

This chapter has described some of the electronic methods used to process sound, and the psychoacoustics behind the processing. There are many other forms of audio processing which rely on the way we hear the sound for their effect such as noise reduction, music coding, etc. In fact psychoacoustics underpins all processing of audio signals because, as in the case of acoustics, the science describes what is happening but the psychoacoustics describes how we hear it and thus forms the means of specifying what we can do. Acoustics and psycho-acoustics in combination are essential to the endeavour of pro-viding good musical experiences and artistic expression to the human race.

References

Bartlett, B. (1970). A scientific explanation of phasing (flanging). *Journal of the Audio Engineering Society*, **18**, (6), 674.

Bohn, D.A. (1986). Constant-Q graphic equalizers. *Journal of the Audio Engineering Society*, **34**, (9), 611.

Bohn, D.A. (1992). Accelerated slope tone control equalizers. *Journal of the Audio Engineering Society*, **40**, (12), 1018.

Bristow-Johnson, R. (1995). A detailed analysis of a time-domain formant-corrected pitch-shifting algorithm. *Journal of the Audio Engineering Society*, **43**, (5), 340.

Dolby, R.M. (1967). An audio noise reduction system. *Journal of the Audio Engineering Society*, **15**, (4), 383.

Dolby, R. (1983). A 20 dB audio noise reduction system for consumer applications. *Journal of the Audio Engineering Society*, **31**, (3), 98.

Greiner, R.A. and Sohoessow, M. (1983). Design aspects of graphic equalizers. *Journal of the Audio Engineering Society*, **31**, (6), 394.

Hartmann, W.M. (1978). Flanging and phasers. *Journal of the Audio Engineering Society*, **26**, (6), 439.

Henriquez, J.A., Riemer, T.E. and Trahan, R.E. Jr (1990). A phase-linear audio equalizer: design and implementation. *Journal of the Audio Engineering Society*, **38**, (9), 653.

Kahrs, M. and Brandenburg, K. (eds) (1998). *Applications of Digital Signal Processing to Audio and Acoustics*. The Kluwer International Series in Engineering and Computer Science, Vol. 437. Boston and London: Kluwer Academic.

Kraght, P.H. (1992). A linear-phase digital equalizer with cubic-spline frequency reponse. *Journal of the Audio Engineering Society*, **40**, (5), 403.

Lee, F.F. (1972). Time compression and expansion of speech by the sampling method. *Journal of the Audio Engineering Society*, **20**, (9), 738.

Moorer, J.A. (1978). The use of the phase vocoder in computer music applications. *Journal of the Audio Engineering Society*, **26**, (1), 42.

Oliveira, A.J. (1989). A feedforward side-chain limiter/compressor/de-esser with improved flexibility (ER). *Journal of the Audio Engineering Society*, **37**, (4), 226.

Orfanidis, S.J. (1997). Digital parametric equalizer design with prescribed Nyquist-frequency gain (P). *Journal of the Audio Engineering Society*, **45**, (6), 444.

Schroeder, M.R. (1962). Natural sounding artifical reverberation. *Journal of the Audio Engineering Society*, **10**, (3), 219.

Stikvoort, E.F. (1986). Digital dynamic range compressor for audio. *Journal of the Audio Engineering Society*, **34**, (1), 3.

Appendix 1:
Solving the ERB equation

To find the centre frequency of the auditory filter whose critical bandwidth is equal to a given f_0, the ERB quadratic equation (equation 2.6), which relates the critical bandwidth to the centre frequency of the filter, can be solved mathematically as follows. The standard formula for solving a quadratic of the form:

$$[a\ x^2] + [b\ x] + [c] = 0 \quad \text{is:} \quad x_{1,\,2} = \frac{-b \pm \sqrt{b^2 - 4ac}}{2a}$$

giving the two solutions x_1, and x_2. To solve equation 2.8 it has to be written in the standard form. Starting with equation 2.8 for the equivalent rectangular bandwidth (ERB):

$$\text{ERB} = \{6.23*10^{-6}\ (f_c)^2 + 93.39*10^{-3}\ f_c + 28.52\}\ \text{Hz}$$

Let: $L = 6.23*10^{-6}$; $M = 93.39*10^{-3}$; $N = 28.52$; and $Y = \text{ERB}$

Substituting: $\quad [L\ (f_c)^2] + [M\ f_c] + [N] = Y$

Subtract Y from each side:

$$[L\ (f_c)^2] + [M\ f_c] + [N - Y] = 0$$

giving:

$$(f_c)_{1,2} \text{ (Hz)} = \frac{-[M] \pm \sqrt{[M]^2 - \{4[L][N-Y]\}}}{2[L]}$$

Replacing L, M, and N:

$$(f_c)_{1,2} \text{ (Hz)} = \frac{-[93.39*10^{-3}] \pm \sqrt{[93.39*10^{-3}]^2 - \{4[6.23*10^{-6}][28.52-Y]\}}}{2[6.23*10^{-6}]}$$

$$= \frac{-[93.39*10^{-3}] \pm \sqrt{[93.39*10^{-3}]^2 - \{[24.92*10^{-6}][28.52-Y]\}}}{[12.46*10^{-6}]} \qquad (A1.1)$$

(Take the positive result)

Appendix 2: Converting between frequency ratios and cents

The cent is defined as one hundredth of an equal tempered semitone, which is equivalent to one twelve-hundredth of an octave since there are twelve semitones to the octave. Thus one cent can be expressed as:

$$\sqrt[1200]{2} \text{ or } 2^{\left[\frac{1}{1200}\right]}$$

The frequency ratio of any interval (F1/F2) can therefore be calculated from that interval in cents (c) as follows:

$$\frac{F1}{F2} = 2^{\left[\frac{c}{1200}\right]}$$

and the number of cents can be calculated from the frequency ratio by rearranging to give:

$$\log_2\left[\frac{F1}{F2}\right] = \left[\frac{c}{1200}\right]$$

therefore: $c = 1200 \log_2\left[\frac{F1}{F2}\right]$ (A2.1)

For calculation convenience, a logarithm to base 2 can be expressed as a logarithm to base 10. Suppose:

$$\log_2[x] = y \quad\quad\quad\quad (A2.2)$$

Then by definition:

$$x = 2^y$$

Taking logarithms to base 10:

$$\log_{10}[x] = \log_{10}[2^y] = y \log_{10}[2]$$

Substituting in Equation A2.2 for y:

$$\log_{10}[x] = \log_2[x] \log_{10}[2]$$

Rearranging:

$$\log_2[x] = \left[\frac{\log_{10}[x]}{\log_{10}[2]}\right] \tag{A2.3}$$

Substituting Equation A2.3 into Equation A2.1:

$$c = 1200 \left\{\frac{\log_{10}\left[\dfrac{F1}{F2}\right]}{\log_{10}[2]}\right\} = \left[\frac{1200}{\log_{10}[2]}\right] \log_{10}\left[\frac{F1}{F2}\right]$$

Evaluating the constants to give the equation for calculating the cents value of a frequency ratio:

$$c = 3986.3137 \log_{10}\left[\frac{F1}{F2}\right] \tag{A2.4}$$

In semitones (s), this is equivalent to:

$$s = \left[\frac{c}{100}\right] = 39.863137 \log_{10}\left[\frac{F1}{F2}\right] \tag{A2.5}$$

Rearranging Equation A2.4 to give the equation for calculating the frequency ratio from a cent value:

$$\left[\frac{F1}{F2}\right] = 10^{\left[\frac{c}{3986.3137}\right]} \tag{A2.6}$$

Appendix 3: Deriving the reverberation time equation

Clearly the length of time that it takes for sound to die is a function not only of the absorption of the surfaces in a room but also is a function of the length of time between interactions with the surfaces of the room. We can use these facts to derive an equation for the reverberation time in a room. The first thing to determine is the average length of time that a sound wave will travel between interactions with the surfaces of the room. This can be found from the mean free path of the room which is a measure of the average distances between surfaces, assuming all possible angles of incidence and position. For an approximately rectangular box the mean free path is give by the following equation:

$$MFP = \frac{4V}{S} \tag{A3.1}$$

where MFP = the mean free path (in m)
V = the volume (in m^{-3})
and S = the surface area (in m^{-2})

The time between surface interactions may be simply calculated from A3.1 by dividing it by the speed of sound to give:

$$\tau = \frac{4V}{Sc} \tag{A3.2}$$

where τ = the time between reflections (in s)
and c = the speed of sound (in ms^{-1}, or metres per second)

Equation A3.2 gives us the time between surface interactions and at each of these interactions α is the proportion of the energy absorbed, where α is the average absorption coefficient discussed earlier. If α of the energy is absorbed at the surface, then $(1 - \alpha)$ is the proportion of the energy reflected back to interact with further surfaces. At each surface a further proportion, α, of energy will be removed so the proportion the original sound energy that is reflected back will go as follows:

$$\text{Energy}_{\text{After one reflection}} = \text{Energy}_{\text{Before reflection}} (1 - \alpha)$$

$$\text{Energy}_{\text{After two reflections}} = \text{Energy}_{\text{Before reflections}} (1 - \alpha)^2$$

$$\text{Energy}_{\text{After three reflections}} = \text{Energy}_{\text{Before reflections}} (1 - \alpha)^3 \tag{A3.3}$$

$$\vdots$$

$$\text{Energy}_{\text{After n reflections}} = \text{Energy}_{\text{Before reflections}} (1 - \alpha)^n$$

As α is less than 1, $(1 - \alpha)$ will be also. Thus Equation A3.3 shows that the sound energy decays away in an exponential manner. We are interested in the time that it takes the sound to decay by a fixed proportion and so need to calculate the number of reflections that have occurred in a given time interval. This is easily calculated by dividing the time interval by the mean time between reflections, calculated using Equation A3.2, to give:

$$n = \frac{t}{\left(\dfrac{4V}{Sc}\right)} = t\left(\frac{Sc}{4V}\right) \tag{A3.4}$$

where t = the time interval (in s)

By substituting Equation A3.4 into Equation A3.3 we can get an expression for the remaining energy in the sound after a given time period as:

$$\text{Energy}_{\text{After a time interval}} = \text{Energy}_{\text{Initial}} (1 - \alpha)^{t\left(\frac{Sc}{4V}\right)} \tag{A3.5}$$

and therefore the ratio that the sound energy has decayed by at that time as:

$$\frac{\text{Energy}_{\text{After n reflections}}}{\text{Energy}_{\text{Before reflections}}} = (1 - \alpha)^{t\left(\frac{Sc}{4V}\right)} \tag{A3.6}$$

In order to find the time that it takes for the sound to decay by a given ratio we must take logarithms, to the base $(1 - \alpha)$, on both sides of Equation A3.6 to give:

$$\log_{(1-\alpha)}\left(\frac{\text{Energy}_{\text{After n reflections}}}{\text{Energy}_{\text{Before reflections}}}\right) = t\left(\frac{Sc}{4V}\right)$$

which can be rearranged to give the time required for a given ratio of sound energy decay as:

$$t = \left(\frac{4V}{Sc}\right)\log_{(1-\alpha)}\left(\frac{\text{Energy}_{\text{After n reflections}}}{\text{Energy}_{\text{Before reflections}}}\right) \tag{A3.7}$$

Unfortunately Equation A3.7 requires that we take a logarithm to the base $(1 - \alpha)$! However we can get round this by remembering that this can be calculated using natural logarithms as:

$$\log_{(1-\alpha)}\left(\frac{\text{Energy}_{\text{After n reflections}}}{\text{Energy}_{\text{Before reflections}}}\right) = \frac{\ln\left(\frac{\text{Energy}_{\text{After n reflections}}}{\text{Energy}_{\text{Before reflections}}}\right)}{\ln(1 - \alpha)}$$

So Equation A3.7 becomes:

$$t = \left(\frac{4V}{Sc}\right)\frac{\ln\left(\frac{\text{Energy}_{\text{After n reflections}}}{\text{Energy}_{\text{Before reflections}}}\right)}{\ln(1 - \alpha)} \tag{A3.8}$$

Equation A3.8 gives a relationship between the ratio of sound energy decay and the time it takes and so can be used to calculate this time. There are an infinite number of possible ratios that could be used. However the most commonly used ratio is that which corresponds to a decrease in sound energy of 60 dB, or 10^6. When this ratio is substituted into Equation A3.8 we get an equation for the 60 dB reverberation time, known as T_{60}, which is:

$$T_{60} = \left(\frac{4V}{Sc}\right)\frac{\ln(10^{-6})}{\ln(1 - \alpha)} = \left(\frac{V}{S\ln(1-\alpha)}\right)\frac{4 \times (-13.82)}{344 \text{ ms}^{-1}}$$

$$= \frac{-0.161V}{S\ln(1 - \alpha)} \tag{A3.9}$$

where T_{60} = the 60 dB reverberation time (in s)

Thus the reverberation time is given by:

$$T_{60} = \frac{-0.161V}{S\ln(1 - \alpha)} \tag{A3.10}$$

where T_{60} = the 60 dB reverberation time (in s)

Equation A3.10 is known as the Norris–Eyring reverberation formula and the negative sign in the numerator compensates for the negative sign that results from the natural logarithm resulting in a reverberation time which is positive. Note that it is possible to calculate the reverberation time for other ratios of decay and that the only difference between these and Equation A3.10 would be the value of the constant. The argument behind the derivation of reverberation time is a statistical one and so there are some important assumptions behind Equation A3.10 These assumptions are:

- that the sound visits all surfaces with equal probability, and at all possible angles of incidence. That is, the sound field is diffuse. This is required in order to invoke the concept of an average absorption coefficient for the room. Note that this is a desirable acoustic goal for subjective reasons as well; we prefer to listen to, and perform, music in rooms with a diffuse field.
- that the concept of a mean free path is valid. Again this is required, in order to have an average absorption coefficient but in addition it means that the room's shape must not be too extreme. This means that this analysis is not valid for rooms which resemble long tunnels; however, most real rooms are not too deviant and the mean free path equation is applicable.

Appendix 4:
Deriving the reverberation time equation for different frequencies and surfaces

In real rooms we must also allow for the presence of a variety of different materials, as well as accounting for their variation of absorption as a function of frequency. This is complicated by the fact that there will be different areas of material, with different absorption coefficients, and these will have to be combined in a way that accurately reflects their relative contribution. For example, a large area of a material with a low value of absorption coefficient may well have more influence than a small area of material with more absorption. In the Sabine equation this is easily done by multiplying the absorption coefficient of the material by its total area and then adding up the contributions from all the surfaces in the room. These resulted in a figure which Sabine called the 'equivalent open window area' as he assumed, and experimentally verified, that the absorption coefficient of an open window was equal to one. It is therefore easy to incorporate the effects of different materials by simply substituting the total open window area for different materials calculated using the method described above for the open window area calculated using the average absorption coefficient in the Sabine equation. This gives a modified equation which allows for a variety of frequency-dependent materials in the room:

$$T_{60(\alpha<0.3)} = \frac{0.161V}{\displaystyle\sum_{\text{All surfaces } S_i} S_i \alpha_i(f)} \tag{A4.1}$$

where $\alpha_i(f)$ = the absorption coefficient for a given material and S_i = its area

For the Norris–Eyring reverberation time equation the situation is a little more complicated because the equation does not use open window area directly. There are two possible approaches. The first is to calculate a weighted average absorption coefficient by calculating the effective open window area, as done in the Sabine equation, and then dividing the result by the total surface area. This gives the following equation for the average absorption coefficient:

$$\alpha_{\text{weighted average}}(f) = \frac{\displaystyle\sum_{\text{All surfaces } S_i} S_i \alpha_i(f)}{S}$$

which can be substituted for α in the Norris–Eyring reverberation time equation, to give a modified equation, which allows for different materials in the room:

$$T_{60} = \frac{-0.161V}{S\ln\left(1 - \dfrac{\displaystyle\sum_{\text{All Surfaces } S_i} S_i \alpha_i(f)}{S}\right)} \tag{A4.2}$$

Equation A4.2 can be used to calculate the effect of a variety of frequency-dependent materials in the room. However, there is an alternative way of looking at the problem which is more in the spirit of the reasoning behind the Norris–Eyring reverberation time equation. This second approach can be derived by considering the effect on the sound energy amplitude of successive reflections which hit surfaces of differing absorption coefficient. In this case the proportion the original sound energy that is reflected back will vary with each reflection as follows:

$$\text{Energy}_{\text{After one reflection}} = \text{Energy}_{\text{Before reflection}} (1 - \alpha_1)$$

$$\text{Energy}_{\text{After two reflections}} = \text{Energy}_{\text{Before reflections}} (1 - \alpha_1)(1 - \alpha_2)$$

$$\text{Energy}_{\text{After three reflections}} = \text{Energy}_{\text{Before reflections}} (1 - \alpha_1)(1 - \alpha_2)(1 - \alpha_3)$$

$$\vdots$$

$$\text{Energy}_{\text{After n reflections}} = \text{Energy}_{\text{Before reflections}} (1 - \alpha_1)(1 - \alpha_2)(1 - \alpha_3) \times \dots \times (1 - \alpha_n)$$

This can be couched in terms of an average α by taking the geometric mean of the different reflection coefficients $(1 - \alpha)$. For example after two reflections the energy is at a level which would be the same as if there had been two reflections from a material whose reflected energy was given by:

$$(1 - \alpha)_{average} = \sqrt{(1 - \alpha_1)(1 - \alpha_2)}$$

After three reflections the average reflection coefficient would be given by:

$$(1 - \alpha)_{average} = \sqrt[3]{(1 - \alpha_1)(1 - \alpha_2)(1 - \alpha_3)}$$

And after n reflections the average reflection coefficient would be given by:

$$(1 - \alpha)_{average} = \sqrt[n]{(1 - \alpha_1)(1 - \alpha_2)(1 - \alpha_3) \times \ldots \times (1 - \alpha_n)}$$

Because there are only a finite number of different materials in the room, but of differing areas, it is necessary only to consider an average based on just the number of different materials but weighted to allow for their differing surface areas. Because logarithms convert products to additions this weighted geometric mean can be simply expressed as a sum of the individual absorption terms and so the Norris–Eyring reverberation time equation can be rewritten in a modified form, which allows for the variation in material absorption due to both nature and frequency, as:

$$T_{60} = \frac{-0.161V}{\displaystyle\sum_{\text{All surfaces } S_i} S_i \ln(1 - \alpha_i(f))} \tag{A4.3}$$

Equation A4.3 is also known as the Millington–Sette equation. Although Equation A4.3 can be used irrespective of the absorption level it is still more complicated than the Sabine equation and, if the absorption coefficient is less than 0.3 it can be approximated very effectively by it, as discussed previously. Thus in many contexts the Sabine equation, Equation A4.1, is preferred.

Index

'A' weightings, 85
Absorption:
  Helmholtz absorbers, 308–9
  materials, 276, 305–10
  porous absorbers, 305–7
  resonant absorbers, 307–8
  reverberant sound, 253–4, 258–62, 264, 272–5,
    280, 287–8
  room mode decay, 298
  sound, 37
  wideband absorbers, 309–10
Acoustic cues, 223–6
Acoustic foldback, 285–7
Acoustic models, 155–98
Acoustic pressure waveforms, 112, 113, 127
Acoustic reflex, 70, 71, 96
Acuity, loss, 91–2
Adaptive inverse filtering, 352, 354
Adiabatic gas law equation, 5
ADT see Automatic double tracking
Aging, presbycusis, 80
Air columns see Bores
Air effects, sound, 5–7, 33–5, 287–8
Airflow, instruments, 166–95, 200
Ambience speakers, 357
Ambisonics, 357–9
American National Standards Institute, 119, 210
Amplitude, 10, 87, 232, 298–9

Amplitude reflection gratings, 314–16
Analogue reverberation techniques, 335–7
Anatomy, human, 65–74, 199
Anechoic rooms, 153
Aperiodic sound waves, 54–6
Artificial reverberation, 335–43
Auditory filters, 77–8, 231
Automatic double tracking (ADT), 343–7
Axial modes, 290, 295–6

Backward masking, 233, 234
Balanced sound mixes, 333–5
Band-pass filters, 57–60, 78–9, 130, 328, 331
Band-reject filters, 57, 58, 328
Bars, percussion, 195–8, 222
Basilar membrane, 71–4, 78–9, 90, 120, 125–6,
  129, 130–4
Bernoulli effect, 189–90, 200
Binary Amplitude Diffuser, 315, 316
Binaural stereo, 107–8, 359–61
Black box model, 152–4, 194–5
Bonello criteria, 293–5
Bores (air columns), 184
Boundary effects
  inverse square law, 31–3
  reflection, 38–40, 42
    bound–unbound boundaries, 39–40, 45–6
    hard boundaries, 38–9, 43–8

Boundary effects (*continued*)
  refraction, 36
  standing waves, 43–8, 288–305
Bowed strings, 160–62
Brass instruments, 189–94, 218, 219
Buffers, 347, 348

'C' weightings, 85
Centre dialogue speakers, 356–7
Centre frequency, 366–7
Cents, 149–50, 368–9
Chiff, 170, 215
Chipmunk effect, 351
Chords, two-note, 137–43
Chorus effects, electronic, 343–7
Clarinet, 187, 188, 217–18, 222
Close miking, 332
Cochlea, 71–4, 92
Coding, perceptual, 234–5
Combination tones, 228–30
Complex sounds, 89–91
Complex waveforms, 51–6
Compression, 2, 170, 171, 172, 173
Compressor/limiters, 361–2
Concert halls, 284–5
Concha, 66
Conical resonators, 185
Consonance, 138–44
Constructive interference, 40–1
Contemporary pitch perception theory,
    133–4
Correlated sound sources, 20–4, 40
CQ *see* Larynx closed quotient
Critical bands:
  complex sounds, 89
  consonance/dissonance, 139–41
  definition, 65
  ERB quadratic equation, 366–7
  hearing system, 74–9, 126
  timbre, 221–3
Critical distance, 258, 262–4
Critical frequency, 300–5
Cut-off, filters, 329, 330, 331
Cut-up, adjustment, 169
Cymbal plates, 198

Damaged hearing, 93
Decay, 61–2, 256, 283–4, 295–300
Deceptive sounds, 228–46
Decibels, 25–8, 81

Delay, 22–3, 103–4, 105, 106, 337–47
Design, room acoustics, 279–88
Destructive interference, 40–1
Difference limen *see* Just noticeable difference
Difference tone, 228–30
Diffraction, 48–50
Diffuse field, 260, 300, 301
Diffusion, 286, 311–16
Digital reverberation techniques, 337–43
Direct sound, 257–8
Direction perception, 66, 96–108
Directivity, reverberant sound, 264–6
Displacement nodes, 157, 186, 187, 204
Dissonance, 138–44
Dolby Stereo™, 356–7
Double bass, 155, 163–4
Duration of sound, 88–9, 90–1, 136

Ear anatomy, 66–74
Early decay time (EDT), 283–4
Early reflections, 249–54, 338, 343
Edgetone, 168
EDT *see* Early decay time
Electronic filters, 129
Electronic sound processing, 327–65
  artificial reverberation, 335–43
  chorus/ADT/flanging effects, 343–7
  equalisation, 331–5
  filtering, 327–30
  loudness processing, 361–4
  pitch processing, 347–54
  sound morphing and vocoding, 354–6
  spatial processing, 356–61
  time modification, 347–54
  tone controls, 331–5
Enclosed space acoustics, 247–325
  absorption materials, 305–11
  critical distance, 258, 262–4
  diffusion materials, 311–16
  direct sound, 247–9
  energy–time considerations, 321–5
  reflected sound, 249–87
  sound isolation, 316–21
  standing waves, 288–305
  see also Room modes
Energy–time considerations, 321–5
Equal loudness contours, 83
Equal-tempered tuning, 149–51, 227
Equalisation, 331–5
Equivalent open window area, 277

Equivalent rectangular bandwidth (ERB), 76, 366–7
Expander/gates, 362–4
Exposure times, noise, 94–5

f_0 *see* Fundamental frequency
'Ff' sound, 167–8
Filters:
    adaptive inverse, 352, 354
    auditory response curves, 77–8
    band-pass, 57–60, 78–9, 130, 328, 331
    band-reject, 57, 58, 328
    centre frequency, 366–7
    electronic, 129, 327–30
    equalisation/tone control, 331
    filter banks, 56
    frequency response, 57–8
    low-pass, 57, 327, 328, 331
    masking, 231
    time responses, 58–60
    types, 57–8
Fingering, instruments, 178, 179, 193
Flanging effects, 343–7
Flanking paths, 320–1
Fletcher–Munson curves, 84
Floating rooms, 321
Flow-control, 177, 184, 202
Flue pipes, 166–80
Flute, 110, 112, 177–80, 217–18, 222
Flutter echoes, 274–5
Focusing surfaces, 250, 251
Foldback, 285–7
Formants, 203–6, 350–1
Forward masking, 233
Four-part harmony, 143–4
Fourier analysis, 51
Free space, 153
Frequency:
    air absorption, 288
    analysis, 66–74, 74–9
    Bonello criteria, 293–5
    definition, 10–11
    diffraction, 50
    domain representations *see* Spectra
    filter responses, 57–8
    interference, 41
    proximity grouping, 238
    ratios, 114–16, 144–51, 368–9
    relative loudness, 84–6
    response, 332–3

reverberation time variation, 275–6
sine waves, 9–12
see also Critical frequency; Fundamental frequency; Modal frequency; Mode frequency
Fundamental frequency (f_0):
    brass instruments, 154, 193
    definition, 110
    flue organ pipes, 169
    instrument/voice ranges, 154
    musical notes, 109–11
    open/stopped pipes, 173–5
    pitch perception, 119–36
    place theory, 121–8
    stringed instruments, 155

Gases, sound velocity, 5–7
Gedacht 8' stop, 176, 177
Golf ball and spring model, 1–3
Grouping illusions, notes, 236–46
Guitar, 164–5

Haas effect, 104–5
Hair cells, 74, 92, 120
Handedness, 236, 241–2
Hard boundaries, 38–9, 43–8
Harmonic additive synthesis, 226
Harmonic number, 113
Harmonics:
    consonance/dissonance, 140–44
    definition, 53
    frequency ratios, 114–16
    long-term average spectra, 212–13
    musical intervals between, 114–19
    musical notes, 111–13
    square waves, 61
    timbre perception, 215, 216, 221–2
Harmony:
    hearing notes, 137–44
    hearing pitch, 119–136
    musical notes, 109–18
    tuning systems, 144–51
Head, sound localisation, 96–108
Head related transfer functions (HRTFs), 359–61
Headphones, 103, 359–61
Hearing:
    anatomical structures, 65–74
    critical bands, 65, 74–9, 89, 126, 139–41, 221–3, 366–7

Hearing: (*continued*)
  frequency sensitivity, 79–82
  loudness perception, 82–91
  noise-induced hearing loss, 91–6
  pitch perception, 119–36
  pressure sensitivity, 79–82
  protection, 95–6
Helmholtz absorbers, 308–9
High-pass filters, 57, 58, 327, 328, 331
HRTFs *see* Head related transfer functions

IID *see* Interaural intensity difference
Impedance, 12–14, 67–70
Impulse sound sources, 195
Incus, 67, 68, 69
Inharmonicity, 158–60
Inner ear, 71–4
Input/system/output model, 152–4, 156, 205, 206
Instruments *see* Musical instruments
Integrated noise doses, 94–5
Intensity:
  direct sound, 249, 258
  early reflections, 250–2, 253
  interaural difference, 101–5
  sound intensity level, 14–16, 18–19, 25, 28–31
Intensity stereo, 105–6
Inter-spike interval histograms, 131, 132
Interaural intensity difference (IID), 101–5
Interaural time difference (ITD), 97–101, 103–5
Interference, 40–2
Intervals, musical, 114–19, 139–44
Inverse square law, 28–33
Isolation, sound, 316–21
ITD *see* Interaural time difference

Just diatonic scale, 147–9
Just noticeable difference (JND), 124–5, 135

Keyboards, 149–51

Languid, 169, 170
Larynx, 200
Larynx closed quotient (CQ), 201
Lateral reflections, 284–5
Leaky bucket model, 259
L_{eq} measurement, 86
Linear superposition, 33, 43
Lips, embouchure, 189–90
Listening, room acoustics, 247–325

Localisation, sound, 66, 96–108
Logarithms, 25–6, 116–17, 118, 163
Long-term average spectra (LTAS), 212–13, 216
Longitudinal waves, 2
Loud sounds, 71
Loudness, 82–91, 85–91, 361–4
Loudspeakers, 249, 251–4, 261–6
Low-pass filters, 57, 327, 328, 331
LTAS *see* Long-term average spectra
Lute, 164–5

Malleus, 67, 68, 69
Masking, 230–6, 333–5
Microphones, 332
Middle ear, 67–71
Millington–Sette equation, 277, 376
Mixed surfaces, 276–9
Mixes, sound balance, 333–5
Modal frequencies, rooms, 289–93
Mode frequencies, 156–8, 184–8, 191, 192, 195–8, 203
Moore's pitch perception model, 133–4
Morphing, 354–6
MPEG systems, 235
Musical instruments:
  'black box' model, 152–4, 194–5
  brass, 189–94, 218, 219
  fundamental frequency ranges, 154
  percussion, 8–9, 194–8, 222
  stringed, 110–13, 129, 155–66, 218, 238, 238–9, 241
  timbre, 210–46
  wind, 110, 112, 166–94, 215, 217, 218, 222
Musical intervals, 114–19, 139–44
Musical notes:
  acoustic pressure waveforms, 110–11, 112
  grouping illusions, 236–46
  harmonics, 109–19
  hearing harmony, 136–44
  pure tones, 228–33

Nerve firing, 120, 129–32, 134
Nicking, flue organ pipes, 169–70
Noise, 91–6, 167–8
Non-harmonic sounds, 123
Non-periodic sounds, 54–6, 118, 127
Non-simultaneous masking, 233, 234
Norris–Eyring reverberation formula, 268–9, 271, 272, 277, 372, 375
Notch filters, 327, 330

Note envelope, 211, 214–15
Note grouping illusions, 236–46
Note offset phase, 211–15, 221
Note onset phase, 211–20, 221

Oblique modes, 292
Oboe, 110, 112, 217–18, 222
Octave illusion, 236, 237
Offset phase, notes, 211–15, 221
Ohm's second (acoustical) law, 121–2
Onset phase, notes, 213–20, 221
Open pipes, 170–7
Open-holes lattice cut-off frequency, 185
Organ, 166–95, 214–17, 226–8, 245
Organ of Corti, 129, 130
Organum, 137–8
Ossicles, 67, 68–9
Outer ear, function, 66–7
Output, 152–4, 156, 163–6, 205, 206
Overblown mode, 169, 179, 215
Overtones, 113

Pain, threshold of, 80–2
Panning, 343
Partials of a waveform, 53
Partitions, sound isolation, 317–20
Perceptual coding systems, 234–5
Percussion instruments, 8–9, 194–8, 222
Performer support, 285–7
Perilymph fluid, 71, 72
Periodic sound waves, 51–4
Perturbation theory, 203
Phase, 10, 54
Phase-locking, 130–1
Phon scale, 84
Piano, 158–60
Piccolo, 179
Pinnae, 66, 102–3
Pipe organ *see* Organ
Pitch perception, 119–36
    basilar membrane, 129, 130, 131
    contemporary theory, 133–4
    duration effect, 136
    fundamental frequency, 134–6
    illusions, 242–6
    non-harmonic sounds, 123
    place theory, 66–79, 120–8
    temporal theory, 128–33
    timbre, 210–11, 221
Pitch processing, electronic, 347–54

Pitch shift phenomenon, 135–6
Place theory of pitch perception, 66–79, 120–8
Plucked instruments, 157–8, 181
Pop concert effect, 35
Popping frequency, 191
Porous absorbers, 305–7
Power level, 16–17, 25
Precedence effect *see* Haas effect
Presbycusis, 80
Pressure:
    complex sounds perception, 89
    flue organ pipes, 171–6
    nodes/antinodes, 177, 178, 186, 187
    ossicular chain, 68–9
    sine waves, 12–14
    sound wave nature, 1–3
    standing wave components, 44–5
    see also Acoustic pressure waveforms;
        Airflow; Compression; Sound pressure
        level
Pressure-controlled valves, 183
Propagation speed *see* Velocity
Protection, hearing, 95–6
Proximity grouping, 237, 238
Psychoacoustics, definition, 65, 66
Pure tones, 228–33
Pythagorean tuning, 145–7
Pythagorus' theorem, 42

Quadratic residue diffuser, 313, 314

R see Room constant
Radiation, sound, 165–6
Rarefaction, 2, 171, 172, 173
Recorders, 177–80
Recordings, 332–3
Reed wind instruments, 166, 180–95
Reflection:
    amplitude reflection gratings, 314–16
    bounded to unbounded areas, 39–40, 45–6
    early, 249–54, 338, 343
    enclosed spaces, 249–88
    Haas effect, 105
    hard boundaries, 38–9, 43–8
    standing waves, 43–8, 156–7, 174, 288–305
Reflection-free zones, 322–5
Refraction, 33–7
Register holes, recorder, 179
Reissner's membrane, 71, 72
Repetition pitch, 136

Resonant absorbers, 307–8
Resonators, 170, 171, 175, 176, 182–3, 184, 185
Reverberant sound
  analogue processing, 335–7
  artificial, 335–43
  digital processing, 337–43
  faults, 272–5
  field, 255–7
  Norris–Eyring formula, 268–9, 271, 272, 277,
    372, 375
  Sabine formula, 271–2, 277, 374–6
  source directivity, 264–6
  steady state level, 258–62
  stereo, 341, 342
  see also Reverberation time
Reverberation time, 267–88, 370–6
Room constant (*R*), 260–3
Room modes, 288–305
Rooms *see* Enclosed space acoustics

Sabine reverberation formula, 271–2, 277, 374–6
Sawtooth waveforms, 162
Saxophone, 187, 188, 217–18, 222
Scala tympani, 71, 72
Scala vestibuli, 71, 72
Scales, musical, 144–51
Scattering, 50–1
Secondary pitch effects, 134–6
Sensitivity, hearing, 79–82, 87, 91
'Sh' sound, 127–8, 168
Shallot opening, 180–2
Shepherd tone effect, 242–4
Signal processing, 327–47
  pitch processing, 347–54
  sound morphing/vocoding, 354–6
SIL *see* Sound intensity level
Similarity grouping, 237, 238
Simple sounds, 86–9
Simultaneous masking, 233, 234
Sine waves, 9–11, 12–14, 22, 52–6, 138–9
Singing voice, 154, 198–208, 219–20
Smoke effects, 287, 288
Snare drum, 118
Snell's law, 34
Solids, sound velocity, 3–5
Sound intensity level (SIL), 14–16, 25, 28–31
Sound isolation, 316–21
Sound level meters, 86
Sound modifiers
  black box model, 152, 153

percussion instruments, 195–8
reed organ pipes, 182–3
singing voice, 199, 202–8
stringed instruments, 162–6
wind instruments, 170–7, 182–3
Sound morphing, electronic, 354–6
Sound power level (SWL), 16–17, 25
Sound pressure level (SPL), 17–20, 25, 27–8, 81,
  82–91
Sound processing, electronic, 327–65
Sound theory, 1–64
Sources of sound:
  boundary effects, 32–3
  components separation, 355
  direction perception, 96–108
  directivity, reverberant sound,
    264–6
  percussion instruments, 194–8
  singing voice, 198–208
  stringed instruments, 155–66
  summation, 20–8
  wind instruments, 166–94
Spatial processing, electronic, 356–61
Spectra:
  analysis, 56–64
  long-term average (LTAS), 212–13, 216
  Shepherd tone, 243
  trombone, 194
  see also Waveforms
Spectrograms, 62–4
Speed *see* Velocity
SPL *see* Sound pressure level
Square waves, 52–3, 55, 58, 59, 60, 61
'Ss' sound, 119, 127–8, 168
Standing waves, 43–8, 156–7, 174, 288–305
  see also Room modes
Stapedius muscle, 70
Stapes footplate, 67, 68, 69, 71
Steady-state, 211–14, 224, 225, 255–62
Stereo, 105–8, 341, 342, 356–61
Stopped pipes, 170–7, 187, 192, 216
Streaming effects, 236, 238–41
Stringed instruments, 155–66
  bowed string, 160–62
  input/system/output model, 156
  plucked string, 157–8
  sound modifiers, 162–6
  struck string, 158–60
  timbre, 238–9, 241
Sub woofer speakers, 357

Subjectivity, timbre, 223
Summation, 20–8
Superposition, 33, 43, 51, 172
Surfaces, diffusion, 311–14
SWL *see* Sound power level
Synthetic sounds, 154, 199, 223, 224, 226–8, 230

Tangential modes, 291–2
Tap resonances, 163
Temperature, 6–7, 33–7
Temporal theory, 128–33
Tensor tympani muscle, 70
Threshold of hearing, 80–2, 87, 232–4
Timbre:
  acoustic cues, 223–6
  acoustics, 211–20
  critical bands, 221–3
  deceiving the ear, 228–46
  definition, 210–11
  modification, 333
  pipe organ, 226–8
  psychoacoustics, 220–6
  spectral shape, 330
  streaming effects, 238–41
Time:
  acoustic system responses, 60–2
  domain effects, 51–6, 329
  early decay time, 283–4
  electronic sound processing, 347–54
  energy–time considerations, 321–5
  filter responses, 58–60
  interaural time difference, 97–101, 103–5
  noise exposure times, 94–5
  see also Reverberation time
Tinnitus, noise-induced, 92
Tone, 90, 331–5
  see also Timbre
Tone-hole lattices, 185
Transverse modes, 195–6
Transverse waves, 2, 7–9
Tristimulus diagram, 224–5
Trombone, 194
Trumpet, 110, 112
Tuba, 194
Tuning systems, 144–51, 227–8
Two note chords, 137–43, 227–30

Two-dimensional standing waves, 47
Tympanic membrane, 67, 69, 70

Unbounded–bounded boundaries, 39–40, 45–6
Uncorrelated sound sources, 20, 24–5
Universal modal frequency equation, 292–3

Velocity components, 3–9, 11–14, 44–5
Vibration:
  strings, 156–66
  vocal folds, 200, 201
Vibrato, 202, 218, 219, 220, 346
Viola, 155, 163–4
Violin, 110–13, 129, 155, 162–3, 218, 238
Virtual pitch, 123
Vocal tract, 199–205
Vocoding, 354–6
Voice *see* Singing voice
Voicing, 169–70
Vowels, 204, 205, 206

Waveforms:
  acoustic pressure waveforms, 110–11, 112, 118
  clarinet, 187, 188
  complex, 51–6
  guitar, 165
  lute, 165
  pitched/non-pitched sounds, 119
  reed organ pipes, 183
  sawtooth, 162
  saxophone, 187, 188
  timbre, 216, 217, 219, 220
  trombone, 194
Wavelength, 9–12, 49–51
Western music, 137–9, 144
White noise, 233
Wideband absorbers, 309–10
Wind, refraction, 36
Wind instruments, 166–94, 215, 217, 218, 219
Windows, 277, 279
Woodwind instruments, 177–80, 183–9, 215, 217, 218

Xylophone, 196–8

Young's modulus, 4, 5, 7

 **Focal Press**

http://www.focalpress.com

Join Focal Press On-line

As a member you will enjoy the following benefits:

- an email bulletin with **information on new books**
- a bi-monthly **Focal Press Newsletter**:
    - o featuring a selection of new titles
    - o keeps you informed of **special offers, discounts and freebies**
    - o alerts you to **Focal Press news and events** such as author signings and seminars
- complete access to **free content** and reference material on the focalpress site, such as the focalXtra articles and commentary from our authors
- a **Sneak Preview** of selected titles (sample chapters) *before* they publish
- a chance to have your say on our **discussion boards** and **review books** for other focal readers

Focal Club Members are invited to give us feedback on our products and services. Email: worldmarketing@focalpress.com – we want to hear your views!

Membership is FREE. To join, visit our website and register. If you require any further information regarding the on-line club please contact:

Emma Hales, Promotions Controller
Email: emma.hales@repp.co.uk
Fax: +44 (0)1865 315472
Address: Focal Press, Linacre House,
Jordan Hill, Oxford,
UK, OX2 8DP

Catalogue
For information on all Focal Press titles, we will be happy to send you a free copy of the Focal Press catalogue:

USA
Email: christine.degon@bhusa.com

Europe and rest of World
Email: carol.burgess@repp.co.uk
Tel: +44 (0)1865 314693

Potential authors
If you have an idea for a book, please get in touch:

USA
Terri Jadick, Associate Editor
Email: terri.jadick@bhusa.com
Tel: +1 781 904 2646
Fax: +1 781 904 2640

Europe and rest of World
Christina Donaldson, Editorial Assistant
Email: christina.donaldson@repp.co.uk
Tel: +44 (0)1865 314027
Fax: +44 (0)1865 315472